Experiments in General Chemistry

Fifth Edition

Steven L. Murov
Modesto Junior College

THOMSON
BROOKS/COLE
Australia • Brazil • Canada • Mexico • Singapore • Spain • United Kingdom • United States

Publisher: David Harris
Acquisitions Editor: Lisa Shaver
Assistant Editor: Sarah Lowe
Marketing Manager: Amee Mosley
Marketing Assistant: Michele Collela
Marketing Communications Manager: Nathaniel Bogson-Michelson

Project Manager, Editorial Production: Alison Throckmorton
Print Buyer: Nora Massuda
Printer: Thomson/West

Printed in the United States of America
1 2 3 4 5 6 7 09 08 07 06 05

Printer: Thomson/West

0-495-12538-5

Thomson Higher Education
10 Davis Drive
Belmont, CA 94002-3098
USA

Asia (including India)
Thomson Learning
5 Shenton Way
#01-01 UIC Building
Singapore 068808

Australia/New Zealand
Thomson Learning Australia
102 Dodds Street
Southbank, Victoria 3006
Australia

Canada
Thomson Nelson
1120 Birchmount Road
Toronto, Ontario M1K 5G4
Canada

UK/Europe/Middle East/Africa
Thomson Learning
High Holborn House
50–51 Bedford Row
London WC1R 4LR
United Kingdom

Latin America
Thomson Learning
Seneca, 53
Colonia Polanco
11560 Mexico
D.F. Mexico

Spain (including Portugal)
Thomson Paraninfo
Calle Magallanes, 25
28015 Madrid, Spain

EXPERIMENTS IN GENERAL CHEMISTRY

Table of Contents

^{C} indicates that a centigram balance will suffice for this experiment.
^{M} indicates that a milligram balance is recommended for this experiment.

EXPERIMENTS IN GENERAL CHEMISTRY

PREFACE

A collection of the best short works of the late, great physicist, Richard Feynman, is entitled *The Pleasure of Finding Things Out.* The goal of this text is to give students the opportunity to experience the pleasure of finding things out. Curiosity is a wonderful attribute that motivates humans to use creative talents to probe, explore and explain the mysteries of our universe. The search for explanations for the mysteries provides an adventure that helps to make life exciting and rewarding. The pathways to the solutions often lead to valuable insights and unexpected significant discoveries. This text has been designed to provide a stimulating environment that will promote curiosity and motivate students to seek solutions to chemical mysteries. To accomplish this goal, an extensive effort has been made to develop experiments that maximize an **inquiry or discovery oriented approach** and **minimize personal hazards and ecological impact.** Simulating a research environment within the constraints imposed by a college course is a demanding challenge. Time limits, safety, chemical toxicity, chemical disposal, chemical costs and equipment costs are a few of the constraints. The preparation of this text has involved a serious effort to overcome these constraints.

In addition to having a positive attitude towards the laboratory portion of the chemistry course, it is very important that the student understands the role of the experience. There is a tendency for students to treat each experiment as a separate unit without sufficient consideration of the connections it makes to other experiments or to the chemical principles being introduced in the lecture portion of the course. During the preparation of this fifth edition, a strong effort has been made to enable students to understand the reason for each experiment's place in the lab and how it adds to the sum of the student's chemical experiences. To help students understand the reasons for doing each experiment and to place the experiments into a broader perspective, this text includes ***Prelaboratory Problems*** with answers to some of the questions, some ***Postlaboratory Problems*** and a set of ***Review Exercises***. Most chemistry laboratory courses do not have final exams so the ***Review Exercises*** and ***Postlaboratory Problems*** strive at having the students reflect upon the course and organize the techniques and principles according to applicability and limitations.

A correlation chart between the experiments in this book and concepts in Thomson general chemistry books has been prepared and follows this preface. It should be noted that the sequence of experiments can be changed (for instance, ***Experiments 29***, ***30*** and ***31*** can be run much earlier in the course) without an interruption in continuity with a few exceptions. The vanillin prepared in ***Experiment 2*** is used in ***Experiment 3***. The absorption spectrum of triiodide is determined in ***Experiment 26*** and used again in one of the options in ***Experiment 28***. Copper(II) glycinate is synthesized in ***Experiment 29*** and analyzed in ***Experiment 31***.

Significant improvements are included in the fifth edition of this text. ***Experiment 5***, in addition to the determination of the empirical formula of a hydrate now also contains the synthesis and empirical formula determination of zinc iodide. If the entire experiment is going to be performed, two laboratory periods should be allotted. Otherwise, one period will suffice for *Part A* or a combination of *Parts B, C* and *D*. The Internet study of the properties of elements and compounds (***Experiment 13***) has been updated and improved. The exercise now includes less duplication of material directly covered in textbooks and new material on nuclear stability. In addition to ***Experiment 13***, Internet exercises are also included in ***Experiments 12*** and ***35***. Finally, a question has been added at the end of the *Results and Discussion* section of each experiment that asks the student if the *Learning Objectives* of the experiment have been achieved.

Another extremely important feature is that this fifth edition like the earlier ones, **has taken the lead out** along with the chromates and barium, mercury and nickel salts. For chemical reactions, the only transition metals used are iron, copper, manganese, zinc and cobalt (trace amounts of silver are used in the optional qualitative test in the distillation experiment). While some might consider cobalt(II) to be too toxic, the LD_{50} for cobalt(II) ion is about five times higher than the values for barium and mercury. Because of the selected use of transition metals, **disposal problems and costs should be minimized by the use of this text.**

Selection of experiments has been based on a search for strengths. While two experiments (*9, 27*) introduce approaches to qualitative analysis concepts and techniques, this text does not cover the classical qualitative analysis scheme. The cursory coverage that would have been afforded to the scheme due to page limitations would have been weak. However, we do find that students learn the important concepts of qualitative analysis by performing ***Experiments*** *9* and *27*. Whenever possible, unknowns have been incorporated to promote interest and a research environment. As part of the responsibility of the laboratory portion of the course is to teach technique, most commonly used techniques are studied and applied including organic chemistry techniques such as distillation, recrystallization, and vacuum filtration.

The experiments have been written assuming that a milligram balance will be standard equipment for the course. ***Experiments 2, 3, 4, 12, 14, 17, 18, 19, 25, 29*** and ***34*** can be performed with a centigram balance without significant loss of information. However, use of anything less than a milligram balance for ***Experiments 5, 7, 11, 21, 22, 24, 30, 31*** and ***33*** will result in a decrease in the ability to interpret the results.

This text is designed for the instructor who takes the laboratory portion of the course seriously. Because of the exploratory nature of many of the experiments, students will frequently need guidance. Students may also need to have some of the techniques demonstrated and the discussion extended. Chemistry has been described as "the central science" and the laboratory is the center of chemistry. Students need to approach the laboratory with a positive attitude and be curious, alert observers who are prepared for the unexpected and willing to use their imaginations to seek insights into the mysteries of our universe.

Acknowledgments. I thank my editors, Sarah Lowe and David Harris, for giving me the opportunity to further improve this general chemistry laboratory text. My coauthor on another lab text, Brian Stedjee, was the original developer of many of the experiments included here and I am grateful for his willingness to test the experiments, proofread this document and prepare the accompanying instructor's manual. I would also like to thank the many Modesto Junior College students who often had to test rather primitive forms of the experiments. Many of the images that have been added to the 5th edition in the upper right hand corner of each experiment were obtained from clipart available at http://www.scs-intl.com/trader/. In some cases, you might want to challenge yourself to determine the connection between the image and the experiment. My wife, Carolyn Ann Murov, sacrificed valuable time to prepare the figures for this text and for several months approximately every three years has to listen to me cursing at the computer while a new edition is prepared. For this edition, Carolyn also had to patiently try to take pictures of our newest kitten curiously exploring a molecule of nepetalactone (catnip). One of her pictures demonstrating the three c's, cats, curiosity and chemistry, was used on page x of this ***Preface***. Finally, I want to thank my three young grandchildren, Dylan (3), Hope (3) and Carson (1) for continually reminding me of the importance of curiosity and the pleasure of finding things out. I sincerely hope students will learn from their chemical laboratory experience and more importantly open their minds to the mysterious and explore it.

Steven L. Murov
Modesto Junior College

Textbook Correlation Table

Murov Expt. #	Key Concepts	Key Techniques	Kotz[1], et. al., Chapter	Masterton[2], et. al., Chapter	Moore[3], et. al., Chapter	Whitten[4], et. al., Chapter
1	scientific method	observation, glassworking	1	1	1	1
2	separation of mixtures	evaporation, filtration, recrystallization	1, 13	1	1, 11	1, 13
3	measurements, sig. figs., density	determination of melting points, densities	1	1	1	1
4	accuracy, precision, density	density determination, graphing, piptetting	1	1	1	1
5	mole, empirical formulas	pyrolysis, flame tests	3	3	3	2
6	classification of chemical reactions, stoichiometry	observation of chemical reactions	4, 5	3, 4	5	3, 4
7	stoichiometry	synthesis, filtration	4, 5	3, 4	4	3
8	conductivity, electrolytes, bonding	measurement of electrical conductivity	5, 9	4, 7, 10	5, 7	4, 7, 18
9	double replacements, net ionic equations	qualitative analysis	5	4	5	4
10	single replacements, activities	observational skills	5, 21	4, 18, 20	5, 19	4, 21
11	concentrations, stoichiometry	titration	5	3, 4	5	4
12	specific heat, enthalpy, Hess's law	calorimetry	6	8	6	15
13	periodicity, LD_{50}	navigating Internet	8	6	7	6
14	spectroscopy	dilutions, spectroscopy	7	6	7	5
15	Lewis structures	modeling	9, 10	7	8, 9	8, 9
16	polarity, chromatography	chromatography	9	7	9	8
17	gas laws	data analysis, graphing	12	5	10	12
18	phase changes	cooling curves	13, 14	9, 10	11, 15	13, 14
19	stoichiometry	distillation, titration	13	1, 4, 9	11	13
20	organic compounds	modeling	11	22	12	27, 28
21	acid and base reactions, stoichiometry	reactions, titration	17, 18	13, 14	16, 17	18, 19

Murov Expt. #	Key Concepts	Key Techniques	Kotz[1], et. al., Chapter	Masterton[2], et. al., Chapter	Moore[3], et. al., Chapter	Whitten[4], et. al., Chapter
22	acid and base stoichiometry	titration	17, 18	13, 14	16, 17	18, 19
23	pH, pK_a	pH meter, titration curves	17, 18	13, 14	16, 17	18, 19
24	acid and base stoichiometry	back titration	17, 18	13, 14	16, 17	18, 19
25	solubility product, free energy	data analysis	18, 19	16, 17	15, 17, 18	17, 20
26	equilibrium constant	spectroscopy, data analysis	18, 19	15, 17	14, 18	17, 20
27	complex formation	qualitative analysis	18	15	17	20, 36
28	kinetics, mechanisms	data analysis, spectroscopy	15	11	13	16
29	coordination compounds	synthesis	22	15, 20	22	25
30	redox, stoichiometry	titration	5, 20	4, 18	5,19	11
31	redox, stoichiometry	titration	5, 20	4, 18	5, 19	11
32	reduction potentials	observational skills	20	18	19	21
33	electrochemistry	electrochemical techniques	20	18	19	21
34	complex formation	spectroscopy	18, 22	15	17, 22	20, 36
35	polymers, material selection	observational skills	11, 13+	9, 22	11, 12, 22	22, 27

1. Kotz, J. C., Treichel, Jr., P., Harman, P. A., *Chemistry & Chemical Reactivity*, 6th ed., Harcourt, Philadelphia, 2006.
2. Masterton, W. L., Hurley, C. N. *Chemistry: Principles and Reactions*, Brooks-Cole, 5th ed., Monterey, 2006.
3. Moore, J. W., Stanitski, C. L., Jurs, P. C., *Chemistry: The Molecular Science*, 2nd ed., Harcourt, Philadelphia, 2005.
4. Whitten, K. W., Davis, R. E., Peck, M. L., Stanley, G. G., *General Chemistry*, 8th ed., Harcourt, Philadelphia, 2006.

Photograph (July, 2005) by Carolyn Murov of our kitten attacking a molecular model of nepetalactone (catnip).

SAFETY FIRST

and Prudent Laboratory Practice

While many people think that the chemical industry must rank high on a list of frequency of accidents, the reverse is actually true. The days lost per person due to accidents in the chemical industry is one of the lowest for all professions. This is not because of fewer potential hazards. To the contrary, the chemical laboratory is full of accidents waiting to happen. However, proper precautions and patience prevent the accidents from happening.

1. Come to the laboratory prepared. Read the experiment and study it. Think about manipulations where special care is needed.

2. Wear eye protection at all times. Your eyes are a very valuable part of your body but are also one of the most vulnerable parts. And don't forget, you are not the only student in the lab. Sometimes chemical accidents spread chemicals throughout the lab.

3. Do not perform unauthorized experiments. Sometimes there is a strong temptation to just mix chemicals together. Do not succumb to this temptation. On the other hand, exploration is strongly encouraged. If you want to try your own experiment, first try to predict its outcome. Then discuss it with the instructor and obtain his/her approval and guidance. Finally for previously untested reactions, use very small quantities and run the reaction in the hood behind a safety shield.

4. When you first enter the laboratory, memorize the locations of the fire extinguishers, eye wash, safety shower, first aid kit and exits from the laboratory. The instructor should discuss the use of each type of safety equipment. If, despite your eye protection, something should get into your eye, thoroughly wash your eyes for several minutes with the eye wash. Don't stop after a brief rinsing.

5. Conduct all experiments that evolve gases or unpleasant odors in the hood and make sure the hood is on and functioning.

6. Never take food or drinks into the laboratory.

7. Do inform the instructor of all accidents, even the most minor of cuts or burns.

8. Pay special attention to the names and concentrations of chemicals and read bottle labels carefully. If an experiment calls for addition of hydrogen peroxide to bleach to generate oxygen and hydrochloric acid is added instead, lethal chlorine gas will be generated that can endanger the lives of the whole class.

9. Treat glass tubing with great respect. Chemical accidents such as the one described in #8 above fortunately seldom occur but minor accidents with glass tubing occur too frequently. In ***Experiment 1***, be very patient and let the hot glass cool before you touch it. Firepolish the ends of all pieces of glass tubing to avoid cuts. Most important, when inserting a piece of glass tubing through a hole in a rubber stopper:

   a. lubricate the hole or glass tubing
   b. hold the tubing very close (within 1 cm) to the part of the tubing that is about to enter the stopper
   c. wrap the tubing in a cloth towel
   d. push slowly while slowly rotating and be patient.

10. If you touch a lab reagent bottle and find that it is wet, put it down and wash your hands.

11. Be sure whenever you are working in the lab that another person familiar with lab safety guidelines is present and that the instructor is within shouting distance.

12. Wear appropriate clothing that covers most of your body. A lab apron is also helpful. Do not wear floppy, loose clothing such as neckties or scarves. Do not wear sandals or open toed shoes.

13. Do not ever leave a heated experiment unattended and do not ever heat a closed system such as a stoppered flask or test tube. Never heat very volatile liquids such as ether or pentane with a flame and in fact don't even use them in rooms with flames burning.

14. Keep your section of the lab clean and treat all equipment, especially the balance as though it is part of your very own Porsche.

15. Carefully clean up all spills and breakages. If you break a mercury thermometer or spill a corrosive chemical, inform your instructor immediately.

16. Keep your books away from sinks especially those that are equipped with aspirators. Occasionally drain hoses jump out of sinks and drench everything near the sink.

17. Do not put chemicals into the sink without first obtaining the instructor's permission.

18. When taking a chemical from a bottle, try not to take more than you need. If you exceed your need, do not return it to the bottle. Give it to another student or properly dispose of it.

19. Do not put spatulas, stirring rods, droppers or pipets into chemical bottles. Transfer the chemical to your own container first. If the chemical will not pour out, seek help from the instructor.

20. Do not put bottle stoppers down on the desk top. This could contaminate the contents of the bottle.

21. If, when using a reagent bottle, you get the bottle wet, rinse off its outside and dry the bottle.

22. Remember that all chemicals have some degree of toxicity. Every effort has been made in the design of this lab text to use chemicals that minimize toxicity and disposal problems. Recognize, however, that the chemicals included here are toxic and need to be treated with care and respect.

23. Do wear eye protection, think, be patient, explore, learn, enjoy and show your enthusiasm.

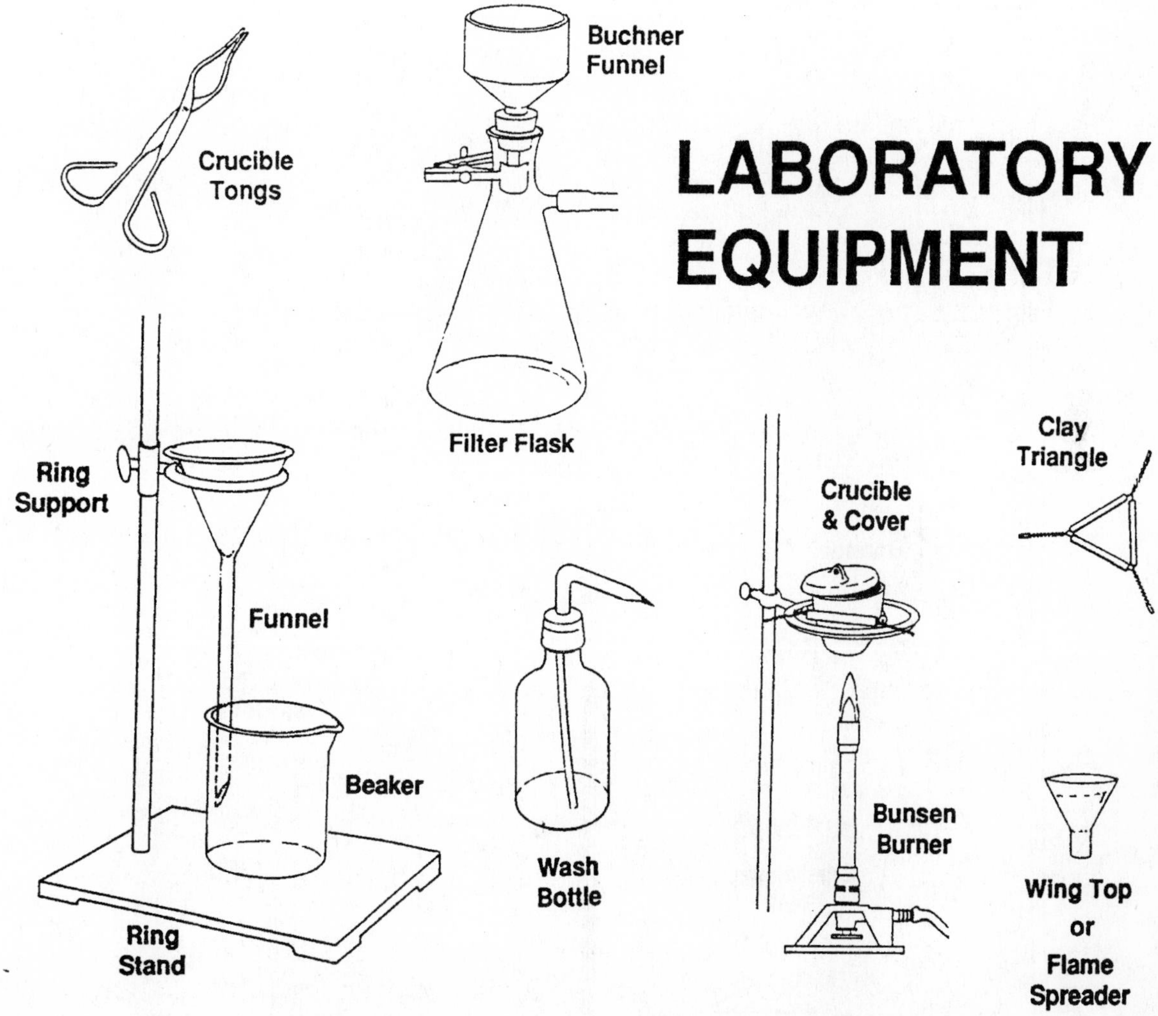

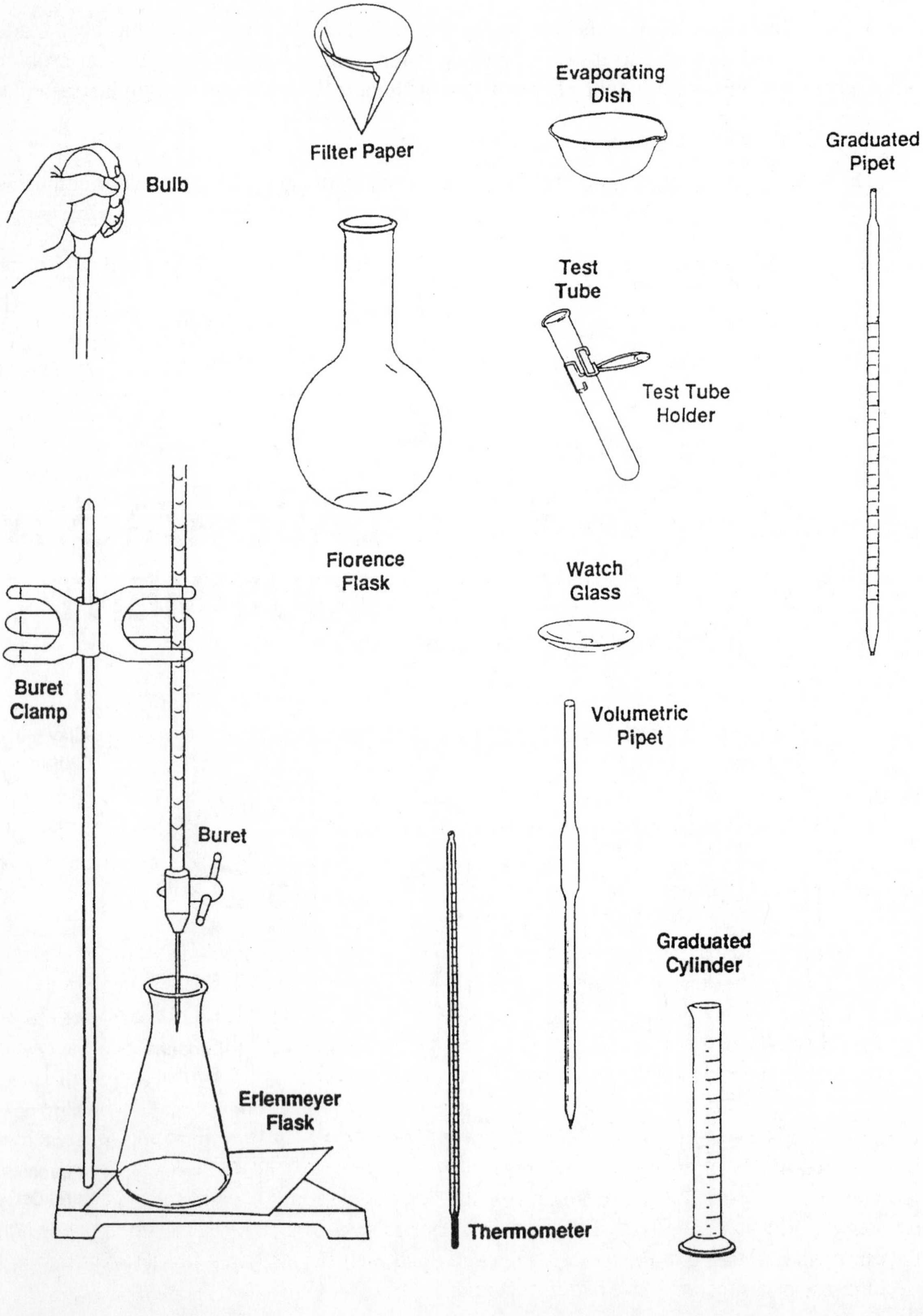
Filter Paper
Evaporating Dish
Graduated Pipet
Bulb
Test Tube
Test Tube Holder
Florence Flask
Watch Glass
Buret Clamp
Volumetric Pipet
Buret
Graduated Cylinder
Erlenmeyer Flask
Thermometer

Experiment 1

EARLY EXPLORATIONS AND TERMINOLOGY

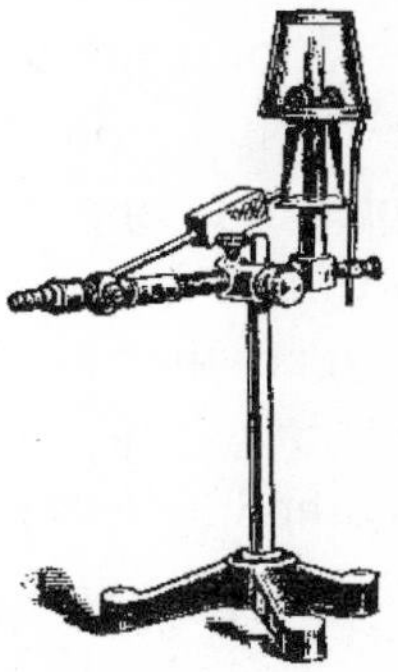

1865 Bunsen burner

Learning Objectives

Upon completion of this experiment, students will have experienced:
1. Practice with techniques of observation and the scientific method.
2. The use of the Bunsen burner.
3. Basic glassworking.
4. Some basic chemistry terminology.

Text Topics

Scientific method, physical properties, chemical reactions (see page ix).

Notes to Students and Instructor

This experiment will probably take more than one laboratory period. It can be started on the first lab day which usually includes check-in procedures and a safety presentation and then finished on the second lab day. Alternatively, selection of only certain sections should shorten the experiment. ***Most sections of this experiment should be thoroughly discussed in class after completion of the experiment.***

Discussion

Pretend for a moment that it is a very hot day. Use your imagination to visualize a glass of ice water. Have you noticed that the ice is floating in the water and not resting on the bottom of the glass? Has this observation ever puzzled you? ***Observation and imagination are two of the keys to good science.*** Mysteries that arise from inconsistencies between expectations and observations often contain clues that lead to exciting and wonderful discoveries.

Consider the ice-water system. What should happen to the density (mass/volume) of a liquid as it is cooled? We might expect a contraction of volume and an increase in density as the temperature decreases (resulting in sinking). And yet, contrary to this expectation, the ice is less dense than liquid water and floats. Included in the observation of the ice-water system then should be the unexpected floating of the ice and questions about why the ice floats and if it is common for the solid phase of a substance to float in its liquid.

Learn to make careful, complete and unbiased observations and include as part of these observations, questions on any inconsistencies that arise from them. Ideally observations should not depend on the observer as we are trying to **record facts in an understandable way for other people**. It is important that scientific observations be reproducible. The observation section of a report should not include interpretations or explanations because explanations might differ from one observer to the next and there may even be more than one possible conclusion. Ice sometimes forms with small air bubbles and in these cases one should record that there are air bubbles in the ice. But to say that the ice floats because of air bubbles is not an observation but in this case an inadequate explanation.

It is very important to record all observations as the act of disregarding or ignoring is actually a conclusion that an observation is not important. Some very important observations have been overlooked only to be found by later investigators to have significance (penicillin and nuclear fission are two examples). Discoveries of teflon and aspartame were made serendipitously by careful observers who did not overlook the unexpected. When doing science, pay heed to the words of Ralph Waldo Emerson, ***"God hides things by putting them near us.,"*** and Louis Pasteur, ***"In the fields of observation, chance favors only the mind that is prepared."***

Observation is the first part of a process commonly called ***the scientific method***. Although its emphasis in some textbooks sometimes gives the misleading impression that scientists operate according to a schedule, the scientific method does describe the process that occurs in scientific exploration. It starts with the puzzling observation and resulting questions. Next with the use of imagination, explanations (or hypotheses) are suggested. Fortunately in science (and this is what makes science easier than most fields), explanations are testable. If experiments support an explanation, the explanation becomes a theory. The theory is always subject to further testing which can result in modification or even discarding of the theory.

As you do your laboratory experiments, remember to stay alert and record all observations including questions about anything curious to you. Be sure that your records are written clearly and concisely in a way that can be understood and tested by others.

Procedure

This exercise has been designed to help you develop your observational skills, distinguish between observations and explanations, and to learn to carefully record all observations. Remember that complete observations often lead to questions.

A. The meniscus. Add water to about the half way point of a 50 mL graduated cylinder and study the features of the water surface. Describe and draw your observations. The phenomenon that you observe is called a meniscus. Write down a question about the meniscus. [Comment: Volumetric glassware has been calibrated to give correct volume measurements when you read the **very** bottom of the meniscus.]

B. The candle flame. Light a candle and study the flame. Write down all your observations and questions about the flame. Try to include observations on states of matter and physical and chemical properties and changes. Be very careful to distinguish observations from explanations. What do you think is actually burning? Write your answer down before you read or experiment further. You will not lose points for an incorrect answer.

Be sure to record observations from the following tests. Put a beaker over the burning candle almost but not all the way down and carefully observe the inside of the beaker. Blow the candle out and immediately put a glass stirring rod on top of the extinguished wick in the region where the flame had been. Inspect the rod. Relight the candle, and, with a burning match in one hand, again blow the candle out. Immediately bring the burning match to the region where the flame had been, moving the match slowly towards the wick for the last 2 cm. Save the candle for Part L.

C. The Bunsen burner. One of the important tools of the laboratory chemist is the Bunsen burner. This exercise has been designed to familiarize you with the burner and introduce you to glassworking. Study *Figure 1-1* and compare it to your Bunsen burner.

Fig. 1-1

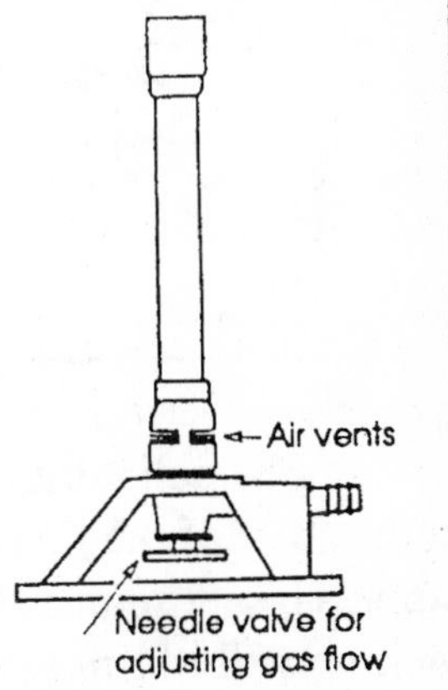

[Comment: The instructor should demonstrate use of the burner.] Close the air control and make sure the gas valve is off. Turn the gas valve on the bench on full, light a match, open the burner gas valve and light the gas. What color is it? Increase the gas flow until the flame is about 8 cm. high and open the air control until the yellow color is gone. Draw and describe the flame. What do you think is the hottest region? Take a wood splint and insert it quickly into the flame right over the top of the burner (*Figure 1-2*). Hold it there until it ignites and observe where it burns. Now hold a wire gauze vertically in the flame so that about 1 cm. of the gauze extends beyond the far edge of the burner (*Figure 1-3*). Heat it until it glows and record your observations about the position and pattern of glow. If significant amounts of colored flame leap from the gauze, rinse the corner of the gauze that you are heating with deionized water and repeat the test.

Fig. 1-2

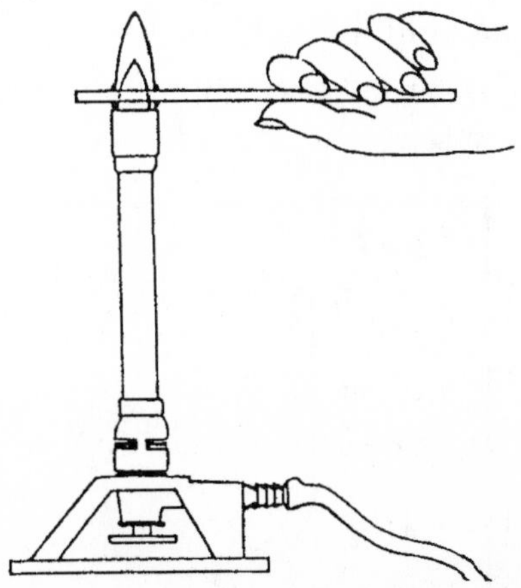

Fig. 1-3

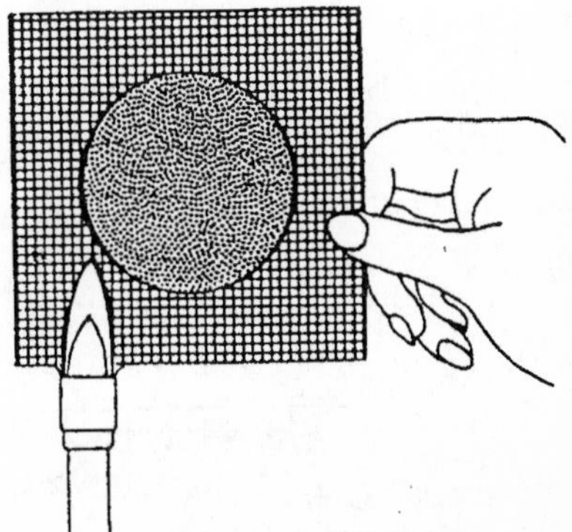

Glassworking. Take a file and make a deep scratch on a piece of 6 mm glass tubing 20 cm from the end. Be sure not to make more than one groove. Holding the scratch away from you with thumbs on either side of the scratch, push your thumbs forward, pulling the two pieces apart (*Figure 1-4*). With a decent scratch, the glass will almost split by itself. However, the ends will still be sharp enough to cut you. Anytime glass tubing is cut, the ends should be firepolished to round off the sharp edges.

Fig. 1-4

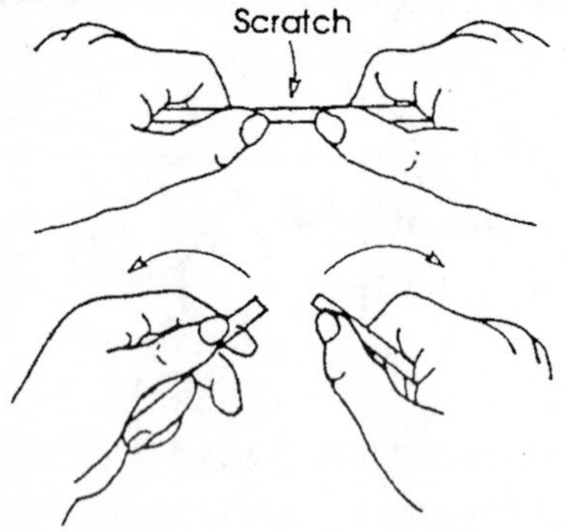

Fig. 1-5

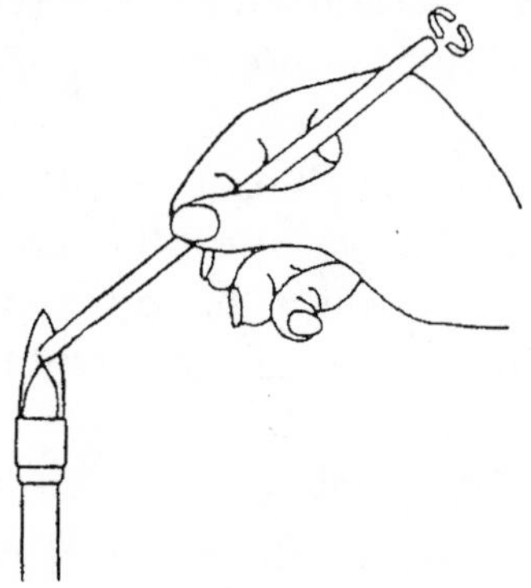

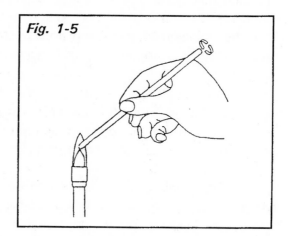

Firepolishing. Following *Figure 1-5*, hold the tubing at about 30° to the horizontal in the hottest part of the flame. Rotate the tubing and observe it carefully. As it approaches its melting point, a bright sodium flame will be observed. Continue to rotate it until it barely melts. Too much melting will begin to constrict the tube opening. Put the tubing down on a wire gauze until it has cooled and firepolish the other end (***Caution: One of the most common lab accidents is the burning of fingers and hands on glass that has been picked up without sufficient cooling.***)

Bending tubing. Chemists often have to make their own specialized pieces of glassware and it is very useful to have some experience with bending glass. Turn off the Bunsen burner and insert the flame spreader (often called a wing tip). Relight just as you did without the flame spreader and adjust the flame as in *Figure 1-6*. Hold both ends of the tube, place it in the hot region of the flame and rotate rapidly. A relatively even yellow glow indicates even heating. After the tube has softened, remove it from the flame, bend it to a right angle and hold it steady for a few seconds. Place it on the gauze for cooling. Do not be disappointed if your bend is not too aesthetically appealing as good glassworking takes many hours of practice.

Fig. 1-6

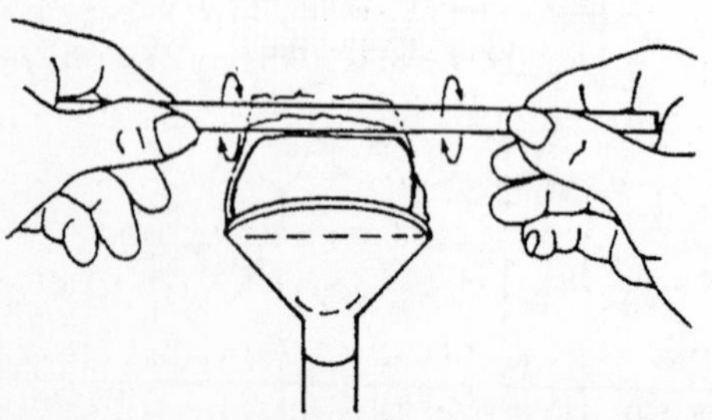

Fig. 1-7

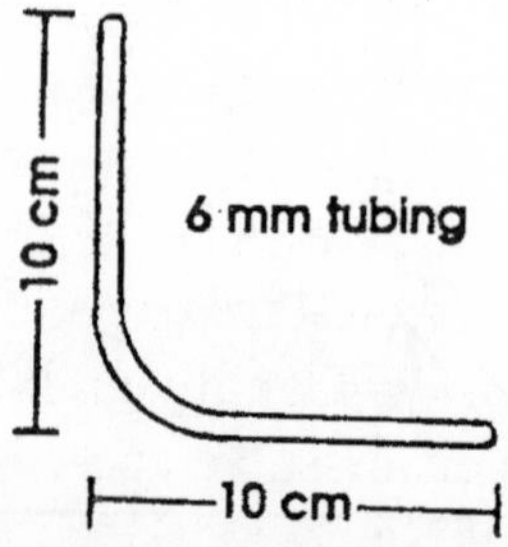

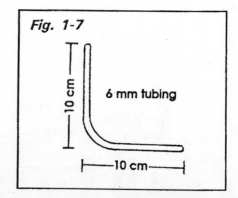

D. Rate of mixing. Mount a 250 mL beaker half full of water on a wire gauze above a Bunsen burner. Heat the water to boiling. Using beaker tongs, pour 50 mL of the hot water into a 150 mL beaker. Add 50 mL of room temperature water to a second 150 mL beaker. Use a dropper bottle of food coloring to add a drop of food coloring to each beaker. Hold the dropper over the lip of each beaker, squeeze out one drop and allow it to fall into the water. Try to add the drop to each beaker in the same way. Do not disturb the beaker in any way but carefully observe the beakers for several minutes and then look at them again in about a half hour.

E. Tearing paper. Try to tear a piece of paper (newspaper is preferable) first vertically and then horizontally. Attempt to make the tears fairly rapidly as opposed to doing it in very small sections with your fingers restricting the direction. Describe and explain your observations.

F. Chemistry reference. This exercise involves the first use of the *Handbook of Chemistry and Physics* in the laboratory portion of this course. Before using the *Handbook*, write down what you think are the three most abundant gases in the atmosphere (not counting water vapor which varies from 0% to 4%). Now look up the atmospheric content (do not change your original answer as you will not lose points for being wrong) in the *Handbook* and record the names of the four most abundant gases and their percentages. [Comment: there is a relationship between this section (F) and a later section (K)].

G. Terminology. Terminology and nomenclature are extremely important in chemistry. An understanding of the language of chemistry makes it much easier to communicate with other chemists and to understand their observations and results. In fact without a working ability with chemical terminology and nomenclature, the transfer of knowledge is close to impossible. This exercise will provide an experience with some of the descriptive terms used routinely by chemists. Early in your chemistry textbook and perhaps in the glossary there should be a discussion of each of the terms used here and in the Prelaboratory problem number 7. For this exercise, find at least one example of each term in your laboratory and describe the example. If appropriate, give its use and location. It is permissible to use observations of the exercises in other parts of this experiment as examples.

H. Solutions. In this procedure, you will apply terminology to some observations on solutions of sodium tetraborate and on the temperature dependence of the solubility of this compound. Add about 200 mL of water to a 400 mL beaker. Place the beaker on a hot plate or on a wire gauze supported above a Bunsen burner and heat the water to boiling. *[Caution: Always use beaker tongs when manipulating beakers of hot liquids. Do not use crucible tongs for beakers.]* While waiting for the water to boil, add about 0.05 gram of sodium tetraborate [$Na_2B_4O_7{\cdot}10\,H_2O$] to about 5 mL of water in a 18 ×150 mm test tube. Mix the contents of the tube by firmly grasping the test tube between your thumb and forefinger of one hand and striking the bottom of the test tube vigorously and frequently with the forefinger of your other hand. ***Never*** put a thumb over the mouth of the test tube to avoid spilling when shaking. If the method above does not achieve adequate mixing, insert a cork or rubber stopper into the tube and shake as you hold the stopper down with your thumb. Continue mixing until changes are no longer observable. Does the $Na_2B_4O_7{\cdot}10\,H_2O$ completely dissolve and what terminology can now be applied to the mixture? Add an additional 0.05 g of $Na_2B_4O_7{\cdot}10\,H_2O$ and repeat the mixing and observing process. Now add an additional 2 grams of $Na_2B_4O_7 \cdot 10\,H_2O$ to the solution and attempt to repeat the dissolving process. If it doesn't dissolve, put the test tube in the beaker of boiling water and stir the mixture in the test tube with a stirring rod until the sodium tetraborate dissolves. Place the test tube in a rack or beaker and allow it to cool to room temperature. After several minutes, report your observations or, if nothing happens, scratch the inside of the tube with a glass rod and let it sit for several minutes and then report your observations.

I. Colors. When an atom loses or gains electrons it becomes an ion. Positive ions are called cations and negative ions are called anions. In this experiment, you will study dilute solutions of compounds which have ionic bonds. This means that when the compounds dissolve in water, dissociation into cations and anions occurs. Ions can be colored or colorless and generally the color of an ion does not depend on its partner [sodium chloride and sodium nitrate solutions are colorless but copper(II) sulfate and copper(II) nitrate solutions are blue because sodium, sulfate, and nitrate ions are colorless and copper(II) is a blue ion]. If we had solutions available for you containing all of the commonly encountered ions, you would notice that most of the common ions are colorless. ***This makes it very useful for identification purposes for you to observe and remember the names and colors of the colored ions.*** Often you can suspect upon visual inspection that an unknown blue solution contains copper(II) ions.

Two sets of samples will be provided in sealed vials. Each sample will be an ionic compound dissolved in water with the concentrations all about the same (0.1 M). For the first set, you should focus your attention on the colors of the cations. Assume that the anions (either chloride, nitrate or sulfate) are colorless and look for a correlation between color and position in the periodic chart. For the second set, focus your attention on the colors of the anions (sodium and potassium ions are coloress).

Set 1:

aluminum nitrate	iron(III) chloride	potassium chloride
barium chloride	lead(II) nitrate	silver nitrate
calcium chloride	lithium chloride	sodium chloride
cerium(III) nitrate	magnesium chloride	strontium chloride
cerium(IV) sulfate	manganese(II) chloride	tin(II) chloride
chromium(III) chloride	mercury(I) nitrate	tin(IV) chloride
cobalt(II) chloride	mercury(II) nitrate	zinc nitrate
copper(II) chloride	nickel(II) chloride	

Set 2:

sodium acetate	potassium ferricyanide	potassium permanganate
sodium bromide	potassium ferrocyanide	sodium phosphate
sodium carbonate	sodium hydroxide	sodium sulfate
sodium chlorate	sodium iodate	sodium sulfite
sodium chloride	sodium iodide	sodium thiocyanate
sodium chromate	sodium nitrate	sodium thiosulfate
sodium dichromate	sodium oxalate	

J. Chemical reactions. Upon mixing two solutions, the four common observations that indicate that a chemical change has occurred are: formation of an insoluble product (precipitate), bubbles (or evolution of a gas), heat, a color change. The absence of these observations often but not always means that there has not been a chemical reaction as a result of the mixing. Negative results are very common and just as important as positive results and must be appropriately recorded. You will prepare 5 different mixtures, make observations and determine if a reaction has occurred. Pour **about** 2 mL of the first solution into a test tube and add about 2 mL of the second solution, mix and observe. Notice that there are times when accuracy and precision are extremely important in chemistry and other situations where the results do not change over a wide range of amounts. For the latter cases, it is a waste of time to spend long amounts of time carefully measuring the amounts. These reactions are examples of cases where 2 mL means between about 1 mL and 3 mL and therefore can be estimated. It might be wise to measure two mL once with a 10 mL graduated cylinder and transfer it to a test tube to give you an idea of the volume.

System	Solution 1	Solution 2
A	calcium chloride (0.1 M)	sodium carbonate (0.1 M)
B	hydrochloric acid (3 M)	sodium hydroxide (3 M)
C	calcium chloride (0.1 M)	potassium nitrate (0.1 M)
D	sodium carbonate (1 M)	hydrochloric acid (3 M)
E	iron(III) chloride (0.1 M)	potassium thiocyanate (0.1 M)

K. Mystery flask. Add the solutions below to a 125 or 250 mL Erlenmeyer flask, swirl until mixed and allow to stand without agitation.

15 mL of a dextrose (glucose) solution (80 g/L)
15 mL of a potassium hydroxide solution (64 g/L)
10 drops of a methylene blue solution (0.4 g/L)

Record all of your significant observations. Vigorously swirl the solution for several seconds and again record your observations. Repeat the sequence as often as you desire but focus your attention on the change that occurs when you swirl the mixture. Save this solution in case you decide later that you want to experiment further with it.

L. Classic Burning Candle Experiment. Light a candle that is at least 6 cm long, drip some melted wax into the middle of a 100 mm x 20 mm Petri dish and set the candle into the melted wax (*Figure 1-8*). After the wax solidifies, fill the Petri dish with water and ignite the candle. Place an inverted gas collecting bottle over the candle and stand it in the Petri dish. Observe the flame and the water level as soon as the bottle has been put in place. Measure the height of the water in the bottle after no more change is apparent. Also devise a method to determine the percentage of the volume filled by the water and perform the measurement.

One of the most important criteria for the testing of a scientific hypothesis is **reproducibility**. One hypothesis that was suggested many years ago to explain the observations in this experiment is that the burning of the candle uses up the oxygen in the bottle and the water rises to replace the used up oxygen. Repeat the experiment a number of times until you are confident the results are either reproducible within experimental error or not reproducible within experimental error. In the event that you come to the latter conclusion, try to come up with a new hypothesis to explain your observations. If possible, devise experiments to test your new hypothesis and with the permission of the instructor, perform the experiments.

Fig. 1-8

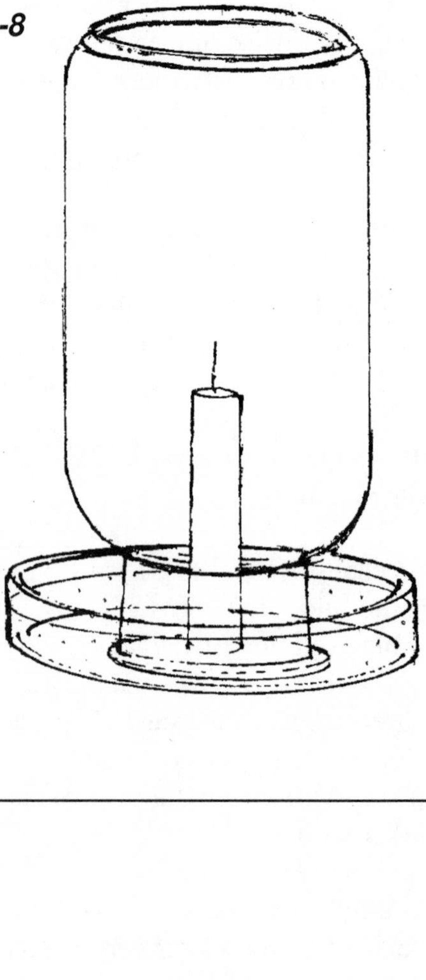

Name________________________________Date_________Lab Section_________

Prelaboratory Problems - *Experiment 1* - Early Explorations and Terminology

The solutions to the starred problems are in ***Appendix 4***.

1. What color is water? ___________

Problems 2 - 5 describe some observations you have probably made. But have any of the observations stimulated you to the point where you asked a question about them? Try to come up with a question now and suggest an explanation.

2. Popcorn pops when heated sufficiently.

3.* Vinegar and oil do not mix.

4. Your laboratory drawer has Erlenmeyer flasks and beakers that hold similar volumes.

5.* The necks of volumetric flasks and pipets have very small diameters.

6. Give observations for *Figure 1-9*.

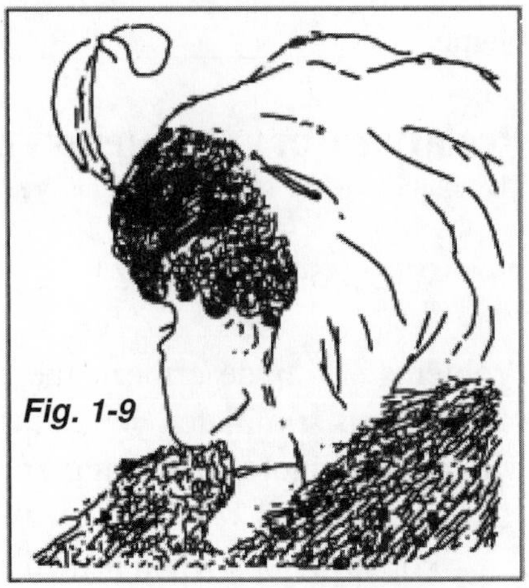

Fig. 1-9

7. Classify the following items a - r using the number codes for the terms below. In some cases, more than one term might be applicable.

1 substance	4 homogeneous mixture	8 intensive physical property
2 element	5 heterogeneous mixture	9 extensive physical property
3 compound	6 saturated solution	10 physical change
	7 unsaturated solution	11 chemical change

a.* gold	________	g. density	________	m. rusting	________
b.* vinegar	________	h. water	________	n. iodine	________
c.* vinegar & oil	________	i. sodium chloride	________	o. orange juice	________
d.* volume	________	j. evaporation	________	p. vodka.	________
e.* melting point	________	k. smog	________	q. coal burning	________
f.* dilute salt water	________	l. freezing of water	________	r. carbonated soda	________

8. Compounds are composed of two or more elements. Are the properties of compounds something like an average of the properties of its component elements? Consider for example sodium chloride (NaCl) and iron(III) oxide [commonly called rust (Fe_2O_3)]. Explain your answer.

Name________________________________Date_________Lab Section_________

Results and Conclusions - *Experiment 1* - Early Explorations and Terminology

A. The meniscus.

1. Description:

2. Drawing

3. Question about the meniscus:

B. The candle flame.

1. Observations (include comments on states of matter, physical and chemical changes):

2. What do you think is sustaining the flame (e.g., burning of the wick, the solid wax, the liquid wax or wax vapor)? Explain your answer.

3. Further observations:

   a. beaker partially over burning candle -

   b. stirring rod next to extinguished wick -

   c. burning match approaching extinguished wick -

4. Was your first explanation (question 2) correct or do you want to modify it or suggest a new one? Explain your answer.

C. The Bunsen burner.

1. Draw a picture of the flame. Based on your results with the splint and the gauze, indicate the hottest and coolest regions of the flame.

D. Rate of mixing.

1. Initial observations:

    a. room temperature water

    b. hot water

2. Observations about ½ hour later:

    a. room temperature water

    b. hot water

3. When two aqueous solutions are introduced into the same container, is stirring needed to achieve a homogeneous system? Explain your answer.

4. How would you stir a solution freshly prepared in a volumetric flask?

E. Tearing paper

1. Describe your observations when you attempted to tear the paper vertically.

2. Describe your observations when you attempted to tear the paper horizontally.

3. Suggest an explanation for any differences observed between vertical and horizontal tearing.

F. Chemistry reference.

Most abundant gases in the atmosphere (excluding water vapor)

Number	Guesstimate	Handbook (edition , page)	%
1	_______	_______	___
2	_______	_______	___
3	_______	_______	___
4	_______	_______	___

Explain why the value for the percentage of carbon dioxide is related to the possibility that the earth could experience a Greenhouse Effect?

G. Terminology (Describe at least one example of each of the following in your laboratory.)

1. substance

2. element

3. compound

4. homogeneous mixture

5. heterogeneous mixture

6. saturated solution

7. unsaturated solution

8. chemical change

9. Shortly after the Chernobyl nuclear reactor accident, some people took potassium iodide tablets to dilute the radioactive iodide in their bodies and diminish its retention in the thyroid gland. Although many experts questioned this practice, little harm was probably caused because of the relatively low toxicity of KI. However, on May 12, 1986, *Newsweek* incorrectly captioned a photo that showed a child apparently receiving KI with "On Alert: Administering iodine to Polish children." Critically evaluate the mistake and possible consequences.

H. Solutions. [Be sure to use appropriate terminology from the section above for all of these questions.]

1. Describe the test tube contents after you have added 0.05 gram of sodium tetraborate decahydrate and stirred..

2. Describe any significant changes after an additional 0.05 gram of $Na_2B_4O_7 \cdot 10\,H_2O$ is added and mixed.

3. How can the sodium tetraborate - water system be distinguished from a compound?

4. Describe your observations when an additional 2 grams of $Na_2B_4O_7 \cdot 10H_2O$ is added to the solution.

5. What happens to the mixture in *#4* when it is heated?

6. Describe what happens when the solution from *#5* is allowed to cool. Are your observations consistent with your expectations? Explain your answer.

I. Colors.

1. List the colored cations and their colors.

cation	color	cation	color	cation	color
______	______	______	______	______	______
______	______	______	______	______	______

2. Can you make any generalizations about color versus position in the periodic chart?

3. List the colored anions and their colors.

anion	color	anion	color	anion	color
______	______	______	______	______	______
______	______	______	______		

J. Chemical Reactions.

Reaction	Observations
A	
B	
C	
D	
E	

K. Mystery flask.

1. Observations upon standing:

2. Observations upon swirling:

3. Suggest an explanation for the change that occurs upon swirling.

4. How could you test your explanation (it might be possible with your instructor's approval to actually perform the test).

L. Classic Burning Candle Experiment.

1. Trial 1 water level (distance above dish water level) ________
2. Volume of water that moved into bottle ________
3. Volume of bottle ________
4. Percentage of bottle filled by water ________
5. Additional trial water levels ______ ______ ______ ______
6. Based on the percentage of the bottle filled by the water, is it likely that the water simply displaced used up oxygen? Explain your answer.
7. Based on the reproducibility of the results, does the hypothesis that the water is displacing used oxygen fit the observations? Explain your answer
8. If your answer to number 7 was negative, suggest another hypothesis to explain your observations.
9. Suggest and if possible perform experiments to test your hypothesis in number 8. Describe the experiments and their results. Do the experiments support or refute your new hypothesis?
10. Some of the *Learning Objectives* of this experiment are listed on the first page of this experiment. Did you achieve the *Learning Objectives*? Explain your answer.

Experiment 2

SEPARATION OF MIXTURES

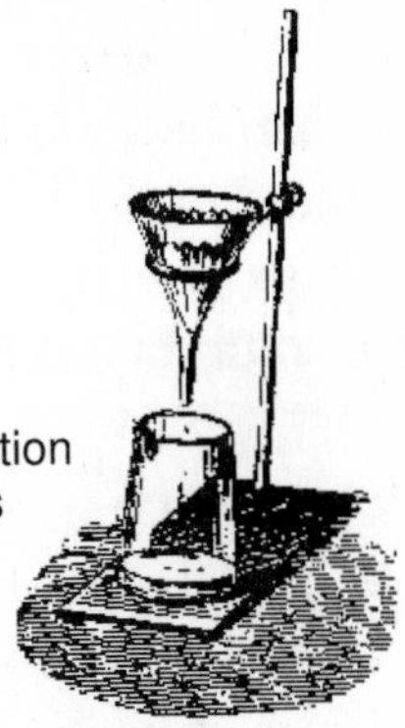

Learning Objectives

Upon completion of this experiment, students will have experienced:
1. The evaporation of water from salt water.
2. The separation of a solid from a liquid using gravity filtration.
3. Recrystallization and use of vacuum filtration.

Text Topics

Physical properties, separations of mixtures using physical properties, temperature effects on solubility (for correlation to some textbooks, see page ix).

Notes to Students and Instructor

The vanillin recrystallized in this experiment will be saved for ***Experiment 3*** where its percent recovery and purity will be determined.

Discussion

Imagine yourself by a river in the mountains. You are thirsty but is it safe to drink the water? Probably not because the river is not just water but is a mixture of several substances and probably contains bacteria that can cause intestinal problems. Almost everything that we encounter in nature is a mixture and this adds a challenge to the work of chemists. Analytical chemists may need to know what substances are present (e.g., are pesticides in a water supply?) and this usually involves separation and purification before some type of identification test can be performed. Most physical and chemical tests on mixtures do not give results that can be easily interpreted. The chemist may also need a substance for use as a starting material in a chemical reaction. The presence of impurities may affect or even eliminate the desired reaction. Even chemicals obtained from the stockroom are not absolutely pure with grades of purity generally ranging between 90 and 99.99%. Sometimes small amounts of impurities can be tolerated and in other cases, they must be removed. Would you want as much as 0.1% of a toxic substance in a medicine?

Many separation and purification techniques have been developed. You will use several during this course and three important ones in today's experiment. Try to consider the applicability of each technique as you use it. Can it be used on large samples or only small ones? Will it work to remove large amounts of impurities or only trace constituents? Will it work on any of the three common states of matter or only one? The three techniques you will use today all take advantage of differences in physical properties of the substances being separated. The evaporation of water from salt water to leave behind the desired sodium chloride utilizes the differences in boiling points of water and sodium chloride. The gravity filtration to collect calcium carbonate takes advantage of the very low solubility of calcium carbonate in water and the ability of filter paper to allow passage of a liquid but not a solid. Finally the recrystallization of vanillin relies on the observation that solubility of a solid in a liquid often increases with increasing temperature.

Procedure

A. Evaporation: Determination of the mass percent of sodium chloride in a saturated sodium chloride solution. The amount of sodium chloride dissolved in a saturated solution can be determined by evaporating a weighed amount of a saturated solution to dryness. Dryness is confirmed by repeated heating and weighing cycles until a constant mass is achieved. Decant (be careful not to agitate) about 6 mL of a saturated sodium chloride solution into a 10 mL graduated cylinder. Weigh a clean, dry, evaporating dish to the nearest 0.01 g (preferably 0.001 g) and add the 6 mL of NaCl solution to it. Reweigh the dish and its contents.

Put a 400 mL beaker about half full of water on a wire gauze above a Bunsen burner. Suspend the evaporating dish in the 400 mL beaker. Boil the water (add water as needed to the beaker to maintain a reasonable level) until the evaporating dish attains apparent dryness (about 20 minutes). Using beaker tongs, remove the beaker and evaporating dish from the flame. Holding it with crucible tongs, place the evaporating dish on a wire gauze and gently flame it for about 3 minutes and allow the dish to cool. Weigh the dish to the at least the nearest 0.01 gram. Again flame the evaporating dish gently for several minutes, cool, and weigh. Repeat the process until successive weighing differences are less than 0.01 g. Calculate the mass percent of NaCl in the saturated solution.

Fig. 2-1

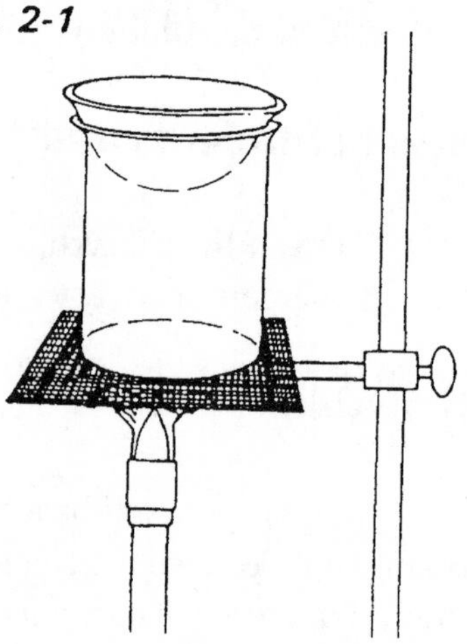

B. Filtration: Collection of calcium carbonate. The addition of aqueous calcium chloride to aqueous sodium carbonate results in a double replacement reaction or an exchange of the positive and negative partners of two ionic compounds.

$$Na_2CO_3(aq) + CaCl_2(aq) = 2\,NaCl(aq) + CaCO_3(s)$$

As calcium carbonate is not soluble in water, it precipitates out of the solution. The solid calcium carbonate can be separated from the solution by gravity filtration.

Pour about 10 mL of 1.0 M sodium carbonate solution into your graduated cylinder and transfer this solution to a 150 mL beaker. Add 10 mL of 1.0 M calcium chloride solution to the beaker, stir, and report your observations. Save for the filtration.

Fig. 2-2

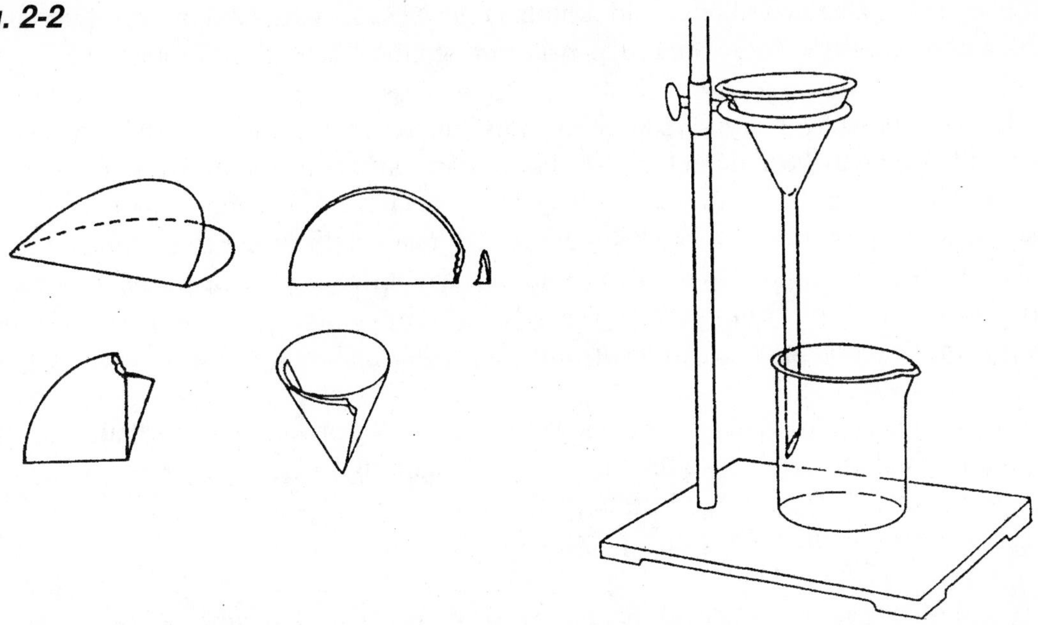

Select a piece of filter paper that is the appropriate size for your long stem funnel. After being folded and opened into a cone, it should fit slightly below the glass rim of the funnel. For the most common size funnel, this paper will have a 12.5 cm diameter. Fold a piece of filter paper in half. Tear off about a half centimeter piece from one corner then fold it into quarters. Open up one pocket of the filter paper so that it forms the shape of a complete cone. Put it into your funnel and wet it thoroughly with deionized water from your wash bottle so that it adheres uniformly and seals to the inside wall of the funnel. Swirl the reaction mixture and transfer part of it to the filter. Make sure the liquid does not get closer than 0.5 cm to the top edge of the filter paper. Continue adding the mixture until all of it has been added to the funnel. With a wash bottle, squirt some water into the beaker and transfer this wash water to the funnel. Repeat this process two more times. Allow the funnel to drain. When the dripping has virtually stopped, scrape the precipitate onto a watch glass with a stirring rod. Using a medicine dropper, rapidly add about 10 drops of 6 M HCl to the precipitate. Describe your observations.

Whenever a chemical synthesis is attempted (in this case calcium carbonate), it is necessary to verify that the intended product has been isolated. The addition of an acid to a carbonate yields carbon dioxide gas. The observation that a gas evolves provides some evidence but certainly not proof that a carbonate has been synthesized.

Try adding a few drops of 6 M HCl to a small piece of egg shell. Report your observations and conclusions concerning the composition of the egg shell.

C. Recrystallization: Purification of vanillin. The solubility of many solids increases dramatically as the solvent temperature increases. A barely saturated solution of the solid in a hot solvent is prepared and allowed to cool. The solubility decreases as the temperature drops and causes recrystallization to occur. The crystals are collected by vacuum filtration hopefully in a purer condition than before they were dissolved in the solvent. Insoluble impurities are removed by filtration of the hot saturated solution and soluble impurities stay dissolved in the solvent and do not crystallize upon cooling. These impurities pass through the final cold filtration.

In the following recrystallization of vanillin, recognize that the difficult part of the experiment has already been done for you. For a recrystallization, a solvent must first be found which will dissolve the vanillin when hot but not when cold. Appropriate amounts must also be chosen. Water turns out to be a good solvent for the recrystallization of vanillin and reasonable amounts have been determined by experimentation and reported to you. *Be sure to save the vanillin that you purify today for* ***Experiment 3*** *where you will determine the percent recovery from the recrystallization and the success of the attempted purification.*

Weigh into a 125 or 250 mL Erlenmeyer flask about 2 g of vanillin to the balance limit. Add about 60 mL of deionized water to the flask and stir vigorously. Using a Bunsen burner, heat the solution just to the boiling point and stir until all the solid dissolves. Allow the solution to cool on your desk for several minutes and then put the flask into an ice bath.

Crystallization should occur on its own but occasionally supersaturation occurs and crystallization needs a little assistance. If this happens, try rubbing the inside of the flask beneath the liquid line with a glass rod hard enough to make a grinding sound. If this does not work, seeding with a small crystal of impure vanillin or better yet, a crystal of another student's recrystallized vanillin should initiate the process. After crystallization is complete and the flask has been cooled to about 5°C, isolate the vanillin using vacuum filtration. Obtain a Buchner funnel and filter flask and assemble them as in *Figure 2-3*. Select a piece of filter paper that has been precut to just fit in the bottom of the Buchner funnel.

Connect the filter flask to an aspirator (Note: It is good lab practice to include a trap between the aspirator and the filter flask as water sometimes backs up from the aspirator into the filter flask. However, in this case you are interested in collecting the precipitate only and the filtrate will be discarded. If the filtrate is needed, be sure to use a trap to prevent contamination.) and wet the paper in the filter with deionized water from a wash bottle. Turn the aspirator on full and transfer the contents of the Erlenmeyer to the Buchner funnel. Rinse the Erlenmeyer flask out once or twice with 5 mL of ice-cooled water and use this water to wash the crystals in the funnel. Weigh a piece of filter paper. Empty the crystals onto the weighed filter paper and place them in your desk for drying, being sure they cannot spill. You will weigh the product and check its purity next week.

Fig. 2-3

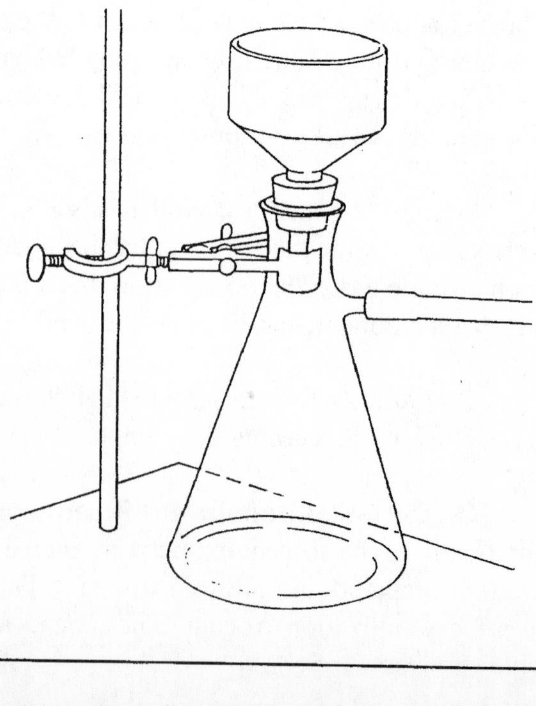

Name________________________________Date__________Lab Section__________

Prelaboratory Problems - *Experiment 2* - Separation of Mixtures

The solutions to the starred problems are in ***Appendix 4***.

1.* 6.7 mL of a potassium chloride solution was added to a 54.730 g evaporating dish. The combination weighed 61.945 g. After evaporation the dish and contents weighed 55.428 g.

a. What was the mass percent of potassium chloride in the solution? ____________

b. If the actual mass percent of potassium in the above solution was 10.00%, what was the percentage error of the above measurement?

____

c. Why was evaporation used to determine the mass percentage of potassium chloride in the solution rather than filtration or recrystallization? Explain your answer.

d. Compare the mass percent calculated in # 1-c above to the value for a saturated solution in ***Appendix 3***. Was the solution prepared above saturated? Explain your answer.

2. 8.5 mL of a sample of sea water solution was added to a 44.317 g evaporating dish. The combination weighed 52.987 g. After evaporation the dish and contents weighed 44.599 g.

a. What was the mass percent of dissolved solids in the sea water?

____

b. The actual sodium chloride content in the sea water was 2.69%. If it had been assumed that the dissolved solid(s) consisted only of sodium chloride, what percentage error would have resulted? [Caution: the answer is not 0.56%]? The numbers included here are for a typical sea water sample. In addition to NaCl, typical sea water contains about 0.56% by mass of compounds (primarily chlorides) of magnesium, calcium and potassium.

____

c. Could the dissolved solids have been isolated using either filtration or recrystallization? Explain your answer.

3. Aspirin (acetyl salicylic acid) can be synthesized by reacting acetic anhydride with salicylic acid. Assume the reaction was run and 1.75 grams of a solid product was obtained. Recrystallization of the 1.75 g mixture yielded 1.50 g of aspirin.

   a. What was the percentage recovery of aspirin from the recrystallization?

   ____

   b. Why was recrystallization used for the purification of aspirin rather than filtration or evaporation? Explain your answer.

4. The reaction of aqueous solutions of barium chloride and sodium sulfate results in the formation of a precipitate of barium sulfate. Sodium chloride is also formed but is soluble in water. Why would you use filtration to isolate the barium sulfate rather than evaporation or recrystallization? Explain your answer.

$$BaCl_2(aq) + Na_2SO_4(aq) = BaSO_4(s) + 2\,NaCl(aq)$$

Name________________________________Date_________Lab Section_________

Results and Discussion - *Experiment 2* - Separation of Mixtures

A. Evaporation: Determination of the mass percent of sodium chloride in a saturated sodium chloride solution.

1. Mass of evaporating dish _________

2. Mass of dish and sodium chloride solution _________

3. Mass of dish and sodium chloride (1st heating) _________

4. Mass of dish and sodium chloride (2nd heating) _________

5. Mass of dish and sodium chloride (3rd heating if necessary) _________

6. Mass of saturated sodium chloride solution _________

7. Mass of sodium chloride _________

8. Mass percent of sodium chloride in saturated solution _________

9. Calculate the percentage deviation between your value and an accepted literature value of 26.4%. _________

10. Is sea water close to being a saturated sodium chloride solution? Explain your answer [Hint: See ***Prelaboratory Problem*** #2-b in this experiment.]. _________

11. How could you tell that the original salt solution was saturated?

B. Filtration: Collection of calcium carbonate.

1. What did you observe when you mixed solutions of sodium carbonate and calcium chloride? ______________

2. What did you observe when you added HCl to the product of your reaction? ______________

3. What conclusion did you draw from your observations in #2?

4. What did you observe when you added HCl to a piece of egg shell? ______________

5. What conclusion did you come to about the egg shell from your observations in #4?

6. Buildings and statues contain significant percentages of carbonates. Based on your observations in #2 and #4, what effects could acid rain have on these structures?

C. Recrystallization: Purification of vanillin.

1. Mass of crude vanillin (also enter on page 41) ________

2. Mass of paper used for storing vanillin (also enter on page 41) ________

3. Why is the vanillin solution cooled in an ice bath before vacuum filtration?

4. What additional step could be added to this procedure to remove impurities insoluble in the solvent? ______________

5. Some of the *Learning Objectives* of this experiment are listed on the first page of this experiment. Did you achieve the *Learning Objectives*? Explain your answer.

Name___________________________________Date__________Lab Section__________

Postlaboratory Problems - *Experiment 2* - Separation of Mixtures

(Note: It might be useful to refer to the solubility chart in ***Appendix 3*** for some of these questions.)

A. Evaporation: Determination of the mass percent of sodium chloride in a saturated sodium chloride solution.

1. What criteria do you think were used is the selection of this system?

   a. Why was water chosen for the solvent?

   b. Why was sodium chloride chosen for the solute?

   c. Suggest reasons why the following were not chosen for the solute:

   calcium carbonate ($CaCO_3$)

   sodium cyanide (NaCN)

   gold chloride ($AuCl_3$)

   hydrogen chloride (HCl)

2. For the evaporation of NaCl, why is the heat-cool-weigh cycle repeated until constant mass is attained?

3. Would evaporation be a very useful separation technique for solutions that contain more than one solute? Explain your answer.

B. Filtration: Collection of calcium carbonate.

Why was gravity filtration used instead of evaporation to isolate the calcium carbonate (consider question *A-3* above)?

C. Recrystallization: Purification of vanillin.

1. What criteria do you think were used to select the solute and solvent for this experiment?

2. a. Suggest any limitations to the use of recrystallization as a purification method for a solid.

   b. Suggest reasons why recrystallization from water would probably not be a suitable procedure for purifying the following substances:

   sodium chloride ($NaCl$) (Hint: Look up its solubility and also the dependence of the solubility on temperature in your textbook.)

   calcium carbonate ($CaCO_3$)

   ethanol (C_2H_6O)

3. Gravity and vacuum filtration separate insoluble solids from a liquid phase. The choice depends on conditions. Suggest criteria you would apply to choose between them.

4. What happens to the salt concentration in a saturated solution if the water is allowed to evaporate (assume constant temperature is maintained)? Explain your answer.

5. Suggest any ways you can think of to improve any part(s) of this experiment.

Experiment 3

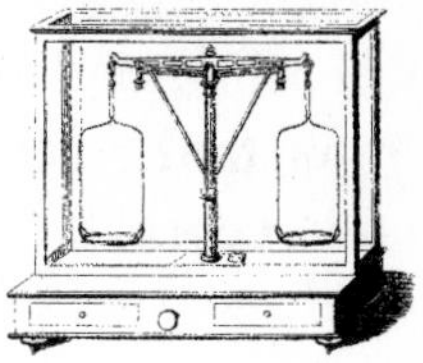

MEASUREMENTS AND IDENTIFICATION TECHNIQUES

Learning Objectives

Upon completion of this experiment, students will have experienced:

1. The determination of the percent recovery and purity of recrystallized vanillin (from ***Experiment 2***).
2. The determination of melting points using the capillary technique and the effects of impurities on melting points.
3. The determination of the density of a solid using two methods.

Text Topics

Significant figures, melting points, density (for correlation to some texts, see p. ix).

Notes to Students and Instructor

With proper preparation, this experiment and part of the next one can be performed in one laboratory period.

Discussion

The importance of careful and complete descriptive observations cannot be overstated. Often, however, it is desirable to seek quantitative relationships between variables. To accomplish this, we need to perform measurements using a consistent set of units. We know that the volume of a gas will increase as the temperature is increased but can we predict how much it will increase? By measuring the volume of a gas as a function of temperature, it is possible to develop an empirical relationship between volume and temperature. Application of the kinetic theory yields the same equation providing support for the validity of the theory.

For units, scientists have joined most of the rest of the world in adopting the convenient and rational metric system. Prefixes relate different length, mass and volume systems by powers of ten (rather than factors of 4, 12, 16, 36 or 5280). Volume is defined in terms of distance cubed (1 L = 1 dm^3 therefore 1 mL = 1 cm^3) instead of independently (how many gallons in 1 ft^3?). The mass scale has been established so that at 3.98°C, water has a density of exactly 1 g/cm^3 or 1 g/mL (What is the density of water in lbs/ft^3 or lbs/gal?). The metric temperature scale is in Celsius degrees with the freezing and boiling points of water logically defined as 0°C and 100°C respectively.

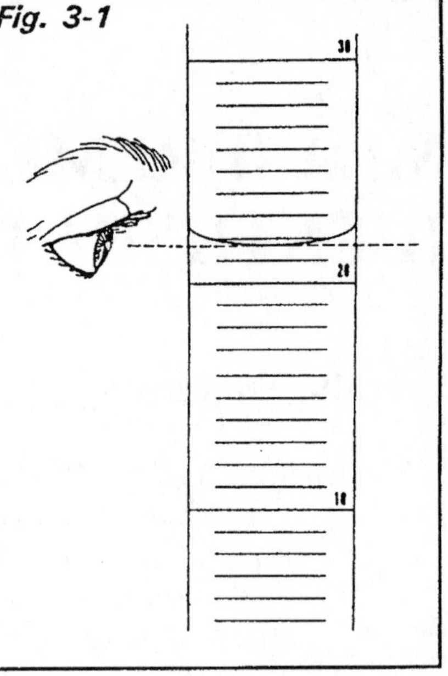

Fig. 3-1

This experiment will focus on techniques of reading, recording, and utilizing measurements. **One of the most important rules to remember is that you should estimate and record 1 digit beyond the last set of graduations on your measuring instrument** (this rule doesn't apply to many of the new high tech instruments that have digital readout although sometimes fluctuation in the last digit is observed and you then do estimate the most probable value of that digit). The estimated digit does have significance and is therefore counted as a ***significant figure***. In *Figure 3-1*, the reading should be recorded as 21.7 mL and not 22 mL. The bottom of the meniscus (see ***Experiment 1***) is obviously below 22.0 mL and above 21.5 mL. To write down 22 mL would be communicating that the value is between 21 and 23 mL. But it is clear that the reading provides more information than this. The value 21.7 mL communicates to the reader that the value is probably between 21.6 and 21.8 mL. A second rule is that ***when the estimated digit is a zero*** (this should happen about 1 out of every 10 readings), ***the zero must be recorded***. The thermometer below reads 23.0°C. 23°C is an inadequate recording of the temperature as it states that the reading is between 22 and 24°C but observation indicates that it is between 22.9 and 23.1°C

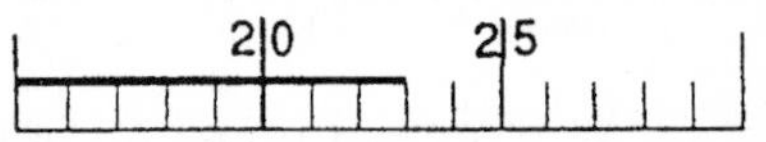

Fig. 3-2

Be especially careful with zeros when a decimal is not showing. If you say you are driving 90 km/hr, do you mean between 89 and 91 or between 80 and 100 (the former should be written 9.0×10^1 km/hr and the latter either 90 km/hr or preferably 9×10^1 km/hr). For more information on measurements, significant figures and rounding off, refer to your textbook.

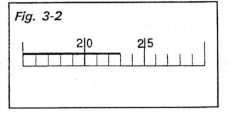

Intensive Physical Properties. Whether dealing with an unknown chemical extracted from a plant or the product of a chemical synthesis, chemical identification can often pose a significant challenge to the chemist. For previously characterized compounds, a comparison of properties of the unknown with those of knowns will usually make an identification possible. To be useful for identification purposes, a property must not depend on the amount of substance. Extensive properties such as mass and volume do depend on the amount of substance and are not useful for identification purposes. However, intensive properties such as melting points and densities do not depend on the amount of material and are and very useful for identification purposes. Melting points of substances are significantly depressed by many additives or impurities. Thus, in addition to being useful for identification purposes, melting points are useful in determining purity.

The density (mass/volume) of a substance is also an intensive property. 1.00 mL of mercury has a mass of 13.6 grams and a density of 13.6 g/mL. 10.0 mL of mercury has a mass of 136 grams and a density of 13.6 g/mL. Although density does require two measurements, mass and volume, both are relatively easy to measure. Comprehensive tables of densities and melting points have been compiled and both are commonly used for identification purposes.

Procedure

A. Percent recovery. Determine the percent recovery of vanillin by weighing the vanillin obtained from the recrystallization in ***Experiment 2***, dividing the amount by the original amount of crude vanillin and then multiplying by 100%. When you are finished with the vanillin, in *Parts A and B,* return your recrystallized vanillin as the vanillin can be used again in ***Experiment 2***.

B. Melting points. The melting point of a pure substance can be measured quickly using the capillary method (described later). The method is accurate to within a few tenths of a degree and requires only a small amount of sample. For identification purposes, the determination of the melting point usually narrows down the number of possible compounds to two or three. When a contaminated compound is melted, we should notice two distinguishing features. First, as with ice water contaminated by salt, the melting point is lowered by additives. Second, the melting process occurs over more than a few tenths of a degree range and is more properly called a melting range than a melting point. **It is very important that the total range be reported; that is the temperature at which the first minute amount of liquid appears to the temperature at which the sample is totally liquid.** The extent of depression and broadening of the melting range serves as a useful measure of the purity of a sample. Very roughly each 1% of impurity (up to about 10%) will depress the high value of the range about 1° below the melting point of the pure compound. Utilizing the melting point depression phenomenon, it is even possible to distinguish between two possible compounds with the same melting points if labeled samples of the two are available (See ***Prelaboratory Problem 5***).

The capillary technique is illustrated in *Figures 3-3* and *3-4*. Select a thermometer that covers the range -10°C to 110°C. Capillary tubes are very convenient sample holders as they are inexpensive and hold very small amounts of sample that can be easily observed during the melting process. Fill the capillary tube by pressing the open end onto the powdered sample until there is about a 0.5-1 cm length of sample in the tube. Now drop the capillary tube, sealed end down, through a 1 meter piece of 6 mm glass tubing that is being held on a hard surface. The impact of the capillary with the hard surface seldom results in breaking and causes the sample to drop to the bottom of the tube. Repeat the dropping procedure until the sample is packed in the bottom of the tube. Attach the capillary tube with a rubber ring (cut off a piece of rubber hose) to a thermometer with the sample even with the mercury bulb of the thermometer. Place the thermometer in a 250 mL beaker half full of water mounted above a Bunsen burner. Support the thermometer using a ***split*** rubber stopper supported by a clamp on a ring stand. Gently heat the water with **continual** stirring and observe the sample. As you approach the melting range, slow the heating rate to around 2°C/min. Heating the sample too rapidly near its melting range will result in large errors. At the first indication of sample melting, record the temperature to the nearest 0.1°C. Continue to slowly heat until the sample has totally liquified and record the high end of the melting range.

Using the technique described above, determine the melting ranges of the three samples below. [Hint: The melting ranges should occur between 60°C and 85°C. As the melting range is approached, the heating rate should be kept as slow as possible.] For the mixed sample, add approximately equal amounts of each compound to a mortar and grind them together with a pestle before inserting into a capillary tube. Don't forget to recycle your vanillin.

a. Impure vanillin from ***Experiment 2.***
b. Recrystallized vanillin from ***Experiment 2.***
c. 50% recrystallized vanillin + 50% phenyl carbonate

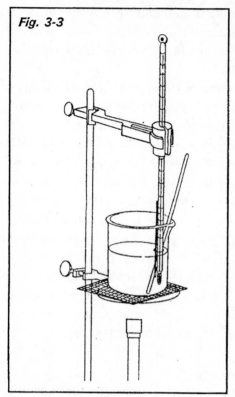

Fig. 3-3

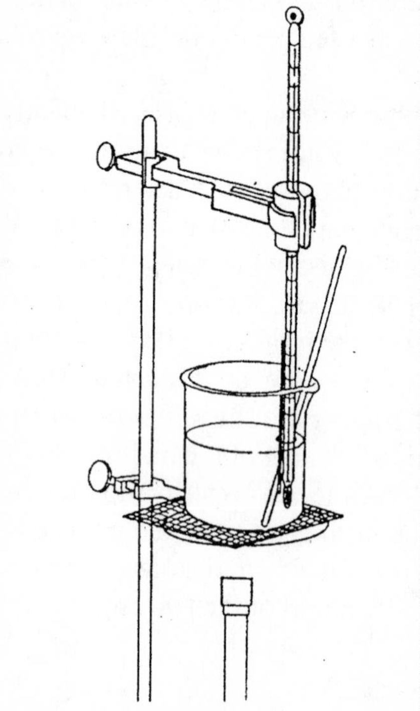

Fig. 3-4

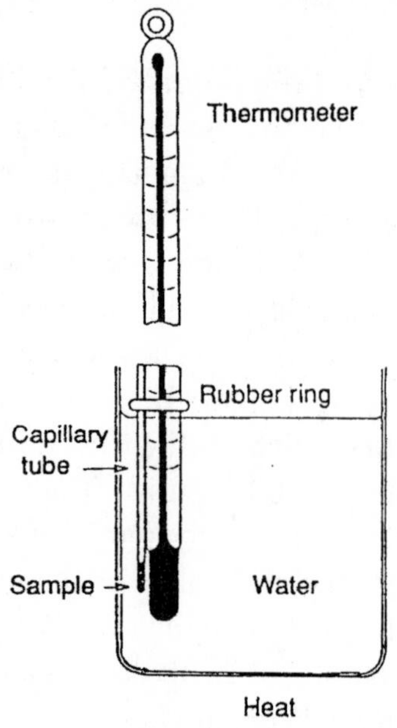

C. Density. The density, or mass/volume, of a metal cylinder will be calculated twice using two different methods to determine its volume. Then you will use the density to attempt to identify the metal. First use your balance to determine the mass of the metal cylinder to the limits of the balance (probably either 0.01 g or 0.001 g.).

1. Volume from linear measurements: For regularly shaped objects such as a cylinder, it is usually possible to measure the dimensions of the object and use a formula to calculate the volume of the object. The volume of a cylinder is calculated by multiplying π times the square of its radius times its length ($V = \pi r^2 L$). Using a metric ruler or preferably vernier calipers, measure the diameter and length of the cylinder and calculate its volume and density.

2. Volume from water displacement: A more general method for volume determination that will work with regular and irregular shaped objects is the method of water displacement. The water level in a graduated cylinder is recorded, an object submerged in the water and the new water level read. The difference in water levels is the volume of the object. Obtain an appropriately sized graduated cylinder for your metal cylinder and fill it about half full with water. Read the water level, tilt the graduated cylinder and slide the metal cylinder gently into the water being sure that it is completely submerged. Read the resulting water level and calculate the volume and density of the object.

3. Assume the metal cylinder is one of the following: aluminum, copper, iron or lead. Look up the density of each metal and determine the composition of your cylinder.

Name__________________________________Date__________Lab Section__________

Prelaboratory Problems - *Experiment 3* - Measurement and Identification

The solutions to the starred problems are in ***Appendix 4***.

1. Math review problems

a.* How many significant figures are in each of the following numbers?

405.0 _____ 0.0789 _____ 2.040 _____ 360 _____ 3.00×10^{10} _____

b. Express each of the following with three significant figures.

0.002537 ________ 12345000 ________ 0.07966 ________ 620.2 ________

c. Perform the following operations: 4.5×10^{3} cm + 1.6×10^{2} cm = _______

$(1.5\times10^{-3}\text{ mole})\dfrac{(6.022\times10^{23}\text{ molecules})}{1\text{ mole}}$ = ________ $\dfrac{1.80\times10^{-6}\text{ g}}{2.00\times10^{-3}\text{ mL}}$ = ________

2. List at least four distinct advantages of the metric system over "our" measurement system.

3. a.* Salt is often spread on icy roads in the winter. What is its function and how does it work?

b. Due to the use of salt on roads, two towns in Massachusetts were having problems with high salt content in ground water supplies. One town chose not to use salt one winter and the automobile accident rate actually went down. Suggest a reason for this unexpected result and comment on the validity of long range conclusions from this study.

c. When preparing home-made ice cream, salt is added to the ice-water mixture. What is its function and why does it work?

d. Antifreeze (commonly ethylene glycol or propylene glycol) is usually mixed with water in the radiator of a car. What is its function and how does it work?

4.* A 2.5 g sample of naphthalene (melting range 73.4-77.1°C) is recrystallized and 1.8 grams of purified naphthalene (melting range 80.5-80.6°C) are recovered.

a. What is the percent recovery? __________

b. Comment on the purity of the original naphthalene sample.

c. Comment on the success of the recrystallization.

5. Cinnamic acid and urea melt at 133°C. The melting point of an unknown is found to be 133°C. What experiments using melting range determinations could be done to determine if the unknown is cinnamic acid or urea?

6.* A 15.00 g metal sphere was found to have a diameter of 1.85 cm. The volume of a sphere is $V = (4/3)\pi r^3$. Calculate the density of the sphere and assuming that the sphere is made out of one of the elements, aluminum, chromium, iron, lead, titanium or zinc, determine its composition.

7. A 55.81 gram irregular object made out of one of the metals in #6 raises the water level in a graduated cylinder from the 17.6 mL level to the 24.7 mL level. What is the density and identity of the metal?

Name________________________________Date__________Lab Section__________

Results and Discussion - *Experiment 3* - Measurement and Identification

A. Percent Recovery

1. Mass of weighed filter paper __________

2. Mass of weighed filter paper plus recrystallized vanillin __________

3. Mass of recrystallized vanillin __________

4. Mass of impure vanillin (from last week) __________

5. Percent recovery __________

6. Suggest how you could have improved your techniques to increase your percent recovery.

B. Melting Point Determinations

1. Melting range of impure vanillin __________

2. Melting range of recrystallized vanillin __________

3. Melting range of 50% recrystallized vanillin + 50% phenyl carbonate __________

4. *Handbook of Chemistry and Physics* value for the melting range of vanillin __________

   and for phenyl carbonate __________
   (edition _______ pages _______)

5. Did your recrystallization significantly purify the vanillin? Explain your answer.

6. How can rapid heating while using the capillary melting point technique give erroneous results?

C. Density

1. Linear measurement method

   a. Identification number of metal cylinder ________

   b. Mass of metal cylinder ________

   c. Length of metal cylinder ________

   d. Diameter of metal cylinder ________

   e. Radius of metal cylinder ________

   f. Volume of metal cylinder ________

   g. Density of metal cylinder ________

2. Water displacement method (cylinder i.d. number and mass same as above)

   a. Initial volume of water in graduated cylinder ________

   b. Final volume of water in graduated cylinder ________

   c. Volume of metal cylinder ________

   d. Density of metal cylinder ________

3. Identity of metal

   Handbook of Chemistry and Physics values for the density of:

   a. aluminum ________

   b. copper ________

   c. iron ________

   d. lead ________

   e. Identity of metal ________

4. For a regular shaped object such as the metal cylinder used in this experiment, which of the two methods do you think is preferable for a density determination. Give reasons for your preference.

5. Some of the *Learning Objectives* of this experiment are listed on the first page of this experiment. Did you achieve the *Learning Objectives*? Explain your answer.

Name________________________________Date_________Lab Section_________

Postlaboratory Problems - *Experiment 3* - Measurement and Identification

1. Why is it important to determine the percent recovery for chemical processes? In addition to considering its importance in the chemical laboratory, consider its importance to a chemical company.

2. Why is it important to determine the melting ranges of the impure and recrystallized samples of vanillin?

3. Melting ranges are helpful for identification purposes and determining purity. Would the melting range of a new (never before reported) compound be worth determining and if so, why?

4. What properties must a solid have for the water displacement method to be useful?

5. For each of the following solids, describe how you would determine its density.

   a. a ball bearing

   b. 10 grams of granular zinc

   c. 10 grams of granular sodium chloride

   d. an irregular shaped piece of maple (about 10 grams)

   e. a lump of sodium (about 10 grams)

6. Another way of determining the volume of solid makes use of the fact that the difference in the mass of an object determined directly and suspended in water is the mass of the water displaced by the object. Division of this mass by the density of water gives the volume of the water displaced and therefore the volume of the object. Would this method have any advantages over the water displacement method used in this experiment?

7. Suggest further tests for distinguishing two metals that have densities within the experimental errors of each other.

8. Suggest any ways you can think of to improve any part(s) of this experiment.

Experiment 4

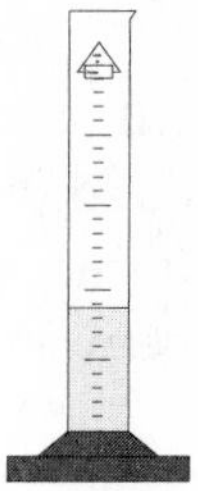

DENSITY, ACCURACY, PRECISION AND GRAPHING

Learning Objectives

Upon completion of this experiment, students will have experienced:

1. The determination of the density of water and a saline solution.
2. A comparison of the accuracy and precision of a graduated cylinder and a pipet.
3. Graphing.

Text Topics

Significant figures, density determinations, accuracy and precision (see page ix).

Notes to Students and Instructor

With proper preparation, it should be possible to perform this experiment and the next one or parts of the previous one in one laboratory period.

Discussion

In the previous experiment, the density of a solid was determined using two different techniques. For a short review of the concept of density, reread the *Discussion* for ***Experiment 3***. Today's experiment will involve the determination of densities of liquids, a comparison of the accuracy and precision of a graduated cylinder and pipet and application of proper graphing techniques to density problems.

Accuracy and Precision. The layperson might not distinguish between accuracy and precision, but to a scientist, they have different meanings. Accuracy is a measure of how close an experimental value is to the actual value. If you measure the circumference of a circle as 25.00 cm and the diameter as 8.000 cm, the value of π that results is 25.00/8.000 = 3.125. The value 3.125 has a percentage error of:

$$\left(\frac{3.1416 - 3.125}{3.1416}\right)(100\%) = 0.53\%.$$

The 0.53% disparity is a measure of the (in)accuracy of the result.

Precision is the reproducibility of the measurement or how closely the measurements agree with each other. Precision is often indicated by the number of significant figures. A measurement of 25.0 cm should be more precise than a measurement of 25 cm as repeated measurements for the former should fall between 24.9 cm and 25.1 cm whereas measurements for the latter should fall between 24 cm and 26 cm. The values between 24.9 cm and 25.1 cm are clearly closer together.

Although the most accurate values are usually the most precise and vice versa, there are exceptions. Very imprecise measurements might by coincidence average out to yield a very accurate value. Or, an incorrectly calibrated device might yield precise but inaccurate measurements.

Graphing. It is very common for scientific experiments to result in tables full of data. While it is sometimes possible to come to meaningful conclusions from tables, it often easier to discern relationships between variables visually using appropriate graphs. A graph shows how one variable changes as another is varied. A graph should be designed to be easily read and interpreted. The guidelines below should help you prepare readable graphs.

1. Select **logical scales** that utilize as much of the graph paper as possible. For many graphs, the two axes will have different scales. Each division should represent **1, 2, 2.5 or 5 units** or some power of ten times one of these units. Avoid the use of 3 or 7 units/division or another number that makes it difficult to locate points on the graph. Be sure the scales are **linear** unless you are doing a log or another function type plot. Unless extrapolating, the graph should cover only the range of measurements on each axis. It is **not** necessary for the point 0, 0 to be on the graph.

2. Label the axes and indicate units used.

3. Locate the points and put dots at the proper locations. **Circle** the dots.

4. If the points appear to fall on a straight line, use a **straight edge** to draw the line that best fits (averages the deviations) the data. Do not sketch straight lines and do not connect the dots with straight line segments. If the points seem to fall on a smooth curve, draw the best curve possible through the data.

Procedure

A. Density of water. The density of three different volumes of water measured with a graduated cylinder will be determined and the mass of the water plotted against its volume.

1. Weigh an empty, **dry** 50 mL graduated cylinder to your balance limits (0.01 g or 0.001 g).

2. Add 10.0 mL of water to the cylinder. Remember, the <u>very</u> bottom of the meniscus should just be touching the 10.0 mL line. Add water up to about the 9 mL mark and use a dropper to reach the 10.0 mL mark.

3. Weigh the cylinder + 10.0 mL of water. You can now calculate the density of the water.

4. Add water up to the 30.0 mL mark and weigh.

5. Add water up to the 50.0 mL mark and weigh.

6. Calculate the densities of the three **total** volumes.

7. As a beginning exercise in graphing, plot the masses of the water on the vertical scale versus the volumes of the water on the horizontal scale.

B. Accuracy and precision. Triplicate determinations of the density of water will be made on 10 mL water samples measured with a graduated cylinder and also with a 10 mL volumetric pipet. Each 10 mL water sample will be weighed and the density calculated. Average densities and average deviations for each measuring device will be calculated and the accuracy and precision of the graduated cylinder and pipet will be compared.

Weigh a 150 mL beaker to the balance limits. Now add 10.0 mL of deionized water to your graduated cylinder. Transfer the 10.0 mL of water to the beaker and weigh the beaker accurately again. Add 10.0 mL more of water from the graduated cylinder to the beaker and weigh again. Repeat this process a third time. Calculate the mass and density of **each** 10.0 mL portion.

Using a pipet bulb, draw deionized water (not from the beaker you just weighed) into a 10 mL volumetric pipet until the water reaches the lower part of the pipet bulge (*Figure 4-1*). Remove the bulb and quickly replace it with your finger. Tip the pipet to a horizontal position and roll the water around being sure to wet all the glass inside the pipet. Drain the pipet out the top. Repeat this rinsing process two more times. Now draw water into the pipet until it is a few centimeters above the fill line in the upper neck. Replace the bulb with your finger and wipe off the outside of the pipet with a towel. Slowly decrease your finger pressure on the pipet opening until the water starts dripping out the tip. Carefully control the dripping until the meniscus just touches the fill line (*Figure 4-2*). Touch the tip to the inside of a glass container to remove the drop or partial drop hanging from the tip. Being sure there are no air bubbles in the tip and the meniscus is still set correctly, drain the pipet into the preweighed beaker allowing the water to drain on its own. Do not use pressure to blow it out. Touch the pipet tip to the inside of the beaker as it completes its draining and hold it there for about 20 seconds after draining is complete. Do not blow out the last half drop as the pipet is calibrated to deliver 10.00 mL from the fill line to the last half drop. Weigh the beaker again and then pipet 10.00 mL more water into it. Repeat this process one more time. Calculate the mass and density of each 10.00 mL portion of water.

Average the three trials for the graduated cylinder and then the three trials for the pipet. For the graduated cylinder, determine the **absolute** value of the difference between each separate graduated cylinder density measurement and the average value. These three values are the deviations of each measurement from the average. Now average the three deviations to arrive at the average deviation. As you should be able to surmise, the average deviation is a measure of the precision of the measurement. Mathematically the average deviation is not as significant as the standard deviation. While it is not difficult to calculate standard deviations, we have chosen to use average deviations here to simplify this treatment. Find average deviations for the pipet measurements in the same way you found them for the graduated cylinder.

Fig. 4-1

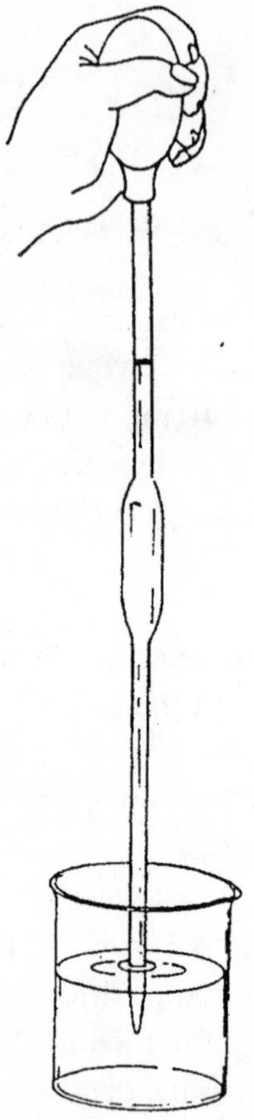

Fig. 4-2

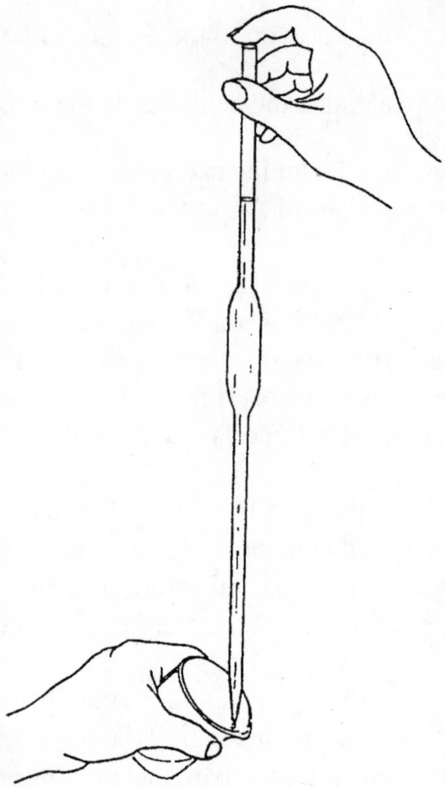

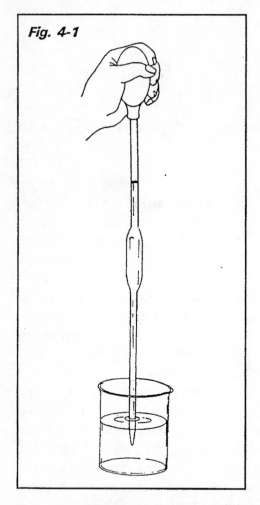

C. Density of a sodium chloride solution. Obtain a salt solution of unknown mass percent salt from your instructor. Rinse a 10.00 mL pipet three times with the unknown solution and then transfer 10.00 mL of the unknown to a preweighed beaker. Weigh the beaker again and calculate the density of the unknown. Use the data in the ***Results and Discussion*** section to plot the density of sodium chloride solutions versus the mass percent of sodium chloride. Use the graph and your experimental density to determine the mass percent of sodium chloride in your unknown. Note that the densities presented of the salt solutions were measured at 20°C. Although densities are temperature dependent, the values are accurate to three significant figures after the decimal between 18 and 24°C. If your solution temperature is not in this range, there will be a slight error.

D. Relative density of liquid and solid phases of a substance. Pour about 5 mL of acetophenone into a 13 x 100 mm test tube. Put the test tube in an ice bath until the acetophenone is frozen solid. Warm the test tube with your hand or a room temperature water bath just sufficiently to be able to slide the frozen acetophenone into an 18 x 150 mm test tube. Add about 3 mL of liquid acetophenone, shake, observe and record your observations and measure the temperature of the system. Return the acetophenone to the used acetophenone bottle.

Add some ice to water and shake and observe (if all the ice melts during shaking, add some more so that the test tube has both liquid and solid water).

Name________________________________ Date__________ Lab Section__________

Prelaboratory Problems - *Experiment 4* - Density, Accuracy, Precision and Graphing

The solutions to the starred problems are in ***Appendix 4***.

1. The mass of a 45.750 g piece of copper is measured three times on two different balances with the following results:

Trial	*Balance 1 (grams)*	*Balance 2 (grams)*
1	45.747	45.76
2	45.745	45.77
3	45.748	45.74

a. Calculate the average deviation for each set of measurements on each balance.

balance 1* __________

balance 2 __________

b. Which balance is more precise? Explain your answer.

c. Which balance is more accurate? Explain your answer.

2. 25.00 mL of heavy water (D_2O where D is a hydrogen with a neutron in its nucleus) at 20°C was pipetted into a 37.234 g beaker. The final mass of the beaker was 64.859 g.

a.* What is the density of heavy water at 20°C? __________

b. The density of normal water (the hydrogens do not have neutrons) at 20°C is 0.9982 g/mL. Calculate the density you would expect for heavy water by assuming that deuterium (^2H or ^2D) is the same size as normal hydrogen (^1H) when it is part of the water.

____

c. Based on your answer to (b), was the assumption in (b) justified? Explain your answer. __________

3. 15.00 mL of radiator liquid (water + ethylene glycol) from a car has a mass of 15.69 g.

   a. Calculate the density of the radiator liquid. ________

   b. Based on the calculation above, is the density of ethylene glycol greater or less than that of water? Explain your answer. ________

   c. Use the *Handbook of Chemistry and Physics* to determine the mass percent of ethylene glycol in the mixture and the freezing temperature of this antifreeze mixture. ________

      ____

Name____________________________________Date__________Lab Section__________

Results and Discussion - *Experiment 4* - Density, Accuracy, Precision and Graphing

A. Density of water

	10.0 mL	30.0 mL	50.0 mL
1. Mass of graduated cylinder + water	________	________	________
2. Mass of empty graduated cylinder	________	________	________
3. Mass of water	________	________	________
4. Density of 10.0 mL sample of water	________		
5. Density of 30.0 mL sample of water		________	
6. Density of 50.0 mL sample of water			________

7. Graph mass (not density) on the vertical axis vs volume on the horizontal axis below:

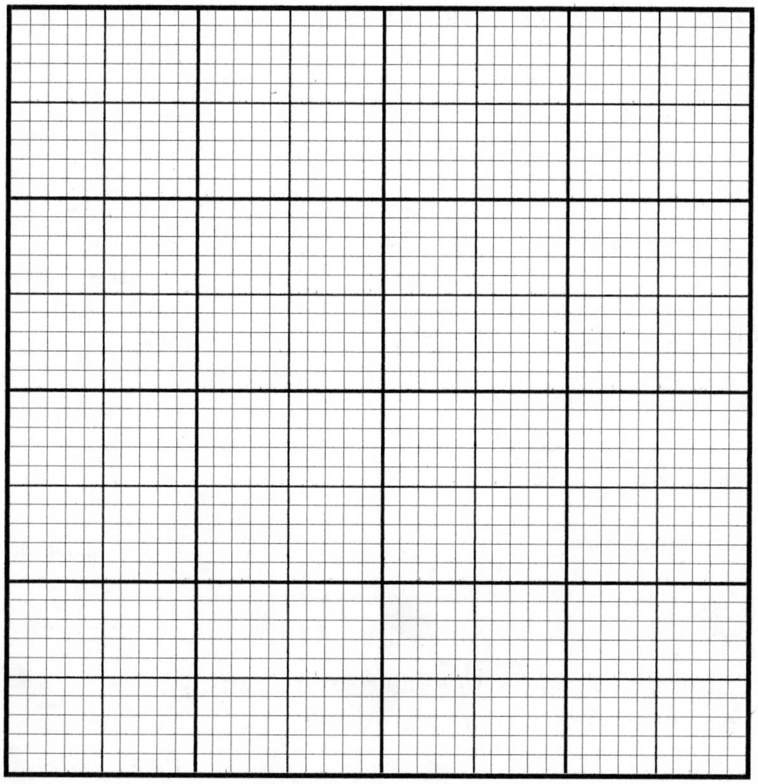

8. The equation for density $d = m/v$ can be rearranged to $m = dv$.
What is the meaning and value of the slope of the line?

slope = ________

B. Accuracy and Precision

	Graduated Cylinder	Pipet
1. Mass of beaker		
2. After first addition		
3. After second addition		
4. After third addition		
5. Mass of **first 10 mL** of water		
6. Mass of **second 10 mL** of water		
7. Mass of **third 10 mL** of water		
8. Temperature of water in beaker		
9. Density of water		
a. Trial 1		
b. Trial 2		
c. Trial 3		
10. Average density		
11. Deviations from average		
a. Deviation of trial 1		
b. Deviation of trial 2		
c. Deviation of trial 3		
12. Average deviation		
13. *Handbook of Chemistry and Physics* value for the density of water at the temperature of your measurement		
14. Percentage error of your average density measurement		

15. Which volume measuring device was more accurate?
    Explain your answer. _________

16. Which volume measuring device was more precise?
    Explain your answer. _________

C. Density of a salt solution.

1. Unknown number _________
2. Mass of beaker _________
3. Mass of beaker + 10.00 mL of unknown _________
4. Mass of 10.00 mL of unknown _________
5. Density of unknown _________
6. The densities of several water-sodium chloride mixtures are reported below. On the accompanying piece of graph paper plot the density on the vertical axis and the % by mass of sodium chloride on the horizontal axis. Density and mass percent are not necessarily linearly related. Draw the best appropriate straight line or curve through the data.

% NaCl (by mass)	density (g/mL)
0.00	0.998
5.00	1.034
10.00	1.071
15.00	1.108
20.00	1.148
25.00	1.189

7. According to your results and the graph, what is
   the mass percent of sodium chloride in your unknown? _________

D. Relative density of liquid and solid phases of a substance.

1. Report your observations on the liquid-solid acetophenone system.

2. Compare the acetophenone system to a liquid-solid system of water. Does one of the systems differ from intuitive expectations? Explain your answer.

3.
   a. Temperature of liquid-solid acetophenone system. ________

   b. What significance if any, does the temperature of the liquid-solid acetophenone system have? Explain your answer.

4. Some of the *Learning Objectives* of this experiment are listed on the first page of this experiment. Did you achieve the *Learning Objectives*? Explain your answer.

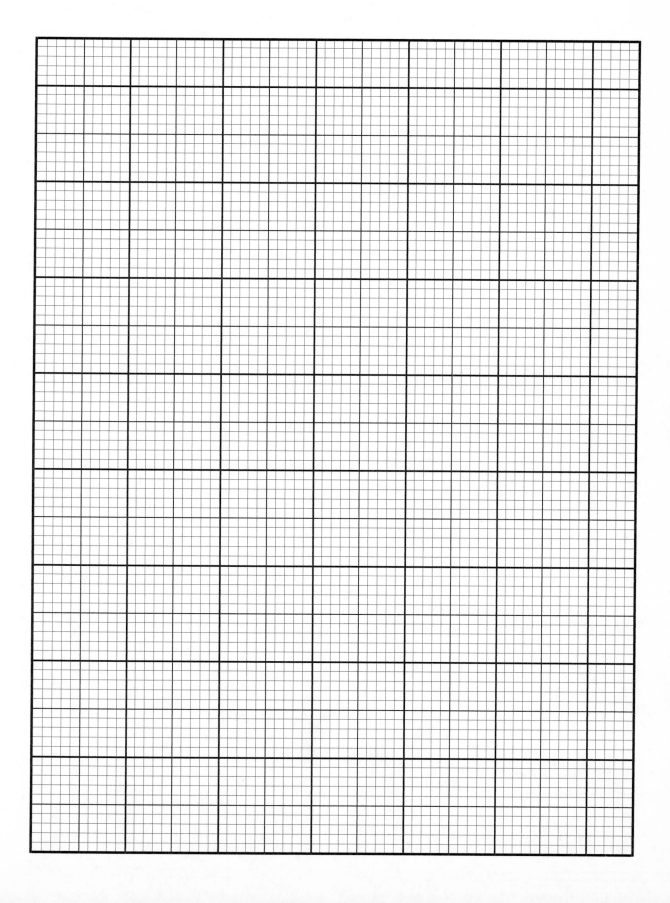

Name_________________________________ Date__________ Lab Section_________

Postlaboratory Problems - *Experiment 4* - Density, Accuracy, Precision and Graphing

1. As density is an intensive property, it is very useful for identification purposes (see *Problem 3* below). Density is also a valuable unit conversion for converting volume to mass or mass to volume. The density of ethanol at 20°C is 0.789 g/mL.

   a. Calculate the mass of 25.0 mL of ethanol. __________

   b. Suggest a situation where weighing may be preferable to a volume measurement.

   c. Calculate the volume of 25.0 g of ethanol. __________

   d. Suggest a situation where a volume measurement may be preferable to weighing.

2. Grams and centimeters in the metric system have been defined so that the density of water at 3.98°C is exactly 1. At 20°C, the density of water is 0.9982 g/mL. What percent error results from the assumption that the density of water is 1.000 g/mL at 20°C?

   ____

3. Suggest simple laboratory tests you could use to distinguish acetone, carbon tetrachloride, cyclohexane, ethyl ether and ethanol. ***Appendix 1*** might be helpful.

4. Describe the technique you would use to determine the density of approximately 5 gram samples of the following at 20°C and 1 atmosphere pressure:

   a. gasoline

   b. granular sugar

   c. a lump of "gold"

   d. mercury

5. In *Part D* of this experiment, the relative densities of liquid and solid acetophenone were studied and compared to the relative densities of liquid and solid water.

   a. Suggest some reasons for the choice of acetophenone for this experiment.

   b. As you undoubtedly noticed, acetophenone has an odor that is somewhat annoying. Use ***Appendix 1*** to see if you can select an alternative substance that could be used in this part of the experiment. Explain your answer.

6. Suggest any ways you can think of to improve any part(s) of this experiment.

Experiment 5

EMPIRICAL FORMULAS

Joseph Proust
1754 - 1844

Learning Objectives

Upon completion of this experiment, students will have experienced:
1. The law of conservation of mass and the concept of limiting reagents.
2. A method for determining the empirical formula of an inorganic salt.
3. A method for determining the identity of an inorganic salt.
4. The determination of the percent water and empirical formula of a hydrate.
5. The use of a platinum wire flame tester.

Text Topics

Percent composition, limiting reagents, empirical formula, hydrates (see page ix).

Notes to Students and Instructor

Two empirical formula determinations are included in this experiment. The first (*Part A*) involves the determination of the formula of a compound formed from the reaction of zinc with iodine. The second (*Parts B, C, and D*) involves the determination of the numbers of waters of hydration in a hydrate. **Performance of both of these experiments will probably require more than one lab period so the instructor needs to decide if one or both of the experiments will be performed.** The instructions for the zinc iodide portion of this experiment (*Part A*) were derived from DeMeo, S., *J. Chem. Ed.*, **1995**, *72*, pp. 836 - 839. When performing *Parts B, C,* and *D, Part D-3* of this experiment should be started first and the remaining parts should be performed when time permits.

Discussion

One of the common techniques used to help determine the identity of a substance involves combustion of very small amounts of the substance. By measuring the amounts of water, carbon dioxide, and other gases produced, it is possible to determine the percentage by mass of each element in the compound. From the percentages, it is possible to calculate the empirical formula of the compound. Suppose, for example, that the combustion experiment for a compound that contains carbon, hydrogen, and oxygen gives the composition as 40.00% carbon, 6.71% hydrogen, and 53.29% oxygen by mass. Next the mass percentage needs to be converted to mole or atom ratios. Whenever dealing with percentages in a calculation of this type, one of the easiest ways to proceed is to assume that you have 100 g of the substance. Thus 100 g of the substance contains 40.00 g of carbon, 6.71 g of hydrogen and 53.29 g oxygen. As is common for many chemical calculations, the calculation of the number of moles is one of the first steps. Thinking ahead, once the number of moles is known, the ratio of moles is easily calculated.

$$40.00 \text{ g C} \left(\frac{1 \text{ mol C}}{12.011 \text{ g C}} \right) = 3.330 \text{ moles C}$$

$$6.71 \text{ g H} \left(\frac{1 \text{ mol H}}{1.008 \text{ g H}} \right) = 6.66 \text{ moles H}$$

$$53.29 \text{ g O} \left(\frac{1 \text{ mol O}}{15.999 \text{ g O}} \right) = 3.331 \text{ moles O}$$

Dividing by the lowest number of moles (3.330 moles C) gives a C:H:O ratio of 1:2:1 so the empirical formula is CH_2O. This means the compound could be glucose ($C_6H_{12}O_6$) but it could also be many other compounds including other sugars or formaldehyde. The determination of the molecular formula from the empirical formula requires an additional measurement of the molecular mass. If a molecular mass measurement for the above compound results in 150 ± 2 g/mole, the molecular formula would be calculated as $C_5H_{10}O_5$. This means the compound could be a five carbon sugar such as ribose.

$$\text{molecular formula} = \frac{\text{molecular mass}}{\text{empirical form. mass}} \text{ x subscripts of empirical formula} = \frac{150}{30} \text{ x } CH_2O = C_5H_{10}O_5$$

The empirical and molecular formulas can go a long way in helping to identify a compound.

Almost all ionic compounds are solids at room temperature. The crystal structure of the ionic solid consists of a repeating array, and it is not straightforward to define the smallest unit (a molecule) of the substance that still would have the properties of the substance. Thus it is better to refer to the formula mass of an ionic substance rather than its molecular mass. For the same reasons, the formula calculations for ionic compounds should be reduced to the lowest whole number ratio or the empirical formula. Except for a few unusual cases, the term molecular formula is not applicable for ionic compounds.

For metals with only one common oxidation state, it is usually possible to predict with a high degree of confidence, the empirical formula that will result from the reaction of a metal with a non-metal. For instance, based on common oxidation numbers, sodium chloride and sodium oxide should be and are NaCl and Na_2O respectively. Aluminum chloride and aluminum oxide should be and are $AlCl_3$ and Al_2O_3. When there are multiple oxidation states of a metal, more than one compound is possible. For instance, copper and oxygen form Cu_2O and CuO and more information is needed to distinguish between the two. When naming a copper oxide, it is very important that the name distinguish between the two possibilities [copper(I) oxide or cuprous oxide and copper(II) oxide or cupric oxide].

Care must be exercised as chemistry often contains wonderful surprises. Suppose the percent by mass experiment for an iron oxide results in 72.36% Fe and 27.64% by mass O. Following the procedure above results in a mole ratio of 1.296:1.727. Division by the lowest number results in a mole ratio of 1:1.33. For cases like this, the next step is to multiply by the lowest whole number that converts each of the subscripts to a whole number. Thus the formula is Fe_3O_4. As iron has two oxidation states (+2, +3), the expected formula is either FeO or Fe_2O_3. Anytime a result that is inconsistent with expectations is obtained, the first and immediate step should be to recheck all the data and calculations. In this case the result is real and Fe_3O_4 is a known compound often called magnetite. Understanding the oxidation state of iron in this compound will be left to you or your instructor but it is results like this that should not be ignored and often lead to important and exciting discoveries.

In ***Part A*** of today's experiment, iodine will be reacted with an excess of zinc. The masses of the starting materials, products and leftover starting material will enable you to calculate the formula of the zinc iodide produced. Since zinc in compounds has only the +2 oxidation state, the product is certainly expected to be ZnI_2. However, as noticed above for iron, expectations do not always agree with results so it is always necessary when performing a synthesis to verify the formula of the product. This will be your primary goal for ***Part A*** of the experiment.

Hydrates. Have you ever noticed the small envelope of chemical that sometimes comes enclosed with water sensitive products such as medicines and electronic equipment? These chemicals are desiccants; that is, they are compounds that absorb water from the air and hopefully keep the other contents of the package dry. When these inorganic compounds absorb water, they form hydrates. Precipitation of many inorganic compounds from water also often results in the incorporation of "waters of hydration" into their crystal structures. Usually the compound incorporates a specific number of water molecules per formula unit of compound. For example, the precipitation of copper(II) sulfate from water yields copper(II) sulfate pentahydrate which is written $CuSO_4 \cdot 5H_2O$. The formula of the salt is written first followed by a dot and finally the number of waters of hydration. Notice that this is the only time a coefficient is included in the formula. While hydrates do have definite formulas, the bonds to the waters are weaker than "normal" ionic and covalent bonds. This is emphasized by the use of the dot in the formula. Since the bonds are relatively weak, it is fairly easy to break them. Mild heating is often all that is necessary to decompose a typical hydrate. Decomposition reactions yield two or more distinct substances. Decomposition of a hydrate is no exception yielding the salt and water. In this experiment, you will explore some of the properties of hydrates, determine the percent by mass of water in a hydrate (simply by heating it and driving off the water) and the empirical formula (the number of waters in the formula) of a hydrate.

Compounds that absorb water from the air and can be used as desiccants are said to be hygroscopic or deliquescent. In some hydrates, the water of hydration is bonded so weakly that it tends to escape even at room temperature when the compound is exposed to the atmosphere. These compounds are efflorescent. In ***Part B*** of this experiment you will study the efflorescence and deliquescence of two compounds. In ***Part C***, you will attempt to verify that $CuSO_4 \cdot 5H_2O$ loses water when heated.

In the *Part D* of the experiment, you will be given an unknown hydrate. Using simple qualitative tests, you will determine the identity of the salt. By quantitatively driving off the water, you will be able to calculate the mass percent of water in the hydrate. Knowing the identity of the salt, it will be possible to determine the number of waters of hydration in the formula.

To help you determine the identity of the salt, you will perform a flame test. In the flame test, a platinum wire is inserted into a solution containing the ions of interest and then inserted into a Bunsen burner flame. For some cations, the energy of the flame will cause electrons to be elevated from the ground state or lowest possible energy levels to higher energy orbitals. As this results in an unstable situation, the electron drops back to the lower energy state and gives off the excess energy often in the form of a characteristic emission of light. The color of the light is useful for the detection of the presence of ions of sodium, barium, calcium, strontium, lithium and potassium. To complete your identification, it may be necessary to attempt one single replacement reaction.

To determine the mass percent of water in the hydrate, the hydrate is weighed, heated and reweighed. The mass percent of water can be calculated from:

$$\frac{\text{weight loss during heating}}{\text{original mass}} \times 100\% = \text{mass \% water in hydrate}$$

Assume that heating of a 3.50 g sample of the hydrate of copper sulfate yields 2.25 g of anhydrous copper sulfate. The mass percent of water in the hydrate would be:

$$\left(\frac{3.50 - 2.25}{3.50}\right)(100\%) = 35.7\% \quad \text{[The theoretical value is 36.1\%]}$$

Now the ratio of the number of water molecules per formula unit of copper sulfate can be determined.

$$(2.25\text{ g } CuSO_4)\left(\frac{1\text{ mol}}{159.6\text{ g } CuSO_4}\right) = 1.41\text{x}10^{-2}\text{ mol } CuSO_4$$

$$(1.25\text{ g } H_2O)\left(\frac{1\text{ mol}}{18.02\text{ g } H_2O}\right) = 6.94\text{x}10^{-2}\text{ mol } H_2O$$

$$\frac{0.0694\text{ moles } H_2O}{0.0141\text{ moles } CuSO_4} = 4.92 \quad \text{[This is within experimental error of the actual number of 5]}$$

This indicates there were 5 moles of water per mole of $CuSO_4$.

$$CuSO_4 \cdot 5H_2O(s) \rightarrow CuSO_4(s) + 5\,H_2O(g)$$

Procedure

<u>Part A - Synthesis and formula of zinc iodide.</u> Be sure to record all masses to the limits of the balance capability. Weigh a 50 mL beaker to the nearest milligram (or if milligram balances are not available, to the nearest 0.01 g) and record the mass. Using a piece of weighing paper, weigh out about 1.2 grams of 20 mesh granular zinc and add it to the 50 mL beaker. Carefully weigh the beaker. Using a second piece of weighing paper, weigh out about 1.2 g of iodine **[Caution: iodine is toxic and should not be touched]**, add it to the same 50 mL beaker and read the mass of the beaker.

Add 3 mL of 0.2 M acetic acid (acetic acid prevents the precipitation of zinc hydroxide) to the solid mixture in the beaker, swirl and observe changes of temperature and color. Continue to swirl until no further changes are evident. The beaker should have cooled back down to room temperature and the iodine and triiodide (some of the iodine and iodide combine in aqueous solution to form I_3^-) color should have disappeared. This process should take about 20 minutes.

Weigh an evaporating dish and record the mass on page 67. Carefully decant (pour off) the aqueous solution into the evaporating dish, being sure that all the leftover zinc remains in the original beaker. Add about 1 mL of 0.2 M acetic acid to the beaker (and contents), swirl and decant again into the evaporating dish. This "washes" the remaining zinc free of residual zinc iodide and transfers the zinc iodide to the evaporating dish. Repeat the washing process two more times. Wash the zinc three more times with 0.2 M acetic acid and discard these additional washings but save the beaker and the remaining zinc.

Mount the evaporating dish in a 400 mL beaker containing about 200 mL of water on a wire gauze above a Bunsen burner as shown in *Figure 5-1*. Heat the beaker until the contents of the evaporating dish appear to be dry. Achieving apparent dryness with steam heating can consume considerable time. Time can be saved if you move to the next step before dryness is completely attained. Carefully remove the beaker with beaker tongs and place the evaporating dish directly on the wire gauze. **Very** gently flame the dish until crackling stops and the solid in the dish turns slightly off-white. Excessive heating results in spattering and/or a dark yellow color indicating that the product is decomposing to iodine. **Both of these problems need to be avoided.** Allow the dish to cool and weigh it. Repeat the direct heating, cooling and weighing cycle until constant mass is achieved (within at least 0.01 grams).

Fig. 5-1

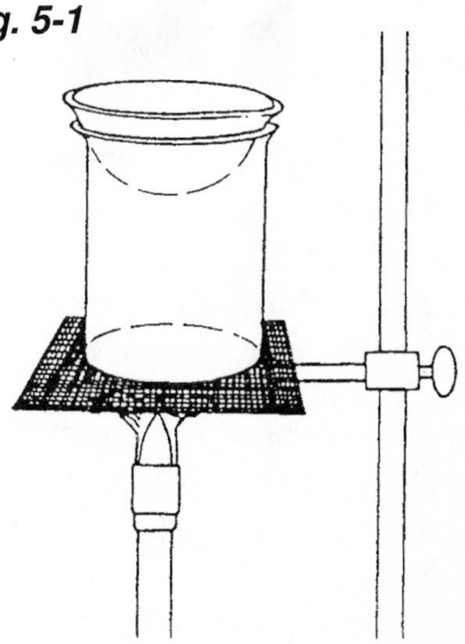

Now flame the beaker containing the zinc until it is dry. Allow it to cool and weigh it. Repeat the heating, cooling and weighing cycle until constant mass is achieved (within at least 0.01 grams).

Optional enhancement. Dissolve about 0.1 gram of the white solid that remains from the aqueous solution in 2 mL of water on a watch glass. Obtain a battery clip (e.g., Radio Shack 270-325) that has had two copper leads attached to the ends of the clip wires. Attach the clip to a 9-volt battery and insert the two copper leads into the solution. Record your observations.

Parts B, C and D are a unit involving hydrates.

B. Deliquescence and efflorescence. Place a few crystals of sodium sulfate decahydrate on a watch glass. Occasionally observe their appearance for about an hour and write an equation for the observed change. On another watch glass, place a few crystals of potassium acetate. As before, observe their appearance over at least a one hour time period.

C. Copper(II) sulfate pentahydrate. Put 1.0 g of $CuSO_4 \cdot 5H_2O$ into a 13 x 100 mm test tube. Stuff a small wad of fine glass wool *(Caution: Minimize contact of the glass wool with your hands - it causes splinters and itching)* into the test tube so that it holds copper(II) sulfate pentahydrate in place when the tube is tilted downward. Clamp the test tube upside down over a watch glass. Holding the base of the burner, heat the $CuSO_4 \cdot 5H_2O$. The blue color should begin to dissipate while the vapors given off condense to a liquid. If any sign of blackening occurs, reduce the heat. Heat until several drops of liquid have been collected and the residue is white.

Using an eyedropper, test a drop of the liquid collected in the watch glass with a piece of blue cobalt chloride test paper (if the paper is pink, pass it quickly high over the burner flame to dry it and turn it blue). Test a drop of water on a piece of blue cobalt chloride paper.

Remove the glass wool plug with a wire hook, break up the white residue with a stir rod or wire and pour it into another watch glass *(caution: avoid letting this powder come into contact with your skin)*. Using a spatula, divide the powder into two piles. Test one pile with a drop of liquid collected in the watch glass. Test the other pile with a drop of water. Write equations for the observed changes.

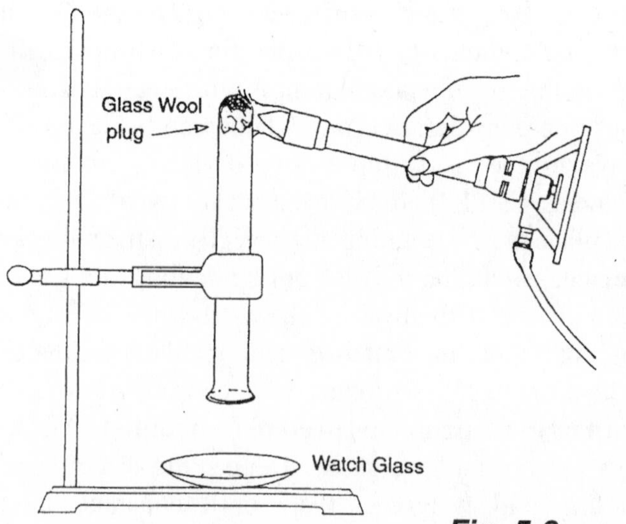

Fig. 5-2

D. Analysis and percent water of an unknown hydrate.

1. Your unknown is a hydrate of strontium chloride, magnesium sulfate or zinc sulfate. Dissolve a small amount (about 0.1 g) in about 5 mL of deionized water in a test tube to use for *#1* and *#2*. Clean a platinum wire by alternately dipping it into 6 M HCl and inserting into the flame until little or no color is observed. Perform flame tests on known solutions of $ZnSO_4$, $MgSO_4$ and $SrCl_2$ and finally the solution of the unknown being sure to clean the wire between each test. This test should enable you to either identify strontium ion or rule it out.

2. If your unknown is not strontium chloride, drop a short piece of magnesium ribbon into your unknown solution. Allow 30 seconds for a reaction to take place. If the magnesium ribbon tarnishes, your unknown is zinc sulfate.

$$Mg(s) + ZnSO_4(aq) = MgSO_4(aq) + Zn(s)$$

If no reaction other than bubbling occurs ($Mg(s) + MgSO_4(aq)$ = no reaction), the unknown is magnesium sulfate. The bubbling is due to a side reaction of the magnesium with water.

$$Mg(s) + 2H_2O(l) = Mg(OH)_2(s) + H_2(g)$$

3. Weigh a clean, dry crucible and cover to at least the nearest 0.01 g (preferably 0.001 g). Place about 4 grams (3.5 - 4.5) of your unknown hydrate crystals in the crucible and weigh to at least the nearest 0.01 g. Suspend the crucible in a clay triangle over a Bunsen burner. Heat gently with the top ***slightly ajar*** for about eight minutes and then vigorously (the crucible should glow a dull orange) for an additional eight minutes. Allow the crucible to cool (about 5 minutes), and weigh it to at least the nearest 0.01 g. Heat the crucible vigorously again for about 5 minutes, cool and weigh again. If the mass difference between the first and second heatings is greater than 0.01 g, perform a third heating, cooling, weighing cycle. Repeat the process until two successive weighings do not differ by more than 0.01 g. Calculate the mass percent of water and the empirical formula of the hydrate.

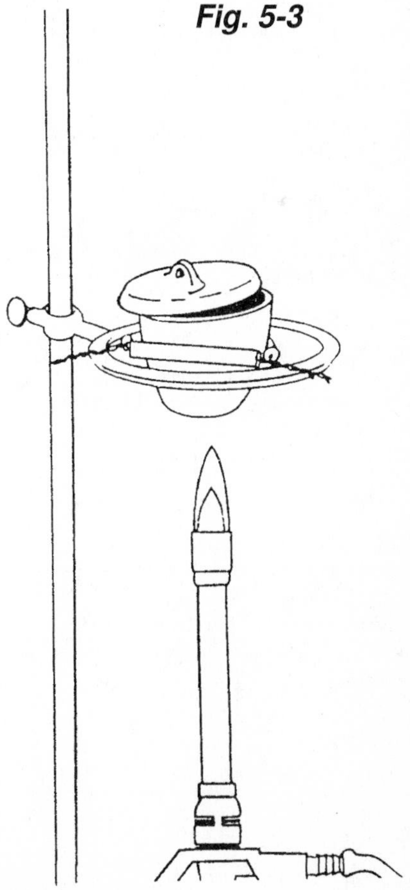

Fig. 5-3

Name_______________________________Date__________Lab Section_________

Prelaboratory Problems - *Experiment 5* - Empirical Formulas

The solutions to the starred problems are in ***Appendix 4***.

1.* 5.0 g of aluminum and 5.0 g of oxygen are reacted and yield the expected product.

a. Write a balanced equation for the reaction.

b. Which one of the two reagents was used in excess and how much of it remains after the reaction is complete (the other reagent is called the limiting reagent).

____

____

2. 2.20 grams of chromium reacts with excess oxygen to give 3.22 grams of a chromium oxide. What is the name and formula of the chromium oxide? Write a balanced equation for the reaction.

____

____

3. Complete and balance the following equations:

a.* $BaCl_2 \cdot 2H_2O_{(s)}$ $-\Delta->$

b. $Co(C_2H_3O_2)_2 \cdot 4H_2O_{(s)}$ $-\Delta->$

4. What is the percent by mass of water in:

a.* $BaCl_2 \cdot 2H_2O$ ____________

b. $Co(C_2H_3O_2)_2 \cdot 4H_2O$ ____________

5.* Determine the empirical formula and the molecular formula of a compound (Freon 11) that consists of 8.74% C, 77.43% Cl, 13.83% F and has a molecular mass of 137 g/mol.

____

____

6. A compound consists of 40.00% carbon, 6.71% hydrogen and 53.29% oxygen and has a molecular mass of 180 ± 1 g/mole. Determine the empirical and molecular formulas of the compound.

____

____

7. The mass percent of water in a hydrate of $MnCl_2$ is 36.41%. What is the empirical formula of the hydrate?

____

8.* A 4.00 gram sample of a hydrate of nickel(II) bromide loses 0.793 grams of water when heated. Determine the mass percent water in the hydrate and the formula of the hydrate.

____

____

9. A 2.500 gram sample of a hydrate of calcium sulfate loses 0.523 grams of water when heated. Determine the mass percent of water in the hydrate and the formula of the hydrate.

____

____

10. An unknown solution containing one of the salts $ZnSO_4$, $MgSO_4$ or $SrCl_2$ gives a negative flame test but tarnishes a piece of magnesium inserted into the solution. Which salt was in the solution?

____

Name________________________________Date_________Lab Section_________

Results and Discussion - *Experiment 5* - Empirical Formulas

A. Synthesis and formula of zinc iodide.

1. Write the expected balanced equation for the reaction between zinc and iodine ______________________

2. Mass of beaker ____________

3. Mass of evaporating dish ____________

4. Mass of iodine ____________

5. Mass of zinc ____________

6. Mass of zinc + iodine ____________

7. Observations after addition of water containing acetic acid

8. Mass of beaker + residual zinc __________ (trial 1) __________ (trial 2) __________ (trial 3 if necessary)

9. Mass of residual zinc ____________

10. Mass of zinc consumed in reaction ____________

11. Mass of evap. dish + zinc iodide __________ (trial 1) __________ (trial 2) __________ (trial 3 if necessary)

12. Mass of zinc iodide ____________

13. a. Mass of zinc iodide + residual zinc ____________

    b. Was mass conserved within experimental error in this reaction? Explain your answer.

14. Moles of zinc consumed in reaction ____________

15. Moles of iodine consumed in reaction ____________

16. Mole ratio of iodine to zinc (use appropriate number of significant figsures) ____________

17. Formula of zinc iodide ____________

18. Does your experimental formula agree with the expected formula? Explain your answer.

19. In terms of the chemistry and energy of the reaction, explain your observations in #7 above.

20. Explain why some zinc was left over at the end of the reaction and why you think the amounts were selected to have zinc left over.

21. (optional) What did you observe in the electrolysis reaction?

22. (optional) Were your observations on the electrolysis consistent with your expectations? Explain your answer.

B. Deliquescence and efflorescence.

System	Observations	Reaction
1. $Na_2SO_4 \cdot 10H_2O$		
2. $KC_2H_3O_2$		$KC_2H_3O_2(s) + n\ H_2O(l) = KC_2H_3O_2 \cdot n\ H_2O(s)$

C. Copper(II) sulfate pentahydrate

System	Observations	Reaction
1. water + $CoCl_2$ paper (forms the hydrate $CoCl_2 \cdot 6H_2O$)		
2. condensate + $CoCl_2$ paper		
3. water + $CuSO_4$		
4. condensate + $CuSO_4$		

5. What evidence supports the conclusion that the condensate is water?

D. Analysis and percent water of an unknown hydrate.

1. Unknown # ________

2. Flame tests

solution	flame color
$SrCl_2$	________
$ZnSO_4$	________
$MgSO_4$	________
unknown	________

Flame test conclusion ______________________________________

3. Mg + unknown → observations ______________________________

   Reaction ____

4. Unknown: $SrCl_2$ or $ZnSO_4$ or $MgSO_4$ ________

5.
   a. Mass of crucible + cover ________
   b. Mass of crucible + cover + unknown ________
   c. Mass of crucible + cover + unknown after first heating ________
   d. Mass of crucible + cover + unknown after second heating ________
   e. Mass of crucible + cover + unknown (after third heating if necessary) ________
   f. Mass of crucible + cover + unknown (after fourth heating if necessary) ________
   g. Mass of original unknown ________
   h. Mass of water lost by unknown ________
   i. Mass percent of water in unknown ________
   j. Mass of unknown salt remaining after heating ________
   k. Formula mass of anhydrous salt ________
   l. Moles of anhydrous unknown salt ________
   m. Formula mass of water ________
   n. Moles of water lost ________
   o. Ratio of moles of water to moles of unknown (use appropriate number of significant figures) ________
   p. Formula of unknown hydrate ________
   q. Unknown identification number ________

6. Some of the *Learning Objectives* of this experiment are listed on the first page of this experiment. Did you achieve the *Learning Objectives*? Explain your answer.

Experiment 6

CLASSIFICATION OF CHEMICAL REACTIONS

phosphorous + oxygen (1850)

Learning Objectives

Upon completion of this experiment, students will have experienced:

1. Five of the common types of chemical reactions.
2. The completion and balancing of chemical equations.
3. The observation of some chemical properties of hydrogen and oxygen.
4. An introduction to concepts of stoichiometry.

Text Topics

The classification of chemical reactions, prediction of reaction products, balancing equations (for correlation with some textbooks, see page ix).

Notes to Students and Instructor

The use of Beral pipets to generate hydrogen and oxygen was derived from an experiment in a text by David Ehrenkranz and John J. Mauch [*Chemistry in Microscale*, Kendall/Hunt. Co., Dubuque (1960)].

Discussion

The process of classification often assists with the simplification and solution of problems. Classification of diseases by cause; viral, bacterial or fungal, facilitates proper treatment. The sciences are frequently classified into disciplines such as physics, chemistry, biology and geology. Then chemistry is often subdivided into organic, inorganic, analytical, theoretical and physical branches. Attempts have been made to classify reactions by type. Later in the course, you will learn how to determine if a reaction is an oxidation-reduction reaction. Today, we will run several reactions and attempt to classify them by the nature of the reaction: combination, decomposition, combustion, single replacement or double replacement.

Combination: the reaction of two substances to form one substance.

$C(s) + O_2(g) = CO_2(g)$

$MgO(s) + H_2O(l) = Mg(OH)_2(s)$

$SO_3(g) + H_2O(l) = H_2SO_4(l)$

$SrCl_2(s) + 6\,H_2O(l) = SrCl_2 \cdot 6H_2O(s)$

Decomposition: The reverse of combination or the breaking down of one substance into two or more substances.

$H_2CO_3(aq) = H_2O(l) + CO_2(g)$

$HgO(s) = Hg(l) + \frac{1}{2}O_2(g)$

$Ca(HCO_3)_2(aq) = CaCO_3(s) + H_2O(l) + CO_2(g)$

$Mg(OH)_2(s) = MgO(s) + H_2O(g)$

$SrCl_2 \cdot 6H_2O(s) = SrCl_2(s) + 6\,H_2O(l)$

Combustion: The rapid reaction of a compound with oxygen. Some combustion reactions are also combination reactions and vice versa.

$2\,CH_4O(l) + 3\,O_2(g) = 2\,CO_2(g) + 4\,H_2O(g)$

$2\,C_8H_{18}(l) + 25\,O_2(g) = 16\,CO_2(g) + 18\,H_2O(g)$

Single Replacement: The replacement of an element in a compound by another element originally in elemental form.

reaction	*observation*
$Cu(s) + 2\,AgNO_3(aq) = Cu(NO_3)_2(aq) + 2\,Ag(s)$	plating of metallic silver on copper and appearance of blue color in solution
$Mg(s) + H_2SO_4(aq) = MgSO_4(aq) + H_2(g)$	gas evolution
$Br_2(aq) + 2\,KI(aq) = 2\,KBr(aq) + I_2(aq)$	loss of red bromine color and appearance of brown iodine color

Double Replacement: An exchange of positive and negative partners by two compounds (most commonly involving two ionic compounds) in aqueous solution. These reactions usually proceed when at least one of the products is a compound insoluble in water (precipitate), a gas or a compound that decomposes into a gas, or a slightly ionized compound.

<u>*reaction*</u>	<u>*observation*</u>
$3\,BaCl_2(aq) + 2\,Na_3PO_4(aq) = Ba_3(PO_4)_2(s) + 6\,NaCl(aq)$	white ppt.
$K_2CO_3(aq) + 2\,HNO_3(aq) = 2\,KNO_3(aq) + H_2O(l) + CO_2(g)$	gas
$H_2SO_4(aq) + 2\,KOH(aq) = K_2SO_4(aq) + 2\,H_2O(g)$	heat

In addition to recording observations about a reaction and classifying the reaction by type, one should write a balanced chemical equation for any reactions that occur. The examples above are all balanced.

The first step in writing a balanced equation is to write the correct formulas for reactants and products. Once this is done, subscripts ***must not be changed*** to balance the equation as this changes the substance. Balance the equation using coefficients (numbers in front of the formula). A coefficient denotes the relative number of moles of the substance whose formula it precedes. Locate the formula with the largest subscript (not subscripts within a polyatomic ion but subscripts that give the number of atoms or ions per formula unit). For the reaction below, 8 is the largest subscript.

$$C_3H_8(g) + O_2(g) \rightarrow CO_2(g) + H_2O(g)$$

Imagine the coefficient 1 in front of the propane (C_3H_8) and balance the hydrogens and then the carbons.

$$1\,C_3H_8(g) + O_2(g) \rightarrow 3\,CO_2(g) + 4\,H_2O(g)$$

Observe that there are now ten oxygen atoms on the right and that there are two oxygen atoms in an oxygen molecule on the left. Divide the number of atoms needed (10) by the number per molecule or formula unit (2) to arrive at the correct coefficient (5).

$$C_3H_8(g) + 5\,O_2(g) = 3\,CO_2(g) + 4\,H_2O(g)$$

Notice that coefficient of 1 is not written in the final equation but is understood. Also notice the very common technique of leaving the O_2 (or H_2) until last if it is present. The coefficient for O_2 affects the amount of one element only whereas the other coefficients change the amounts of at least two elements.

For double replacement reactions, start with the ion with the largest subscript. In case of a tie, choose the ion with the largest oxidation number. For the reaction below,

$$BaCl_2(aq) + Na_3PO_4(aq) \rightarrow Ba_3(PO_4)_2(s) + NaCl(aq)$$

Na^+ and Ba^{2+} have subscripts of 3 but Ba^{2+} has the higher charge. Notice that the subscript 4 is part of the phosphate polyatomic ion and is not part of this consideration. Start with the Ba^{2+} and work from one side of the equation to the other. Balance the barium ions on the left side of the equation by inserting a coefficient of 3.

$$3\,BaCl_2(aq) + Na_3PO_4(aq) \rightarrow 1\,Ba_3(PO_4)_2(s) + NaCl(aq)$$

The coefficient of 3 in front of $BaCl_2$ locked in 6 chlorides so a 6 is now needed in front of NaCl on the right.

$$3\,BaCl_2(aq) + Na_3PO_4(aq) \rightarrow 1\,Ba_3(PO_4)_2(s) + 6\,NaCl(aq)$$

The 6 locks in 6 sodiums so the coefficient in front of Na_3PO_4 is the number needed (6) divided by the number per formula unit (3) resulting in a coefficient of 2. Finally check to see if the phosphates are balanced to be sure you haven't made an error.

$$3\,BaCl_2(aq) + 2\,Na_3PO_4(aq) = Ba_3(PO_4)_2(s) + 6\,NaCl(aq)$$

Procedure

A. Classification of reactions. Carry out the reactions as instructed. Record all your observations (precipitate, gas evolution, heat evolution, or color change). Use your observations, the nature of the reactants and the examples given in the ***Discussion*** section to classify the reactions by type. In particular, be alert for combination reactions when two compounds react to form one, decomposition when one compound decomposes into two or more compounds, combustion when a compound reacts with oxygen, single replacement when an element reacts with a compound to give another element and compound, double replacement when two compounds react to give two new compounds (detectable by formation of a precipitate, a gas, or heat evolution). After mixing, touch the exterior of each test tube to check for heat evolution. Write balanced equations for all observed reactions. If a reaction is not detected, write "NAR" for no apparent reaction. Be sure to give each mixture ample time (at least 5 minutes) before concluding there is no apparent reaction (by vision and/or touch). For sample analyses of reactions, see the ***Prelaboratory Problems***.

1. Mix 3 mL of 0.1 M $CaCl_2$ with 2 mL of 0.1 M Na_3PO_4.

2. Add a few drops of water to a test tube containing about 0.5 g $CuSO_4$ (anhydrous). *[Caution: Anhydrous copper sulfate (white color) is corrosive. Avoid skin contact but wash with copious quantities of water if contact occurs.]*

3. Heat a test tube containing about 0.5 g $Cu(OH)_2$ with a burner.

4. To a test tube containing 3 mL of 6 M HCl, add a 1 cm^2 piece of zinc ribbon. If a gas evolves, quickly insert a lighted splint into the mouth of the test tube. Balance the equations for both reactions.

5. Mix 2 mL of 3M HCl with 2 mL of 1 M Na_2CO_3 (Note that in this case, two types of reactions occur, one right after the other).

6. To a test tube containing 3 mL of 3% H_2O_2 (hydrogen peroxide), add 0.1 g of the catalyst MnO_2. (Note that a catalyst affects the rate of a reaction but is not involved in the overall reaction for the process).

7. Add 2 mL of a saturated calcium acetate solution (about 35 g/100 mL H_2O) to an evaporating dish. To the dish add 15 mL of ethanol and swirl the contents. Pour off any excess liquid and ignite the remaining contents with a match. For an additional effect, sprinkle some boric acid on the mixture. Although the reaction is actually more complex, assume that the reactants in this reaction are only ethanol (C_2H_5OH) and oxygen.

8. Add 3 g of NH_4Cl and 7 g of $Sr(OH)_2 \cdot 8H_2O$ to a 125 mL Erlenmeyer flask and swirl vigorously for about 5 minutes. Be sure to record all observations including odor, sounds and touch (flask, not contents) sensations in addition to visual.

9. Mix 2 mL of 3 M H_2SO_4 with 4 mL of 3 M NaOH.

10. Mix 2 mL of 0.1 M $CaCl_2$ with 2 mL of 0.1 M Na_2CO_3.

11. To a test tube containing 3 mL of 0.1 M $CuSO_4$, add a 1 cm^2 piece of zinc ribbon.

12. To a test tube containing 3 mL of 0.1 M $ZnSO_4$, add a 1 inch long piece of copper wire.

13. Mix 1 mL of 0.1 M $CaCl_2$ and 2 mL of 0.1 M $NaNO_3$.

14. Heat a test tube containing about 2 g of $CuSO_4 \cdot 5H_2O$ with a burner.

15. Mix 2 mL of 6 M HCl with 4 mL of 3 M NaOH.

B. Beral pipet rockets. Reactions similar to those performed earlier (*A-4* and *A-6*) to generate hydrogen and oxygen will be explored further and then used for the propulsion system for rockets made from Beral pipets. Because of the potentially explosive nature of hydrogen - oxygen mixtures, the small volumes provided by Beral pipets are ideally suited for these experiments.

Step 1. Prepare 2 test tube gas generators (see *Figure 6-1*). One will be used in *Steps 1 - 8* and *12, 13* as a hydrogen generator. The second will be used in *Steps 9 - 13* as an oxygen generator.

a. Cut off the stem of a graduated Beral pipet (e.g., Flinn Scientific # AP 1516) about 1 cm from the bulb. The bulb will be referred to as the "collection bulb".

b. Cut the tip of the pipet stem at a point just before the diameter begins to get smaller and insert its tip into a #2 one hole rubber stopper as shown. Place the stopper in a 20 × 150 mm test tube.

Step 2. Add about 200 mL of water to a 400 mL beaker. This will be used as a temperature regulator during the generation of hydrogen and oxygen.

Step 3. Fill the collection bulb with water by holding it (opening upward) under water in a dish pan, squeezing the air out and allowing water to enter.

Step 4. Add about 8 grams of 20 mesh granular zinc to one of the gas generator tubes.

Step 5. Light a candle.

Step 6. Add 25 mL of 3 M hydrochloric acid to the test tube containing the zinc and stopper it with the #2 stopper device. Put the tube in the 400 mL beaker of water and wait 10 seconds before beginning the next part (see *Figure 6-2*). [Note: If gas generation slows during the experiment, replace the hydrochloric acid but do not replace the zinc]

Step 7. Put the collection bulb over the pipet tip (nozzle) and collect hydrogen until all the water has been displaced from the bulb.

Step 8. Remove the bulb from the hydrogen generator and squirt the hydrogen you collected into the candle flame (*Figure 6-3*). Repeat the hydrogen collecting and squirting once or twice and report your observations.

Step 9. Add about 7 grams of manganese metal to the second (oxygen) generator tube, label it and fill another collection bulb as in *Step 3*.

Step 10. Add 25 mL of 3% hydrogen peroxide to the oxygen generating tube and put this tube in the beaker of water next to the hydrogen generator tube. [Note: If gas generation slows during the experiment, replace the 3% hydrogen peroxide but do not replace the manganese metal.

Fig. 6-1

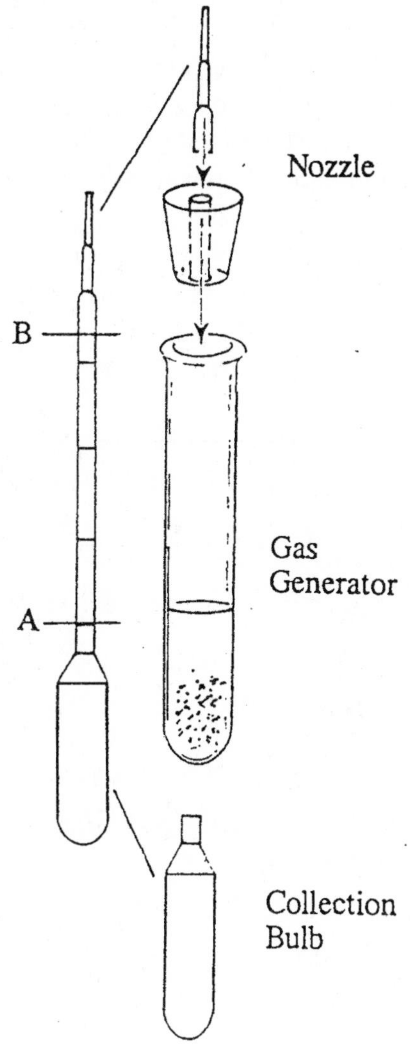

Fig. 6-2

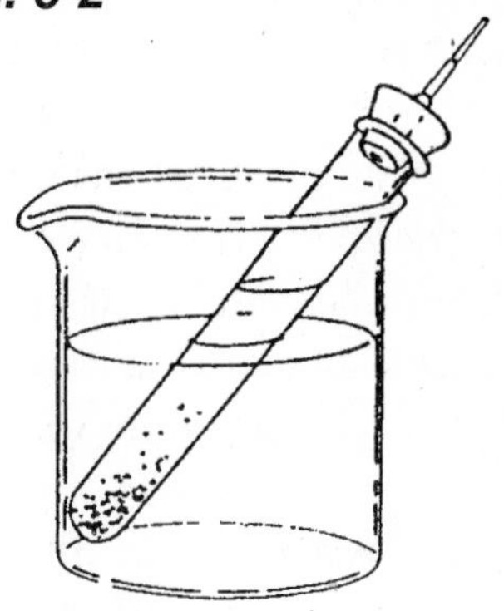

Fig. 6-3

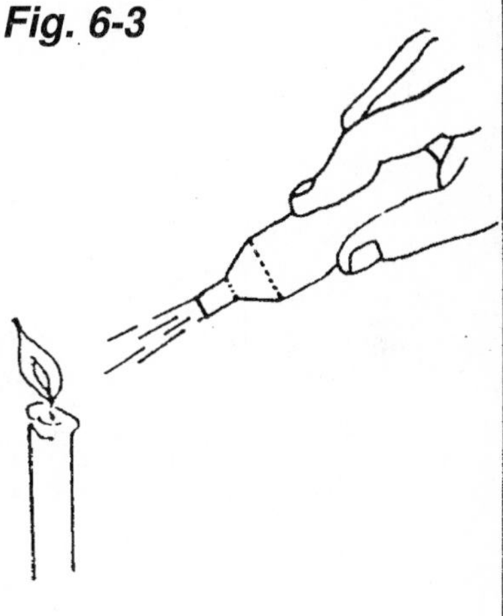

Step 11. Collect a bulb full of oxygen as in *step 7* and squirt the oxygen into the candle flame. Repeat this process a couple of times and report your observations.

Step 12. Now collect half a bulb full of hydrogen, remove the bulb from the hydrogen generator and put it on the oxygen generator until all the water is displaced from the bulb. This should give a 50/50 mixture of hydrogen and oxygen. Squirt the gas mixture into the candle flame and report your observations. Repeat the experiment with several different ratios of hydrogen to oxygen and try to find the optimum mixture.

Step 13. Refill the pipet with the optimum mixture and convert your pipet to a rocket by placing it on the rocket launcher. A rocket launcher can be made by putting a nail through a piece of cardboard, attaching a piece of wire (about 0.5 meters) to the head of the nail (sort of a ground) and putting the "collection bulb rocket" over the point of the nail (*Figure 6-4*). The rocket can be launched by bringing a Tesla coil up to bottom (narrow part) of the rocket so that the spark jumps through the plastic to the nail. [***Caution: Because of potential hazards of electric shock from the coil, the instructor should launch all rockets.***] Have a contest with the other students in the class to determine who can make the farthest flying rocket. Try varying the takeoff angle (nail angle) and the nail diameter (the larger the diameter without creating friction, the better in our experience). **Also try leaving some water in the rocket.**

Step 14. When you are finished with the experiment, place the unreacted manganese and zinc in the appropriate strainers (in the sink). These metals can be reused.

Fig. 6-4

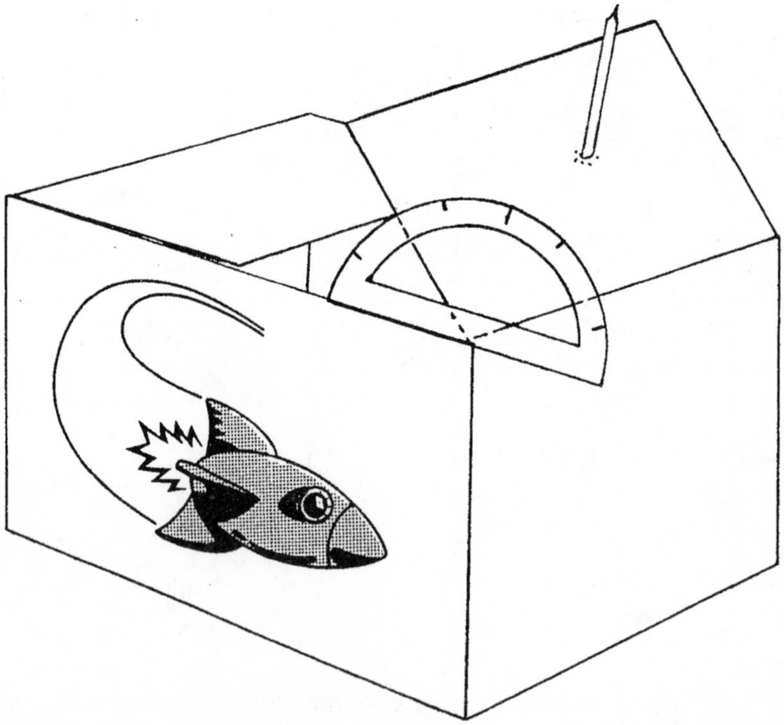

Name________________________________Date__________Lab Section__________

Prelaboratory Problems - *Experiment 6* - Classification of Chemical Reactions

The solutions to the starred problems are in ***Appendix 4***.

1. Classify the following reactions according to combination (CA), decomposition (D), combustion (CU), single replacement (SR) or double replacement (DR) and then balance the equations.

reaction	**classification**
a.* ___$Mg(s)$ + ___$ZnCl_2(aq)$ = ___$MgCl_2(aq)$ + ___$Zn(s)$	_______
b.* ___$AgNO_3(aq)$ + ___$CaCl_2(aq)$ = ___$AgCl(s)$ + ___$Ca(NO_3)_2(aq)$	_______
c.* ___$C_2H_6(g)$ + ___$O_2(g)$ = ___$CO_2(g)$ + ___$H_2O(g)$	_______
d.* ___$Na_2O(s)$ + ___$H_2O_{(l)}$ = ___$NaOH(aq)$	_______
e.* ___$KClO_3(s)$ = ___$KCl(s)$ + ___$O_2(g)$	_______
f. ___$Al(s)$ + ___$HCl(aq)$ = ___$AlCl_3(aq)$ + ___$H_2(g)$	_______
g. ___$C_3H_6O(g)$ + ___$O_2(g)$ = ___$CO_2(g)$ + ___$H_2O(g)$	_______
h. ___$Fe(s)$ + ___$O_2(g)$ = ___$Fe_2O_3(s)$	_______
i. ___$Cl_2(aq)$ + ___$KBr(aq)$ = ___$Br_2(aq)$ + ___$KCl(aq)$	_______
j. ___$Ca(NO_3)_2(aq)$ + ___$K_3PO_4(aq)$ = ___$Ca_3(PO_4)_2(s)$ + ___$KNO_3(aq)$	_______
k. ___$Ca(HCO_3)_2(aq)$ = ___$CaCO_3(s)$ + ___$H_2O(l)$ + ___$CO_2(g)$	_______
l. ___$NaOH(aq)$ + ___$H_3PO_4(aq)$ = ___$Na_3PO_4(s)$ + ___$H_2O(aq)$	_______

2. Complete, balance and classify the following reactions:

a.* ___$BaCl_2(aq)$ + ___$Na_2SO_4(aq)$ =	_______
b.* ___$Fe(s)$ + ___$CuCl_2(aq)$ =	_______
c.* ___$C_6H_6(l)$ + ___$O_2(g)$ =	_______

reaction	**classification**
d.* ____$BaCl_2$(s) + ____H_2O(l) =	________
e. ____$Pb(NO_3)_2$(aq) + ____K_2CrO_4(aq) =	________
f. ____Mg(s) + ____H_2SO_4(aq) =	________
g. ____CaO(s) + ____H_2O(l) =	________
h. ____C_2H_4O(l) + ____O_2(g) =	________
i. ____HNO_3(aq) + ____$Ba(OH)_2$(aq) =	________
j. ____HCl(aq) + ____$KHCO_3$(aq) =	________

3. The Haber process for the preparation of ammonia involves a combination reaction between nitrogen and hydrogen.

   a.* Write a balanced molecular equation for the reaction.

   b.* Give the optimum nitrogen to hydrogen mole ratio for the reaction. ________

   c. Could the following figures be used as a model for the Haber reaction? Explain your answer.

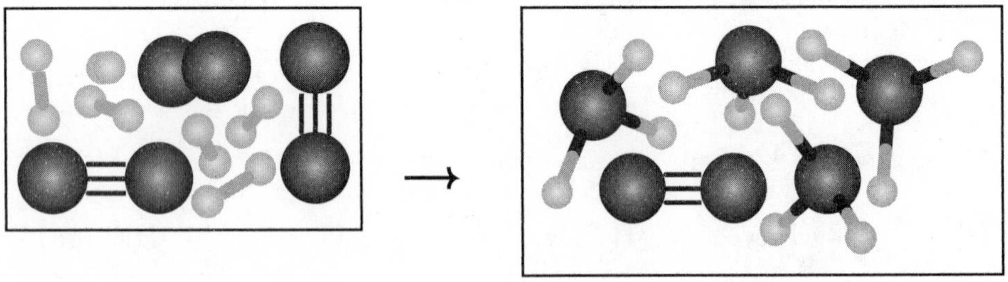

4. The contact process for the preparation of sulfuric acid involves a combination reaction between sulfur dioxide and oxygen to form sulfur trioxide.

   a. Write a balanced molecular equation for the reaction.

   b. Give the optimum sulfur dioxide to oxygen mole ratio for the reaction. ________

Name________________________________Date__________Lab Section__________

Results and Discussion - *Experiment 6* - Classification of Chemical Reactions

A. Classification of reactions. For each system record significant observations (e.g., white precipitate, gas evolution, heat evolution, color change, or *NAR* for no apparent reaction). For numbers 1-9 classify the potential reaction according to combination (CA), decomposition (D), combustion (CU), single replacement (SR), double replacement (DR) and balance the equations. For numbers 10-15, record observations and complete, balance and classify any reactions that occur.

reaction	observations	class.
1.		
____ $CaCl_2(aq)$ + ____ $Na_3PO_4(aq)$ = ____ $Ca_3(PO_4)_2(s)$ + ____ $NaCl(aq)$	________________	______
2.		
____ $CuSO_4(s)$ + ____ $H_2O(l)$ = ____ $CuSO_4 \cdot$ ____ $H_2O(s)$	________________	______
3.		
____ $Cu(OH)_2(s)$ = ____ $CuO(s)$ + ____ $H_2O(g)$	________________	______
4.		
____ $Zn(s)$ + ____ $HCl(aq)$ = ____ $ZnCl_2(aq)$ + ____ $H_2(g)$	________________	______
____ $H_2(g)$ + ____ $O_2(g)$ = ____ $H_2O(g)$	________________	______
5.		
____ $HCl(aq)$ + ____ $Na_2CO_3(aq)$ = ____ $NaCl(aq)$ + $H_2O(l)$ + $CO_2(g)$	________________	______
6.		
____ $H_2O_2(aq)$ = ____ $H_2O(l)$ + ____ $O_2(g)$	________________	______
7.		
____ $C_2H_6O(l)$ + ____ $O_2(g)$ = ____ $CO_2(g)$ + ____ $H_2O(g)$	________________	______
8.		
____ $Sr(OH)_2 \cdot 8H_2O(s)$ + ____ $NH_4Cl(s)$ = ____ $SrCl_2(aq)$ + ____ $NH_3(aq)$ + ____ $H_2O(l)$		
	________________	______

reaction	**observations**	**class.**
9. ____ H_2SO_4(aq) + ____ NaOH(aq) = ____ Na_2SO_4(aq) + ____ H_2O(l)	____________________	______

* * * * * * * * * * * * *

reaction	**observations**	**class.**
10. ____ $CaCl_2$(aq) + ____ Na_2CO_3(aq) =	____________________	______
11. ____ Zn(s) + ____ $CuSO_4$(aq) =	____________________	______
12. ____ Cu(s) + ____ $ZnSO_4$(aq) =	____________________	______
13. ____ $CaCl_2$(aq) + ____ $NaNO_3$(aq) =	____________________	______
14. ____ $CuSO_4 \cdot 5H_2O$(s) =	____________________	______
15. ____ HCl(aq) + ____ NaOH(aq) =	____________________	______

B. Beral pipet rockets

1. Write a balanced chemical equation for the reaction between zinc and hydrochloric acid.

2. What did you observe when the hydrogen was "squirted" into the flame?

3. Write a balanced chemical equation for the reaction of hydrogen in the flame.

4. Write a balanced chemical equation for the decomposition of hydrogen peroxide into oxygen and water (The surface of manganese had air oxidized, prior to its use, to manganese dioxide. MnO_2 catalyzes the decomposition of H_2O_2. But do not include manganese or manganese dioxide in the balanced equation.).

5. What did you observe when the oxygen was "squirted" in to the flame? Explain your observation.

6. Rate the report from each of the hydrogen - oxygen mixtures on a loudness scale of 1 to 10 with 10 being very loud and 1 being no report.

<u>hydrogen/oxygen ratio</u>	<u>rating</u>
________	________
________	________
________	________
________	________
________	________

7. What mixture seemed to give the loudest reports? Is this result consistent with the ratio predicted by the balanced equation (*#B-3*)? Explain your answer.

8. Could the figures below serve as a model for any of the mixtures you tested above? Explain your answer.

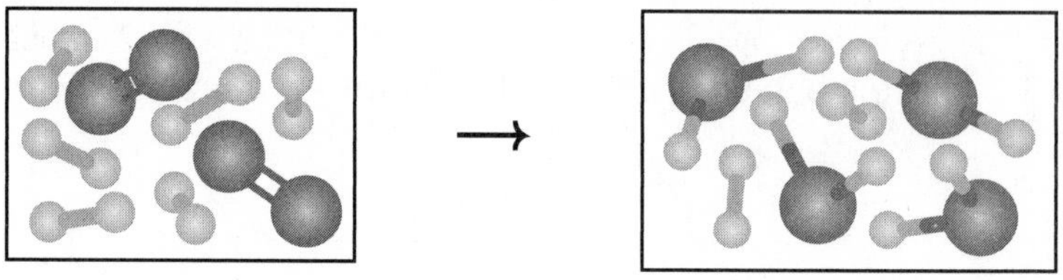

9. Describe the results of your rocket launch. What parameters affect the flight distance?

10. Suggest any ways you can think of to improve any part(s) of this experiment.

11. Some of the *Learning Objectives* of this experiment are listed on the first page of this experiment. Did you achieve the *Learning Objectives*? Explain your answer.

Experiment 7

QUANTITATIVE PRECIPITATION

Antoine Lavoisier
1743 - 1794 (guillotined)

Learning Objectives

Upon completion of this experiment, students will have experienced:
1. Gravity filtration.
2. Stoichiometric calculations.

Text Topics

Formula mass, the mole, hydrates, stoichiometry, experimental yield, theoretical yield, percent yield (for correlation to some textbooks, see page ix).

Notes to Students and Instructor

While the laboratory manipulations of this experiment should not require more than two hours, the best results are obtained if the product is dried several hours and preferably overnight before the final weighing. It is recommended that students return the day after lab for the final weighing.

Discussion

The addition of hydroxide ions to a solution containing copper(II) ions results in the precipitation of copper(II) hydroxide. Subsequent heating *in situ* of the copper(II) hydroxide results in decomposition to copper(II) oxide and water. The CuO can be quantitatively filtered, dried and weighed.

If a weighed amount of a soluble copper(II) compound is used in this procedure, it is possible to calculate the amount of copper(II) oxide that should be formed (the theoretical yield). The experimental yield can be compared with the theoretical yield to calculate the percent yield.

Assuming the procedure is quantitative, the technique can be utilized to determine the mass percent of copper in an unknown copper(II) compound or in a mixture. If the compound or mixture is water soluble and addition of hydroxide ions results in a copper(II) hydroxide precipitate only, the mass determination of copper(II) oxide enables a calculation of the amount of copper in the original sample to be made.

Procedure

Weigh between 1.8 and 2.0 grams of copper(II) sulfate pentahydrate into a 250 mL beaker to the nearest milligram. Add 10 mL of deionized water to the beaker and dissolve the copper salt. Add 10 mL of 6.0 M NaOH to the solution with stirring. Place a watch glass over the beaker and heat the mixture to the boiling point. Try to avoid spattering especially onto the watch glass. If spattering occurs, use a wash bottle to wash all the solid back down into the solution. Heat until all of the blue solid has been decomposed to copper(II) oxide and water (a few minutes). Allow the mixture to cool before filtering.

Fold a 12.5 cm diameter piece of Whatman #1 filter paper (see *Figure 7-1*). Tear off a small outside corner of the paper to improve filtering. Weigh and record the mass of the paper. Place the paper in a long stemmed funnel (see *Figure 7-1*), wet the paper with deionized water from a wash bottle and transfer the previously heated mixture to the filter. Be sure not to overload the filter (the liquid should not rise to less than 0.5 cm from the top of the paper). Be patient and add small portions. While you are filtering, begin the preparation of the second sample for filtration (see bottom paragraph). After all liquid has been transferred to the funnel, use a wash bottle to direct spurts of water at the remaining solid in the beaker and transfer this mixture to the funnel. Continue this procedure until virtually all the solid has been transferred. Now wash the precipitate once more with water and allow it to filter through until dripping ceases. Discard the filtrate which should be colorless and not contain any precipitate. Carefully lift out the paper and unfold it on a watch glass. Place the watch glass in an 105°C oven for a minimum of 3 hours (preferably overnight). Remove and weigh. Calculate the percent yield.

Repeat the above procedure with a 1.8 - 2.0 gram sample of an unknown copper(II) compound supplied by your instructor. Determine the mass percent of copper in the sample.

Fig. 7-1

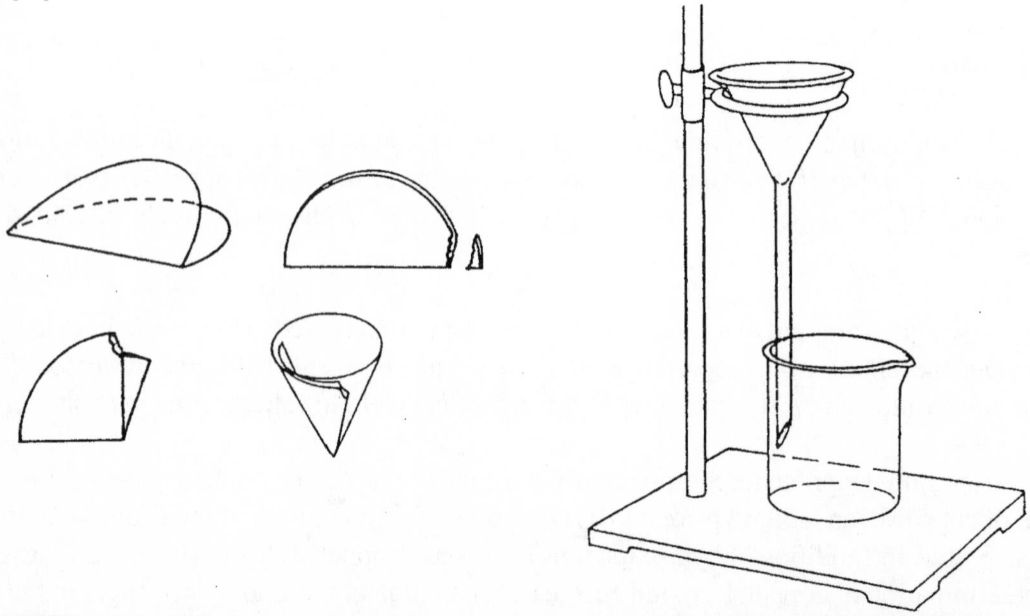

Name____Date___________Lab Section___________

Prelaboratory Problems - *Experiment 7* - Quantitative Precipitation

The solutions to the starred problems are in ***Appendix 4***.

1. Calculate the number of moles in:

   a.* 33 grams of carbon dioxide ________

   b. 12.7 grams of iodine ________

   c. 0.777 grams of cobalt(II) sulfate heptahydrate ________

2. Calculate the mass in grams of:

   a.* 3.45×10^{-2} moles of silver nitrate ________

   b. 8.76×10^{-3} moles of sodium acetate trihydrate ________

3. For the reactions below that are predicted to go (see ***Appendix 3***), write balanced equations. Otherwise write *NAR*.

   a.* magnesium nitrate + sodium hydroxide

   b. barium chloride + sodium hydroxide

   c. zinc chloride + sodium hydroxide

4.* 2.3 grams of sodium reacts violently with excess water and 0.080 grams of hydrogen gas are collected. What are the theoretical and percent yields of hydrogen gas?

5. 1.7 g of benzoic acid ($C_7H_6O_2$) is reacted with an excess of methanol in the presence of a catalyst and 1.7 g of methyl benzoate ($C_8H_8O_2$) is obtained (the mole ratio of benzoic acid to methyl benzoate is 1:1). What are the theoretical and percent yields of methyl benzoate?

6. 2.500 grams of a solid mixture contains barium chloride and sodium chloride. The solid is dissolved in water and an aqueous solution containing an excess of sulfate ions is added. A precipitate is formed, collected, dried and is determined to weigh 2.123 grams. What was the mass percent of barium chloride in the original sample?

7. A 2.50 gram sample of an unknown silver salt (possibly $AgClO_3$, $AgClO_4$, $AgNO_3$ or AgF) is dissolved in water. An excess of HCl solution is added and 1.88 g of AgCl is collected. Determine the identity of the silver salt.

Name_________________________________Date__________Lab Section__________

Results and Conclusions - *Experiment 7* - Quantitative Precipitation

1. Write a balanced molecular equation for the reaction between aqueous copper(II) sulfate and sodium hydroxide.

2. Write a balanced molecular equation for the decomposition of copper(II) hydroxide with heating.

3. Write the overall equation for the process starting with copper(II) sulfate pentahydrate and ending with copper(II) oxide.

4. Calculation of the theoretical yield and percent yield of CuO.

   a. Mass of beaker + $CuSO_4 \cdot 5H_2O$ __________

   b. Mass of empty beaker __________

   c. Mass of $CuSO_4 \cdot 5H_2O$ __________

   d. Formula mass of $CuSO_4 \cdot 5H_2O$ __________

   e. Moles of $CuSO_4 \cdot 5H_2O$ __________

   f. Theoretical number of moles of CuO __________

   g. Formula mass of CuO __________

   h. Theoretical yield in grams of CuO __________

   i. Mass of filter paper + CuO after drying __________

   j. Mass of filter paper __________

   k. Experimental mass of CuO __________

   l. Percent yield of CuO __________

   m. Percent deviation between experimental and theoretical yields __________

n. Show the series of unit conversions that one could use to calculate the theoretical yield in grams of CuO from the grams of $CuSO_4 \cdot 5H_2O$.

5. Calculation of the mass percent of copper in the unknown.

a. Unknown number ________

b. Mass of beaker + unknown ________

c. Mass of beaker ________

d. Mass of unknown ________

e. Mass of filter paper + CuO ________

f. Mass of filter paper ________

g. Mass of CuO ________

h. Moles of CuO ________

i. Grams of copper in CuO (and unknown) ________

j. Mass percent of copper in unknown ________

6. Your unknown was one of the following copper(II) compounds: $CuBr_2$, $CuCl_2 \cdot 2H_2O$, $Cu(NO_3)_2 \cdot 3H_2O$.

a. Calculate the mass percent of copper in each of the possible unknowns.

$CuBr_2$ ________

$CuCl_2 \cdot 2H_2O$ ________

$Cu(NO_3)_2 \cdot 3H_2O$ ________

b. Identify your unknown. Explain your answer. ________

7. Some of the *Learning Objectives* of this experiment are listed on the first page of this experiment. Did you achieve the *Learning Objectives*? Explain your answer.

Name_________________________________Date__________Lab Section__________

Postlaboratory Problems - *Experiment 7* - Quantitative Precipitation

1. Critically evaluate the procedure used in this experiment as a quantitative technique for the analysis of copper(II) ion. Suggest sources of errors and ways of minimizing the errors.

2. Why must the CuO precipitate in the funnel be thoroughly washed with water?

3. Could this procedure be used to determine the mass percent of copper in a sample that also contains magnesium ions? Explain your answer (see *Prelaboratory Problem 3* in this experiment).

4. Several criteria need to be satisfied for quantitative precipitation to be a useful analytical technique. Suggest procedures that might work for the analyses of the samples below or give reasons why the analysis might be very difficult. Assume that at least 1 gram samples are available unless stated otherwise.

   a. the percent of barium in a solid mixture of sodium chloride and barium chloride

   b. the percent of silver in a solid mixture of sodium chloride and silver chloride

   c. the percent of sodium in a solid mixture of sodium chloride and potassium chloride

   d. the percent of copper in 1.00×10^{-3} grams of a mixture containing soluble copper and sodium salts

5. What would the consequences have been in this experiment if you had used 10 mL of 0.1 M NaOH instead of 10 mL of 6.0 M NaOH?

6. Suggest any ways you can think of to improve any part(s) of this experiment.

Experiment 8

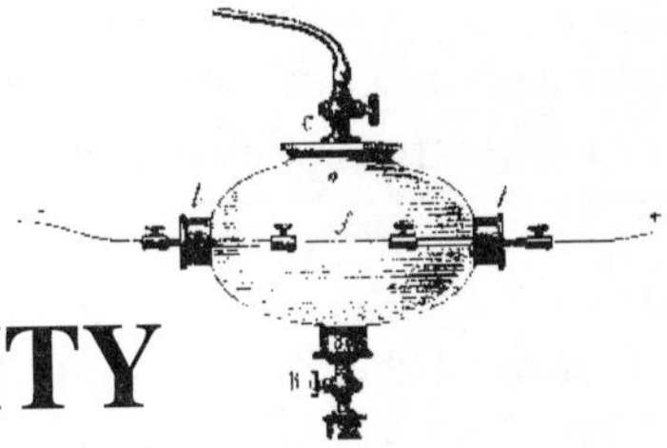

ELECTRICAL CONDUCTIVITY AND ELECTROLYTES

Gas conductance test
W. Magnus, 1850

Learning Objectives

Upon completion of this experiment, students will have experienced:
1. The determination of the electrical conductivity of solids and solutions.
2. A study of the nature of chemical bonding.
3. A study of conductivity change during chemical reaction.

Text Topics

Electrical conductivity, covalent and ionic bonding, double replacement reactions, dissociation of bonds in solution (for correlation to some textbooks, see page ix).

Discussion

Electricity was discovered even before John Dalton set forth the modern concept of the atomic nature of matter (1804). The observation that electricity can travel through matter such as wires (or even your body under unfortunate and shocking circumstances), suggests that ***matter might be electrical in nature.*** In fact, later in the 19th century, scientists probing the nature of electrical beams were able to demonstrate that atoms of all elements are composed of protons and electrons (the discovery of the other particle common to all atoms, the neutron, was not made until 1932). Can you imagine what your life would be like without electricity? 39% of our energy consumption is used for production of electricity (actual conversion to electricity is only about 35% efficient). Have you ever wondered how electricity flows through wires and how your body can serve as an electrical conductor? We will discuss electrical conductance and then use experimental conductance determinations to probe the nature of bonding in several compounds.

For an electrical current there must be a flow of charge. In solid conductors such as copper wire which have metallic bonding, some electrons are held very loosely in the lattice and are relatively free to flow. A current, however, does not consist of continuous movement of the same electrons but is more like a domino effect where one electron moves a short distance before colliding with and transferring its energy to the next electron. Once this electron domino effect is set in motion by an energy source such as a battery or a generator, the flow continues with only a little resistance in metals. Resistance does cause loss and conversion of the electrical energy to heat. Nonmetals do not have loose electrons. They impede electron flow and are electrical insulators.

Passage of a current through a solution occurs in a rather different way. Pure water only conducts under extreme conditions but aqueous solutions containing ions can be good conductors. For conduction to occur in solution, cations must migrate to the negative electrode (cathode) and accept electrons while simultaneously anions migrate to the positive electrode (anode) and deposit electrons. The net result of the double migration is a flow of electrons from the anode to the cathode and completion of the circuit.

The question remains concerning how the solution is provided with a sufficient number of ions to conduct a current. Compounds are composed of two or more elements held together by attractive forces called chemical bonds. Your text will focus attention on two types of bonds, ionic (the electrostatic attraction between cations and anions) and covalent (sharing of a pair of electrons). When ionic compounds dissolve in water, dissociation occurs into hydrated cations and anions that are potential conductors.

$$NaCl(s) \xrightarrow{H_2O} Na^+ + Cl^-$$

If the ionic compound has sufficient solubility in water, then the solution will be conductive and the compound is called an ***electrolyte.*** Except for acids and bases, the dissolving of covalently bonded compounds (molecules) in water does not result in formation of ions. Most molecules are therefore nonelectrolytes. Strong acids with very polar covalent bonds totally dissociate in water and like soluble ionic compounds are strong electrolytes.

$$HCl(g) \xrightarrow{H_2O} H^+ + Cl^-$$

It should be noted that H^+ is a shorthand notation for hydrated protons and H_3O^+ is probably a better representation of the species present. Afterall, H^+ is simply a proton and the concentrated charge certainly attracts solvent molecules to it.

Weak acids and bases only partially dissociate in water and are usually fair to poor electrolytes.

$$HC_2H_3O_2(l) \xrightarrow{H_2O} H^+ + C_2H_3O_2^-$$

$$NH_3(g) + H_2O(l) \longrightarrow NH_4^+ + OH^-$$

The strong bases you will encounter in this course (e.g., sodium hydroxide) are ionic compounds and strong electrolytes.

How can you predict if the bonding is ionic or covalent? As the electronegativity difference between the partners increases, the ionic character of the bond increases. For our purposes here, assume that metal-nonmetal bonds and metal-polyatomic ion bonds are ionic and nonmetal-nonmetal bonds are covalent.

In this experiment, the conductance of several solutions will be studied to determine if the bonding present in the solute is ionic, covalent or covalent with the potential for ion formation.

In addition to studying the conductivities of aqueous solutions containing single compounds, three of the studies will involve the monitoring of conductivity as a second compound is added. For the three examples, double replacement reactions will occur. In double replacement reactions, positive and negative ions switch partners.

$$AgNO_3(aq) + NaCl(aq) = NaNO_3(aq) + AgCl(s)$$

This might result in formation of electrolytes from weak electrolytes or nonelectrolytes from electrolytes.

Procedure

A. Conductivity of solids. A simple device for testing conductivity can be constructed from a 9 volt battery, battery clips, a 1 kiloohm resister and a T 1¾ LED (e.g., Radio Shack 276-041). A light bulb is not satisfactory for testing solution conductance. You will either be supplied with the circuit or the parts to construct one. Test the circuit to check its integrity by touching the probes together. The LED should glow. Now test the conductance of the solids listed in the ***Results and Discussion*** section.

Fig. 8-1

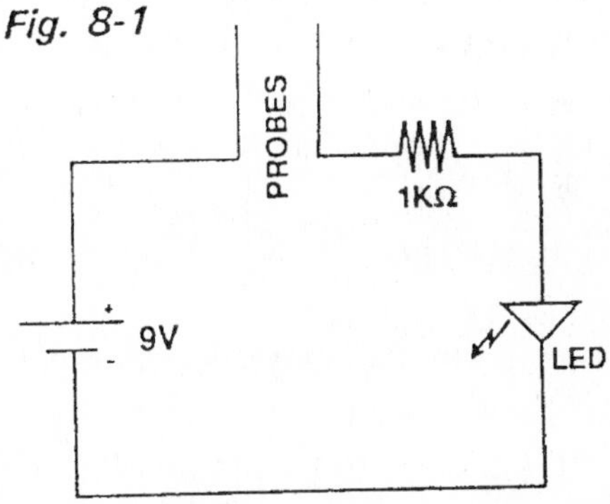

B. Conductivity of solutions. Test the conductance in 50 or 100 mL beakers of the solutions listed in the ***Results and Discussion*** section. Use sufficient solution for conductance measurements with your apparatus (typically about 5 to 25 mL of solution) and attempt to be consistent with the amount of the solution and the technique you apply. For the three samples with a special notation (4, 12 and 14), be sure to follow the additional instructions. For numbers 12 and 14, try to start with volumes of 10 mL or less if possible to minimize the amount of dropwise addition required.

Prelaboratory Problems - *Experiment 8* - Electrical Conductivity and Electrolytes

Before beginning the laboratory manipulations, predict in the ***Results and Discussion*** section which solids and solutions will be strong conductors or electrolytes (SC or SE), fair conductors or electrolytes (FC or FE), weak conductors or electrolytes (WC or WE) or nonconductors or nonelectrolytes (NC or NE). Also predict what your observations will be for the experiments involving the monitoring of the conductivity to follow the progress of a chemical reaction.

Examples: Acetone (CH_3COCH_3) has only covalent bonds. The addition of acetone to water will not add any ions to the water and the solution will not conduct. Sodium nitrate which has ionic bonding is a strong electrolyte and hydrobromic acid (covalently bonded but a strong acid) is also a strong electrolyte. Ant venom (formic acid) is a covalently bonded weak acid and is a fair electrolyte. Barium hydroxide is ionic and sufficiently soluble in water to be a strong electrolyte. If sulfuric acid is added dropwise to a barium hydroxide solution while the conductivity is continuously monitored, the conductivity will gradually decrease.

$$Ba(OH)_2(aq) + H_2SO_4(aq) = BaSO_4(s) + 2H_2O(l)$$

As indicated by the above double replacement reaction, one product is insoluble in water and the other is the nonelectrolyte, water. When the equivalence point is reached, there should not be any ions left in the solution and the solution should be nonconductive. However, continued addition of the strong acid, sulfuric acid, after the equivalence point has been reached should lead to an increase in the number of ions in solution and the conductance should return.

Name__________________________________Date__________Lab Section__________

Results and Discussion - *Experiment 8* - Electrical Conductivity and Electrolytes

A. Conductivity of solids (Mark SC, FC, WC, or NC for strong, fair, weak or non-conductor)

Solid	Prediction	Observed Conductivity
aluminum	________	________
copper	________	________
glass rod	________	________
pencil lead	________	________
plastic strip	________	________
wood splint	________	________

B. Conductivity of solutions (Mark SE, FE, WE, or NE for strong, fair, weak or non electrolyte)

#	Solution	Prediction	Observed Conductivity	
1.	0.1 M NaCl	________	________	
2.	1×10^{-2} M NaCl	________	________	
3.	1×10^{-3} M NaCl	________	________	
4.	crystalline NaCl	________	________	(go to #21)
5.	deionized water	________	________	
6.	tap water	________	________	
7.	0.1 M glucose	________	________	
8.	0.1 M ethanol	________	________	
9.	0.1 M NaOH	________	________	
10.	0.1 M NH_3	________	________	
11.	0.1 M HCl	________	________	
12.	0.1 M $HC_2H_3O_2$ ($\leq$10 mL)	________	________	(go to # 22)
13.	0.1 M H_2SO_4	________	________	
14.	saturated $Ca(OH)_2$ (10 mL)	________	________	(go to # 23)

To answer exercises 15 - 20 below, compare, contrast and explain your observations for the indicated exercises.

15. #1, 4

16. #1, 2, 3

17. #1, 5, 6

18. #1, 7, 8

19. #1, 9, 10

20. #1, 11, 12

For systems 21 - 23, predict what your observations will be when the instructions are followed. Perform the experiment, record and explain your observations utilizing net ionic equations when appropriate.

21. While monitoring conductivity of initially crystalline NaCl, add water dropwise with stirring until all of the NaCl has dissolved.

    Predictions:

    Observations:

    Explanation and net ionic equation:

22. While monitoring the conductivity of 0.1 M $HC_2H_3O_2$ (preferably 10 mL or less), add 0.1 M NH_3 dropwise with stirring until an approximately equal volume of 0.1 M NH_3 has been added.

    Predictions:

    Observations:

    Explanation, molecular and net ionic equations:

23. While monitoring the conductivity of 10 mL of a saturated calcium hydroxide solution, add 40 drops of 0.5 M H_3PO_4, mixing after each addition. Check the conductivity after the addition of each drop.

Predictions:

Observations:

Explanation, molecular and net ionic equations:

24. Suggest any ways you can think of to improve any part(s) of this experiment.

25. Some of the *Learning Objectives* of this experiment are listed on the first page of this experiment. Did you achieve the *Learning Objectives*? Explain your answer.

Experiment 9

IONIC REACTIONS

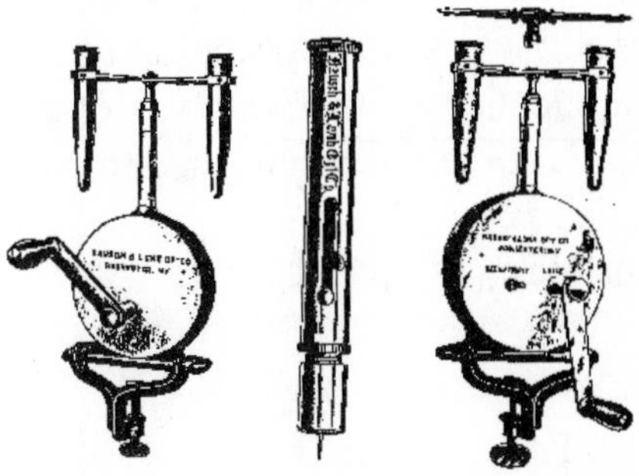

1880 centrifuges

Learning Objectives

Upon completion of this experiment, students will have experienced:

1. The balancing of double replacement equations.
2. The derivation and writing of net ionic equations.
3. The development of schemes for the analysis of unlabeled bottles and identification of cations in a mixture.

Text Topics

Formula equations, net ionic equations, solubility of ionic compounds, double replacement reactions (for correlation to some textbooks, see page ix).

Notes to Students and Instructor

It is very important that students derive schemes for the two challenges in this experiment prior to beginning the laboratory manipulations.

Discussion

In the conductivity experiment, the behavior of ionic and covalently bonded compounds in aqueous solution was explored. You observed and studied double replacement reactions in ***Experiments 1*** and ***6*** and monitored conductivity at different stages of two double replacement reactions in ***Experiment 8***. You should be acquiring the experience needed to enable you to predict if a potential double replacement reaction will go and the observations that should accompany it. This experiment will add to that experience.

When two aqueous solutions are mixed, each containing an ionized or partially ionized compound, a double replacement reaction is possible.

$$MX + NY = MY + NX$$

However, generally only those with at least one of three potential driving forces (formation of an insoluble product, formation of a gas, formation of a weakly ionized compound) actually result in chemical change. Examples of the three common driving forces are given below along with the observation that usually accompanies the reaction.

$$Cu(NO_3)_2(aq) + 2\,NaOH(aq) = \underset{\textbf{blue ppt.}}{Cu(OH)_2(s)} + 2\,NaNO_3(aq)$$

$$2\,HCl(aq) + Na_2CO_3(aq) = 2\,NaCl(aq) + H_2O(l) + \underset{\textbf{gas}}{CO_2(g)}$$

$$HNO_3(aq) + NaOH(aq) = NaNO_3(aq) + H_2O(l) + \textbf{heat}$$

For solubility information, reference to solubility tables (e.g., ***Appendix 3***) is often necessary but it is useful to remember that nitrates, acetates, ammonium compounds and compounds of the Group IA elements are generally soluble. The three most commonly encountered insoluble compounds in textbooks are silver chloride, barium sulfate and calcium carbonate. Except for CO_2 (from H_2CO_3), SO_2 (from H_2SO_3), H_2S and NH_3, gas evolution will generally not be observed because almost all ionic compounds are solids at room temperature. The most commonly encountered weakly ionized compounds are water, acetic acid and ammonia.

An alternative method to determine if a reaction will go is to derive its net ionic equation. To do this recall from the conductivity experiment that ionic compounds and strong acids totally ionize in water.

$$NaCl \xrightarrow{H_2O} Na^+ + Cl^-$$

$$K_2SO_4 \xrightarrow{H_2O} 2\,K^+ + SO_4^{2-}$$

$$HCl \xrightarrow{H_2O} H^+ + Cl^-$$

To arrive at a net ionic equation (until you have considerable experience), start by writing a total ionic equation. Write all **totally** ionized, **dissolved** chemicals in ionic form. Leave insoluble compounds and weakly ionized compounds in molecular form and do not break up polyatomic ions. For example, water (H_2O), acetic acid ($HC_2H_3O_2$) and formulas of insoluble compounds ($BaSO_4$) are written intact but dissolved, ionic compounds such as sodium phosphate are written as ions ($3\,Na^+ + PO_4^{3-}$).

For the reaction between copper(II) nitrate and sodium hydroxide:

$$Cu(NO_3)_2(aq) + 2\,NaOH(aq) = 2\,NaNO_3(aq) + Cu(OH)_2(s)$$

the total ionic equation is:

$$Cu^{2+} + 2\,NO_3^- + 2\,Na^+ + 2\,OH^- = 2\,Na^+ + 2\,NO_3^- + Cu(OH)_2(s)$$

Note that the subscripts that are not part of a polyatomic formula become coefficients. The product, $Cu(OH)_2$, is written in molecular form as it is insoluble in water. When you see an algebraic equation in the form $a + b = a + c$, you would subtract a from both sides of the equation leaving $b = c$. The process going from the total ionic to the net ionic equation is the same as with the algebraic equation. Sodium and nitrate ions are unchanged in the reaction (spectator ions) and are dropped leaving the net ionic equation:

$$Cu^{2+} + 2\,OH^- = Cu(OH)_2(s)$$

This equation and all net ionic equations are of great importance because they enable you to focus on the actual chemical change that occurs. Try to write the net ionic equation for the double replacement reaction between HCl and Na_2CO_3 by first writing the total ionic equation and then eliminating the spectator ions. You should obtain:

$$2\,H^+ + 2\,Cl^- + 2\,Na^+ + CO_3^{2-} = H_2O(l) + CO_2(g) + 2\,Na^+ + 2\,Cl^-$$

$$2\,H^+ + CO_3^{2-} = H_2O(l) + CO_2(g)$$

Make sure when you are finished that the elements and charges are balanced. For cases where no reaction should occur, the algebraic form of the net ionic equation should end up $0 = 0$!

Because it is usually possible to correctly predict observations and products for double replacement reactions, it is often possible to develop schemes for the qualitative analysis of cations and anions and this will be the topic of today's experiment.

Procedure

A. Unlabeled bottles. The challenges that follow will be considerably facilitated if you develop prediction schemes prior to performing the laboratory exercise. The first experiment includes a set of seven bottles labeled only *A to G*. The identities of the solutions they contain are listed in the ***Results and Discussion*** section. (Note: mixtures of any two can undergo only double replacement reactions (possibly followed by decomposition) although several will not react at all). Your challenge is to assign correct names to each bottle based on a comparison of a prediction matrix with an experimental matrix. You are allowed to mix the contents of any of the bottles but no other reagents, test papers or solutions are permitted. To enable you to solve this problem and to develop your ability to write double replacement equations and net ionic equations, you should complete the prediction equations and matrix on pages 109 - 112 before beginning this lab. For example, what should be observed if barium chloride is mixed with sodium sulfate? First write and balance the double replacement reaction:

$$BaCl_2(aq) + Na_2SO_4(aq) = BaSO_4(s) + 2\,NaCl(aq)$$

Next ask if either of the products is insoluble, a weakly ionized compound or a gas or a compound that decomposes into a gas. As both products in this case are ionic and almost all ionic compounds are solids at room temperature, the only question that needs further consideration concerns solubility. Either from memory or a table (for example, see ***Appendix 3***) you should conclude that barium sulfate is insoluble and a reaction will take place. The net ionic equation is:

$$Ba^{2+} + SO_4^{2-} = BaSO_{4(s)}$$

By studying and comparing the pattern of precipitates, gas and heat evolution, you should be able to assign identities to the seven bottles. Pay special attention to unique observations such as heat or gas evolution. These observations can considerably narrow down the choices. Experimentally, one approach is to pour about 2 mL of *A* into 6 test tubes and add 2 mL of *B* to the first tube, two mL of *C* to the second and so on being sure to mix and observe each tube with each addition.

B. Qualitative analysis. The second challenge requires even a little more creativity on your part. You will be provided with a solution containing the cations Li^+, Mg^{2+}, Na^+, and Sr^{2+}. Your goal is to develop a scheme for detecting the presence of each ion in the mixture. In addition to using a solubility table (see ***Appendix 3***) and observing double replacement reactions, you will need to make use of flame tests. Also you will probably find that to test for at least one of the ions, it will first be necessary to remove other ions that would interfere with your proposed test. To do this you might want to use a centrifuge.

Flame tests. The energy provided by a Bunsen burner flame is capable of promoting the ground state electrons of some elements to higher energy orbitals resulting in an excited state. For some cations, the return of the exited electron to the ground state results in the emission of a characteristically colored flame.

Ion	Color
barium	green
calcium	red-orange
lithium	red
potassium	violet (masked by sodium - use cobalt glass)
sodium	yellow-orange
strontium	red
zinc	greenish-white (weak)

To perform a flame test, first adjust the flame to a pure blue. Then clean a platinum wire by alternately inserting it into dilute (6M) HCl and into the flame until the insertion yields little or no flame color. Now dip the wire into a test solution, hold it in the flame and observe the color. Try this with solutions containing each of the possible ions, Li^+, Mg^{2+}, Na^+, and Sr^{2+} and a mixture of all four. Be sure to clean the wire each time you change test solutions.

Centrifuge. Centrifuging is a quick method for separating liquid from solid and is especially useful for small volumes such as you will have in this experiment. Often in qualitative analysis schemes, it is necessary to remove ions that would interfere with other tests by precipitating them out of solution. When a precipitate is formed, the solution is centrifuged. The decantate is poured off into a clean test tube and tested again with the precipitating agent. If further precipitation occurs, centrifuging is required again. Otherwise the decantate is ready for continued testing. The precipitate is washed with water to remove soluble ions, centrifuged and the wash liquid is discarded. Washing should be repeated two times. The presence of the precipitate may be conclusive enough evidence for the presence of an ion. Or the precipitate, now that it is free of other ions, may be redissolved (often by acid) and subjected to confirmatory testing.

Lets consider as an example a solution that could contain Ag^+, Ba^{2+}, Cu^{2+} and Na^+. A blue color should indicate the presence of copper(II) and a colorless solution its absence. Examination of ***Appendix 3*** reveals that of the possible ions, only Ag^+ forms an insoluble chloride. If addition of HCl to the mixture yields a white precipitate, Ag^+ is present. After centrifuging, the precipitate should be properly disposed of and the decantate tested with sodium sulfate solution. If barium ion is present, a precipitate will form. Again the precipitate can be discarded and the decantate tested with sodium hydroxide. A precipitate or absence thereof should confirm the earlier conclusion about copper(II). Now a flame test may be run on the original sample to check for sodium ion. The bright yellow-orange sodium flame should be visible and may mask the flames of barium and/or copper. A flow diagram should be constructed to help keep track of the sequence.

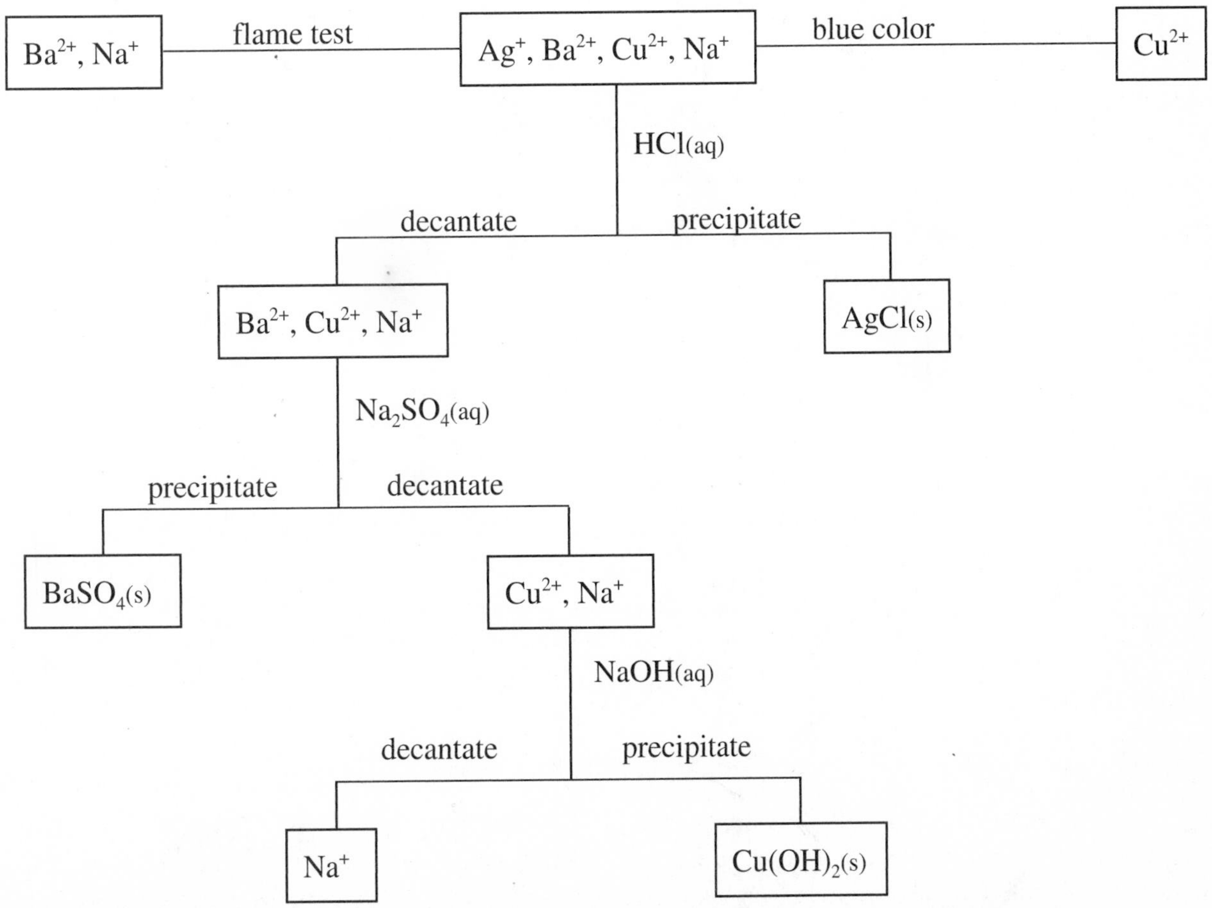

Now, for the ions, Li^+, Mg^{2+}, Na^+ and Sr^{2+}, develop a scheme and write a flow diagram for your scheme. Test the scheme on a mixture containing all four cations. If it works, obtain an unknown from your instructor that could contain 1 to 4 of the cations and analyze it. Otherwise, modify your scheme and try again.

Important and useful tips. Lithium phosphate, while fairly insoluble is slow to form at room temperature. All solutions in **Parts A and B** suspected of containing lithium phosphate should be heated in a hot water bath. This considerably speeds up the precipitation process. Also, strontium hydroxide is marginally soluble and formation of a precipitate is usually slow and dependent on the amounts of reagents used. However, the slight cloudy precipitate that results might be useful for characterizing the precipitate. The mixture of sulfuric acid with potassium phosphate gives a slightly exothermic reaction. The attention given to the nature of this reaction is up to the instructor. Also we have found that commercial potassium phosphate is sometimes contaminated with carbonate and alert students observe some bubbling from the reaction of the contaminant, potassium carbonate, with sulfuric acid.

Name________________________________ Date__________ Lab Section__________

Prelaboratory Problems - *Experiment 9* - Ionic Reactions

The solutions to the starred problems are in ***Appendix 4***.

1.* Patients with gastrointestinal pain are often given a "milk shake" containing barium sulfate to drink and then they are fluoroscoped. Barium salts are very toxic. Why doesn't barium sulfate kill the patient and why is it used?

2. The labels of bottles that contain barium hydroxide, hydrochloric acid and sodium carbonate had fallen off. When the contents of bottle *A* were mixed with *B's* contents, a white precipitate formed, *A* with *C* yielded a gas and *B* with *C* gave heat. Identify *A*, *B*, *C* and write molecular and net ionic equations for the chemical changes observed.

3. These questions deal with the scheme for Ag^+, Ba^{2+}, Cu^{2+} and Na^+ discussed in the ***Procedure*** section.

   a.* Could addition of sodium sulfate be used in the first step? Explain your answer.

   b. Could addition of sodium hydroxide be used in the first step? Explain your answer.

4. Write molecular and net ionic equations and fill in the prediction matrix before coming to lab for *reactions 1 - 21* on pages 109- 112.

5. Develop an analysis scheme and flow diagram before coming to lab and fill it in on page 114.

Name_________________________ Date_________ Lab Section_________

Results and Discussion - *Experiment 9* - Ionic Reactions

A. Unlabeled bottles. Seven unlabeled bottles containing the solutions below in some scrambled sequence will be provided.

1.5 M H_2SO_4	0.1 M $Mg(NO_3)_2$	3 M NaOH
1 M K_3PO_4	1 M Na_2CO_3	0.1 M $SrCl_2$
0.5 M Li_2SO_4		

For the 21 possible mixtures of the seven solutions, write formula (FE), total ionic (TIE) and net ionic (NIE) equations. When no reaction is expected, write *NAR* for no apparent reaction. Based on these equations, fill in the prediction matrix. Based on your experimental observations, fill in the observation matrix. Compare the two and assign identities to *A* - *G*. (For solubilities, see ***Appendix 3***)

1. sulfuric acid + potassium phosphate

FE ____

TIE ____

NIE ____

2. sulfuric acid + lithium sulfate

FE ____

TIE ____

NIE ____

3. sulfuric acid + magnesium nitrate

FE ____

TIE ____

NIE ____

4. sulfuric acid + sodium carbonate

FE ____

TIE ____

NIE ____

5. sulfuric acid + sodium hydroxide

FE ____

TIE ____

NIE ____

6. sulfuric acid + strontium chloride

FE ____

TIE ____

NIE ____

7. potassium phosphate + lithium sulfate

FE ____

TIE ____

NIE ____

8. potassium phosphate + magnesium nitrate

FE ____

TIE ____

NIE ____

9. potassium phosphate + sodium carbonate

FE ____

TIE ____

NIE ____

10. potassium phosphate + sodium hydroxide

FE ______________________________

TIE ______________________________

NIE ______________________________

11. potassium phosphate + strontium chloride

FE ______________________________

TIE ______________________________

NIE ______________________________

12. lithium sulfate + magnesium nitrate

FE ______________________________

TIE ______________________________

NIE ______________________________

13. lithium sulfate + sodium carbonate

FE ______________________________

TIE ______________________________

NIE ______________________________

14. lithium sulfate + sodium hydroxide

FE ______________________________

TIE ______________________________

NIE ______________________________

15. lithium sulfate + strontium chloride

FE ______________________________

TIE ______________________________

NIE ______________________________

16. magnesium nitrate + sodium carbonate

FE ______________________________

TIE ______________________________

NIE ______________________________

17. magnesium nitrate + sodium hydroxide

FE ______________________________

TIE ______________________________

NIE ______________________________

18. magnesium nitrate + strontium chloride

FE ______________________________

TIE ______________________________

NIE ______________________________

19. sodium carbonate + sodium hydroxide

FE ______________________________

TIE ______________________________

NIE ______________________________

20. sodium carbonate + strontium chloride

FE ______________________________

TIE ______________________________

NIE ______________________________

21. sodium hydroxide + strontium chloride

FE ______________________________

TIE ______________________________

NIE ______________________________

Prediction Matrix

	K_3PO_4	Li_2SO_4	$Mg(NO_3)_2$	Na_2CO_3	NaOH	$SrCl_2$
H_2SO_4	1	2	3	4	5	6
K_3PO_4		7	8	9	10	11
Li_2SO_4			12	13	14	15
$Mg(NO_3)_2$				16	17	18
Na_2CO_3					19	20
NaOH						21

Experimental Observation Matrix

	B	C	D	E	F	G
A						
B						
C						
D						
E						
F						

label color ___________

A = _______________ D = _______________ F = _______________

B = _______________ E = _______________ G = _______________

C = _______________

B. Qualitative Analysis.

1. Prepare a flow diagram for the analysis of the cations, Li^+, Mg^{2+}, Na^+, and Sr^{2+}.

Li^+, Mg^{2+}, Na^+, Sr^{2+}

2. List the steps of your procedure for your known (a mixture of all four cations) followed by your observations and molecular and net ionic equations that account for your observations.

Procedure

Observations (for the known) Equations

3. Give your observations and conclusions for your unknown.

4. Cations present in your unknown (unknown # = ____), ______________________

5. Suggest any ways you can think of to improve any part(s) of this experiment.

6. Some of the *Learning Objectives* of this experiment are listed on the first page of this experiment. Did you achieve the *Learning Objectives*? Explain your answer.

Experiment 10

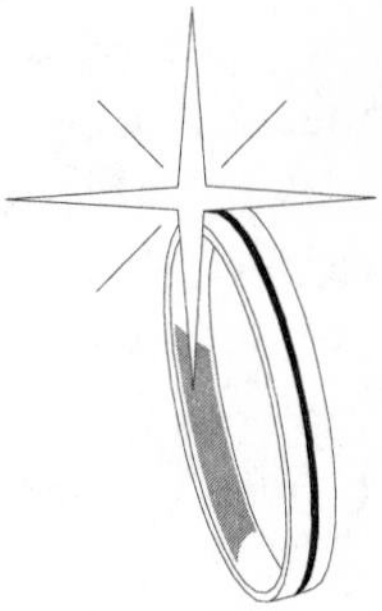

ACTIVITIES OF METALS

Learning Objectives

Upon completion of this experiment, students will have experienced:

1. Observation of simple single replacement reactions
2. The preparation of an activity series of metals

Text Topics

Single replacement reactions, oxidation-reduction reactions, net ionic reactions, activity series for metals (for correlation to some textbooks, see page ix).

Notes to Students and Instructor

This experiment is relatively short and can be done with another short experiment in the same laboratory period. A scheme should be devised prior to the laboratory period for determining the activity series using the minimum number of reactions.

Discussion

In previous experiments, you have observed that elemental zinc reacts with hydrochloric acid to give zinc chloride and hydrogen gas and that zinc reacts with copper sulfate to give elemental copper and zinc sulfate. Both of these reactions are examples of single replacement reactions (reactions where a reactant in elemental form replaces an element in a compound). Since elemental zinc spontaneously replaces copper ion from solution, as might be expected, elemental copper does not spontaneously react with zinc sulfate.

A ball rolls downhill on its own but will only go uphill if there is an external input of energy. Chemical reactions are the same. If they are spontaneous (spontaneous means they can go on their own) in one direction, the reverse reaction will be non-spontaneous.

Single replacement reactions are a type of oxidation-reduction reaction where oxidation means a decrease in the number of electrons and reduction is an increase in the number of electrons. It is possible to rank metals in terms of a decreasing tendency to undergo oxidation. Those that oxidize the easiest will be considered to be the most active so this arrangement is often referred to as an activity list. The activity list is useful as a predictive tool. If we find that A replaces B and B replaces C, the resulting activity list would be A>B>C and we would predict that A should replace C even though the particular reaction has not yet been performed.

Procedure

Your goal will be to devise a series of reactions with the reactants below that utilize the minimum number of reactions necessary for the ranking of the six elements copper, hydrogen, iron, magnesium, sodium and zinc.

available reactants

compounds	***elements***
copper(II) sulfate	copper
iron(III) nitrate	iron
magnesium sulfate	magnesium
sodium sulfate	
sulfuric acid	
zinc sulfate	zinc

The compounds will all be available in aqueous solutions. Each reaction you propose and test will be between an element and a compound. Sodium and hydrogen will not be available in elemental form but you should still be able to devise a scheme that will enable you to rank sodium and hydrogen along with the other four elements. Also you should be able to devise a scheme that will not have to involve testing of every possible combination.

Write balanced net ionic equations for those reactions that go. Predict results for a few reactions that you did not have to try and experimentally test your predictions.

Useful information. In the presence of some salts, magnesium apparently replaces hydrogen in water and produces hydrogen gas. The observation of bubbles then, while important, should not be accepted as evidence of the single replacement reaction that is being investigated.

Name________________________________ Date__________ Lab Section__________

Prelaboratory Problems - *Experiment 10* - Activities of Metals

The solutions to the starred problems are in ***Appendix 4***.

1. Name the following:

a.	$FeCl_3$	____________	d.	$Zn_3(PO_4)_2$	____________
b.*	$Cu(OH)_2$	____________	e.	$Bi(NO_3)_3$	____________
c.	SnO_2	____________	f.	$CoSO_4$	____________

2.* Elemental copper reacts with silver ion spontaneously to give copper(II) ion and elemental silver.

a. Write a net ionic equation for this reaction.

b. Which is more active, copper or silver? Explain your answer. __________

c. In this reaction, is copper oxidized or reduced? Explain your answer. __________

3. a. Given the following information about the elements aluminum, lead, potassium and silver, write an activity series that accounts for the observations. Aluminum does not replace potassium but (theoretically) does replace lead. Silver does not replace lead.

________ > ________ > ________ > ________

b. According to the activity list above, would aluminum replace silver ion? If so, write a net ionic equation for this reaction.

4. Your goal in this experiment will be to devise a series of reactions with the reactants below that utilize the <u>minimum</u> number of reactions necessary for the ranking of the six elements copper, hydrogen, iron, magnesium, sodium and zinc. Luck can play a role in this as each reaction choice should be based on previous results. If you start with certain ones, it will lead you to step through this experiment with fewer reactions than for other choices.

<u>available reactants</u>

compounds	***elements***
copper(II) sulfate	copper
iron(III) nitrate	iron
magnesium sulfate	magnesium
sodium sulfate	
sulfuric acid	
zinc sulfate	zinc

a. Explain why it is unnecessary to run both of the following reactions:

zinc + copper(II) sulfate, copper + zinc sulfate

b. Explain why it would have been inadvisable to supply hydrogen and sodium in elemental form and why it is not necessary.

Name____________________________________Date__________Lab Section__________

Results and Discussion - *Experiment 10* - Activities of Metals

1. List all reactions attempted, complete and balance net ionic equations for those that go and write *NAR* for those that apparently do not go. Remember to try and perform the minimum number [you should not need all the answer spaces (a to t) below] of reactions necessary for the determination of the activity series for the six elements copper, hydrogen, iron, magnesium, sodium and zinc.

	Reactants		Products	Observations
a.	__________	=	__________	__________
b.	__________	=	__________	__________
c.	__________	=	__________	__________
d.	__________	=	__________	__________
e.	__________	=	__________	__________
f.	__________	=	__________	__________
g.	__________	=	__________	__________
h.	__________	=	__________	__________
i.	__________	=	__________	__________
j.	__________	=	__________	__________
k.	__________	=	__________	__________
l.	__________	=	__________	__________
m.	__________	=	__________	__________
n.	__________	=	__________	__________
o.	__________	=	__________	__________
p.	__________	=	__________	__________
q.	__________	=	__________	__________
r.	__________	=	__________	__________
s.	__________	=	__________	__________
t.	__________	=	__________	__________

2. Based on the results of the reactions that you attempted, rank the six elements, copper, hydrogen, iron, magnesium, sodium and zinc from most active (best reducing agent) to least active.

a. ________ d. ________

b. ________ e. ________

c. ________ f. ________

3. Hopefully, you did not have to run the 20 combinations possible with the supplied chemicals to determine the activity series. Predict the outcome of four reactions that you did not attempt and experimentally test your predictions.

	Reactants		Products	Prediction	Observation
a.	________________	=	________________	________	________
b.	________________	=	________________	________	________
c.	________________	=	________________	________	________
d.	________________	=	________________	________	________

4. The literature activity series from most active to least active is most commonly listed as follows: Li>K>Ba>Ca>Na>Mg>Al>Mn>Zn>Cr>Fe>Cd>Co>Ni>Sn>Pb>H_2>Cu>Hg>Ag>Pt>Au

   a. Do your experimental results agree with this ranking? Explain your answer.

   b. Do you notice any correlation between activity and position in the periodic table. Comment especially with regard to periodicity going down groups and across periods [Note: Later in this course (e.g., ***Experiment 32***), you will study standard reduction potentials, a more quantitative method for comparing the activities of chemical species. You will find that "activity" is very complex due to oxidation to different oxidation states and intimately involves interaction with the solvent. These complications may make expectations for correlation unreasonable.].

   c. Besides for appearance, suggest a reason that silver, gold and platinum are used in jewelry.

5. Some of the *Learning Objectives* of this experiment are listed on the first page of this experiment. Did you achieve the *Learning Objectives*? Explain your answer.

Experiment 11

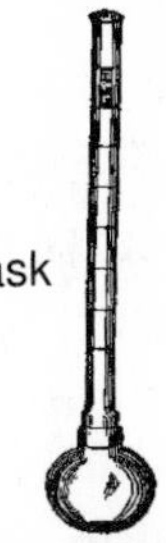

1890 volumetric flask

QUANTITATIVE SOLUTION CHEMISTRY

Learning Objectives

Upon completion of this experiment, students will have experienced:

1. Preparation of standard solutions.
2. Use of volumetric flasks and burets.
3. Determination of the stoichiometry of a reaction in solution.
4. Determination of the concentration of a ferrocyanide solution.

Text Topics

Molarity, standard solutions, titrations, stoichiometry of reactions in solution (for correlation with some textbooks, see page ix).

Discussion

Consider the chemical reactions that you have run in the last few weeks. Most were performed in solution. Why? Reactions between two solids are rare because of the immobility of bound ions. The high degree of ionic mobility available in solution makes solution the preferred medium for most preparative chemical reactions. As chemical reactions and analysis are usually performed in solution, it is imperative that we be able to prepare solutions of known concentrations accurately and be familiar with the techniques for determining unknown concentrations.

As you have probably learned, chemists prefer the use of molarity (moles/liter = M) to express the concentration of a solute in a solution. Make sure you remember that the denominator represents total volume or liters of solution and not just liters of solvent. This means solutions must be prepared by dilution **to** the desired volume, **not with** the desired volume.

Your goals will be to prepare solutions of known concentration of potassium ferrocyanide and zinc sulfate, determine the stoichiometry of the reaction between zinc ion and ferrocyanide and determine the ferrocyanide molarity in a solution of unknown concentration or the zinc ion content in a commercial tablet. Solutions of known concentration are usually prepared by one of three techniques: diluting a measured mass of substance to a certain volume, quantitative dilution of a solution of known concentration, or by preparation of an approximately known concentration followed by a quantitative chemical reaction technique (such as a titration).

Solutions of accurately known concentrations can only be prepared directly by weighing out a solute that can be obtained in pure form (>99.9%) and is relatively nonhygroscopic. Compounds that meet these criteria are called primary standards. Concentrations of other compounds must be determined by a quantitative chemical technique and their solutions are called secondary standards. Titration is the name of a commonly used technique that utilizes a primary standard to standardize a solution whose concentration either is not known or not known accurately enough. Ideally in a titration, you will determine the volume of a solution of accurately known concentration that is required to exactly complete a stoichiometric reaction with a solution of unknown concentration.

For the known, molarity multiplied by volume (in Liters) gives the number of moles used. Now multiplication by the mole ratio from the balanced equation gives the moles of the other reactant or product. The number of moles can then be used depending on what else is known to calculate concentration, molecular mass or mass percent. To determine when a stoichiometric mixture has been achieved, either a chemical indicator or an instrumental technique is utilized. It should be recognized, however, that although indicators are selected because they change color when the desired stoichiometric reaction is complete, the color change does not occur at the exact equivalence point. Thus there are small but real errors inherent in an indicator monitored titration that can be minimized by careful indicator selection.

Procedure

A. Stoichiometry of the reaction. The titration of ferrocyanide with zinc ion will be studied. The usual goal is to use one solution of known concentration to determine the concentration of the second solution. Today however, both reactants will be considered to be primary standards and you will use the titration to determine the stoichiometry of the reaction. The four most reasonable reactions are:

$$K_4Fe(CN)_6(aq) + ZnSO_4(aq) = K_2ZnFe(CN)_6(s) + K_2SO_4(aq)$$

$$K_4Fe(CN)_6(aq) + 2\,ZnSO_4(aq) = Zn_2Fe(CN)_6(s) + 2\,K_2SO_4(aq)$$

$$2\,K_4Fe(CN)_6(aq) + ZnSO_4(aq) = K_6Zn[Fe(CN)_6]_2(s) + K_2SO_4(aq)$$

$$2\,K_4Fe(CN)_6(aq) + 3\,ZnSO_4(aq) = K_2Zn_3[Fe(CN)_6]_2(s) + 3\,K_2SO_4(aq)$$

The products for 3 of the 4 reactions are called mixed salts because they contain 2 different cations, potassium and zinc. The experiment will enable you to calculate the molar ratio of potassium ferrocyanide to zinc sulfate which should enable you to select among the four most reasonable reactions. Once this has been accomplished, you will perform additional titrations to determine the concentration of potassium ferrocyanide in an unknown or the amount of zinc ion in a commercially available tablet.

0.025 M Potassium ferrocyanide solution. Calculate the amount of potassium ferrocyanide trihydrate [$K_4Fe(CN)_6 \cdot 3H_2O$] needed to prepare 250 mL of 0.025 M solution. *The stockroom should provide $K_4Fe(CN)_6 \cdot 3H_2O$ that is as pure as possible.* Weigh out about the calculated amount in a beaker to at least the nearest 0.001 g. Transfer it using a funnel to a 250 mL volumetric flask. Rinse the beaker with deionized water and add this to the funnel. Repeat the washing process until all of the solids in the beaker and the funnel have been washed into the volumetric flask. Add water until the bulb is about ⅔ full and swirl until the solids are totally dissolved. Add water up to the mark using a dropper as you approach the mark. Stopper, invert and swirl and repeat several times. Calculate the concentration of $K_4Fe(CN)_6 \cdot 3H_2O$ to the appropriate number of significant figures.

0.050 M Zinc sulfate soln. Calculate the mass of zinc sulfate heptahydrate [$ZnSO_4 \cdot 7H_2O$] needed to prepare 250 mL of 0.050 M solution. *The $ZnSO_4 \cdot 7H_2O$ should be as pure as possible.* Weigh out about the calculated amount in a beaker to a least the nearest 0.001 g. Transfer it to a funnel in a 250 mL volumetric flask as above for $K_4Fe(CN)_6 \cdot 3H_2O$, dilute to the mark with deionized water and swirl. Calculate the concentration of $ZnSO_4 \cdot 7H_2O$ to the appropriate number of significant figures.

Additional solutions. Also needed for the titrations are freshly prepared aqueous 1% potassium ferricyanide and a solution of 1% diphenylamine in concentrated sulfuric acid. The latter solution should be prepared by the instructor. To prepare 200 mL of diphenylamine indicator, dissolve 2.0 grams of diphenylamine in 100 mL of concentrated sulfuric acid. Weigh 100 grams of ice into a 400 mL beaker. Add the diphenylamine in sulfuric acid to the ice with stirring. Cool the mixture in an ice bath.

Filling a buret with zinc sulfate solution. To rinse a 25 mL buret (if water beads in the buret, it needs further cleaning), add about 5 mL of the zinc sulfate solution. Run a little through the stopcock, then tip it horizontal and rotate it to wet all of the inside with the solution and pour it out the top. Repeat two more times and load the buret with zinc sulfate solution. Mount the buret in a buret clamp and open the stopcock and allow the solution to flow out until the air bubbles in the tip are gone. The reading at the top should be at or below 0.00 mL. Remember one digit beyond the last set of graduations is read and this means **you estimate the reading to the nearest 0.01 mL.**

Pipeting and titration. Pipet (see *Procedure* in ***Experiment 3***) 25.00 mL of the $K_4Fe(CN)_6$ solution into a 250 mL Erlenmeyer flask. Add ~5 mL of 3.0 M sulfuric acid, 3 drops of freshly prepared 1% potassium ferricyanide (oxidizes the diphenylamine to its colored form in the absence of ferrocyanide) and 4 drops of diphenylamine indicator. Set the flask on a white piece of paper directly below the buret with the buret tip about 1 cm into the flask. Read the buret and add the zinc sulfate solution at a moderate rate with swirling until you approach the end point. At this time, you should add titrant in half drop quantities pausing and checking for the end point each time after swirling. Half drops are easily delivered by quickly turning the closed stopcock 180°. The end point will be observed with the appearance of a tinge of blue-violet color which will intensify slightly upon standing. Repeat the titration two more times and determine the stoichiometry of the reaction.

B. (Option 1) Concentration of ferrocyanide in an unknown. Your instructor will give you a potassium ferrocyanide solution of unknown concentration. Perform three titrations on the unknown using the same procedure as in the previous paragraph (except the endpoint probably won't be near 18 mL) and determine the concentration of the ferrocyanide in the solution by using the mole ratio of potassium ferrocyanide to zinc sulfate determined in the first part of this experiment.

C. (Option 2) The amount of zinc ion in a commercially available tablet. Pipet 25.00 mL of the 0.025 M $K_4Fe(CN)_6$ solution into a 250 mL Erlenmeyer flask. Grind up a commercial zinc tablet containing 50 mg (according to the label) of zinc in a mortar and transfer the solid quantitatively to the ferrocyanide solution. Add ~5 mL of 3 M sulfuric acid, 3 drops of freshly prepared 1% potassium ferricyanide and 4 drops of diphenylamine indicator. If the solution is milky white, it is ready for titration. Be aware that the titration will only take a few mL so titrate slowly from the start (Note: If the solution is blue, pipet 10.00 additional mL of the ferrocyanide into the flask. The solution should turn white and can now be titrated with the zinc ion.) Repeat the titration at least once.

Fig. 11-1

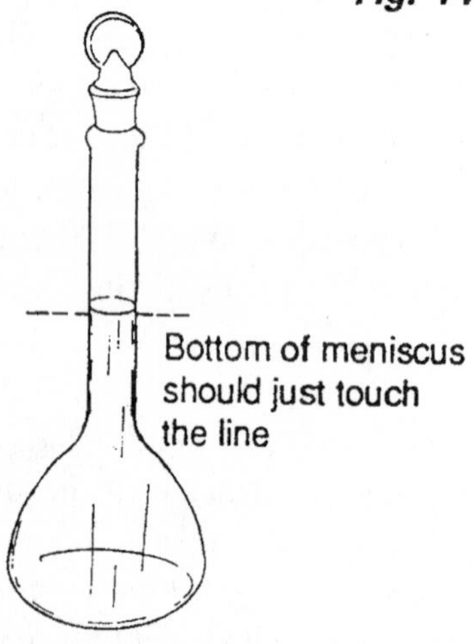

Fig. 11-2

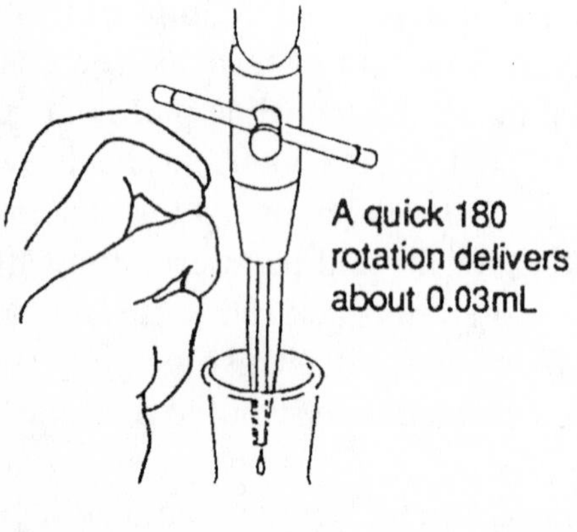

Fig. 11-3

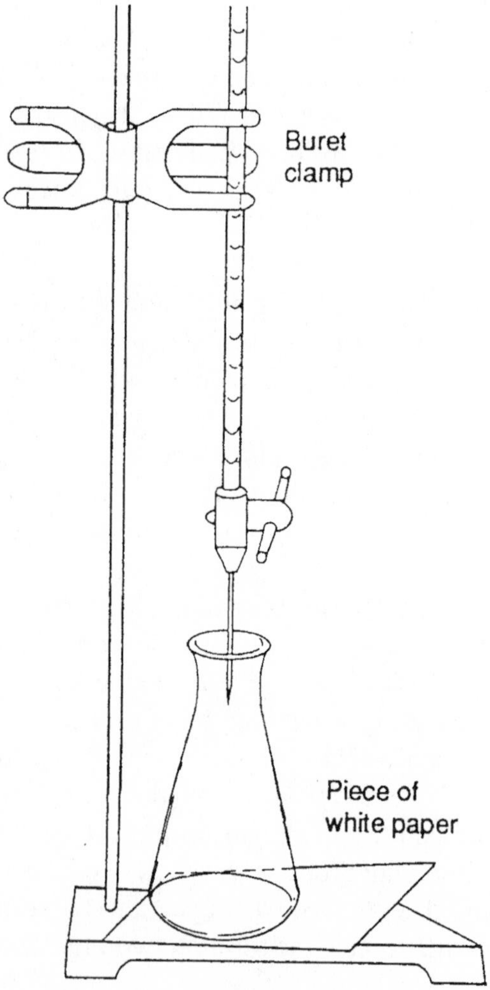

Name__________________________________Date__________Lab Section_________

Prelaboratory Problems - *Experiment 11* - Quantitative Solution Chemistry

The solutions to the starred problems are in ***Appendix 4***.

1. Molecules are mobile in a gas and some reactions can be run in the gas phase. Give several reasons solutions are preferable for synthetic reactions.

2. Sodium hydroxide comes in pellet form in bottles from laboratory supply companies but it cannot be used as a primary standard. Why not?

3. a.* How many grams of $CuSO_4 \cdot 5H_2O$ are needed to prepare 50.0 mL of a 0.75 M $CuSO_4$ solution?

   b. How many grams of $NiSO_4 \cdot 6H_2O$ are needed to prepare 25.0 mL of a 15.0×10^{-2} M $NiSO_4$ solution?

4.* 10.00 mL of a silver nitrate solution is titrated to the equivalence point with 8.50 mL of a 0.1100 M HCl solution. What is the concentration of the silver nitrate solution?

5. 15.50 mL of a 0.1075 M NaOH solution are required to neutralize 15.00 mL of a sulfuric acid solution. What is the concentration of the sulfuric acid solution?

6. 22.87 mL of a 0.1075 M NaOH solution are required to neutralize 0.300 grams of an unknown monoprotic acid. What is the molecular mass of the acid?

7.* What is the function of an indicator in a titration?

8. Suggest at least two different ways an indicator could work in this experiment (***Experiment 11***).

Name______________________________ Date__________ Lab Section__________

Results and Discussion - *Experiment 11* - Quantitative Solution Chemistry

A. Stoichiometry of the reaction

1. Formula Mass of $K_4Fe(CN)_6 \cdot 3H_2O$ ________
2. Mass of $K_4Fe(CN)_6 \cdot 3H_2O$ needed for 250 mL of 0.025 M soln. ________
3. Mass of beaker + $K_4Fe(CN)_6 \cdot 3H_2O$ ________
4. Mass of beaker ________
5. Mass of $K_4Fe(CN)_6 \cdot 3H_2O$ ________
6. Moles of $K_4Fe(CN)_6 \cdot 3H_2O$ ________
7. Concentration of $K_4Fe(CN)_6$ solution ________
8. Formula mass of $ZnSO_4 \cdot 7H_2O$ ________
9. Amount of $ZnSO_4 \cdot 7H_2O$ needed for 250 mL of 0.050 M solution ________
10. Mass of beaker + $ZnSO_4 \cdot 7H_2O$ ________
11. Mass of beaker ________
12. Mass of $ZnSO_4 \cdot 7H_2O$ ________
13. Moles of $ZnSO_4 \cdot 7H_2O$ ________
14. Concentration of $ZnSO_4$ ________
15. Volume of $K_4Fe(CN)_6$ solution for each titration ________
16. Moles of $K_4Fe(CN)_6$ titrated ________

Titration with $ZnSO_4$	1st	2nd	3rd
17. Final buret reading	________	________	________
18. Initial buret reading	________	________	________
19. Volume of $ZnSO_4$	________	________	________
20. Moles of $ZnSO_4$	________	________	________

21. Average number of moles of $ZnSO_4$ ________
22. Mole ratio of $K_4Fe(CN)_6$ to $ZnSO_4$ (Also enter on line 9 on next page) ________

23. Which of the four possible equations (page 124) agrees most closely with the mole ratio in # 22? Write down the equation and explain your choice.

24. Balanced net ionic equation for the reaction:

B. Concentration of ferrocyanide in an unknown

1. Unknown identification number ________

2. Concentration of $ZnSO_4$ ________

3. Volume of unknown $K_4Fe(CN)_6$ solution ________

Titration with $ZnSO_4$	1st	2nd	3rd
4. Final buret reading	________	________	________
5. Initial buret reading	________	________	________
6. Volume of $ZnSO_4$	________	________	________
7. Moles of $ZnSO_4$	________	________	________

8. Average number of moles of $ZnSO_4$ ________

9. Mole ratio of $K_4Fe(CN)_6$ to $ZnSO_4$ ________

10. Moles of $K_4Fe(CN)_6$ ________

11. Concentration of $K_4Fe(CN)_6$ in unknown solution ________

[Note: Zinc concentrations can be determined by addition of an excess of standard potassium ferrocyanide and back titration with a standard zinc solution as in Part C below.]

C. (Option 2) The amount of zinc ion in a commercially available tablet.

1. Amount of zinc ion in tablet according to label ________

2. Concentration of $ZnSO_4$ ________

3. Volume of 0.025 M $K_4Fe(CN)_6$ solution ________
(If not 25.00 mL, then the calculations below need to be modified)

Titration with $ZnSO_4$	1st	2nd	3rd
4. Final buret reading	________	________	________
5. Initial buret reading	________	________	________
6. Volume of $ZnSO_4$	________	________	________

8. Average volume of $ZnSO_4$ ________

9. Average volume of $ZnSO_4$ used in Part A ________

10. Volume of $ZnSO_4$ equivalent to amount of zinc in tablet (#9 - #8) ________

11. Moles of $ZnSO_4$ equivalent to amount of zinc in tablet ________

12. Grams of zinc ion experimentally determined in tablet ________

13. Percent difference between label and experiment ________

14. Critically comment on the claim on the bottle that it contains 50 mg of zinc.

15. Suggest any ways you can think of to improve any part(s) of this experiment.

16. Some of the *Learning Objectives* of this experiment are listed on the first page of this experiment. Did you achieve the *Learning Objectives*? Explain your answer.

Experiment 12

THERMOCHEMISTRY

Herman von Helmoltz
1821 - 1894 (ΔH)

Learning Objectives

Upon completion of this experiment, students will have experienced:

1. The determination of the enthalpy of a reaction.
2. Endothermic and exothermic processes.
3. Hess's law using experimental data and Internet data.

Text Topics

Endothermic and exothermic processes, enthalpy, Hess's law, (see page ix).

Notes to Students and Instructor

This experiment can be done with a thermometer and manual graphing or more conveniently with a temperature probe connected to a computer interface coupled with the appropriate software.

Discussion

Combustion is probably the most commonly run chemical reaction. The controlled burning of coal, oil and gasoline to produce energy has enabled us to progress quickly into today's technologically based society. Unfortunately, our haste to progress has not always been accompanied by enough consideration for the environmental impact of our actions. If our population continues to increase and if we continue to use nonrenewable resources at our present rate, we could jeopardize our future quality of life. As part of our priorities, we need to develop alternative sources of energy that should have a lower environmental impact such as nuclear fusion and solar voltaic cells. For the short term, combustion will continue to be our primary source of energy and it is important that we understand the energetics of combustion and other chemical reactions.

A consideration of the overall energy balance of a reaction enables us to determine if a reaction can go (is it spontaneous?). The overall energetics of a reaction can often be accurately calculated from a rather short compilation of parameters available in most chemistry handbooks. This provides an extremely powerful technique for determining optimum conditions for a reaction. Generally the most important part of the overall energetics is the enthalpy or heat content of the reaction. We call reactions that consume thermal energy, endothermic (positive enthalpy or the sign of ΔH is +) and those that evolve thermal energy, exothermic (negative enthalpy or the sign of ΔH

is −). This experiment will focus on the determination of the enthalpy changes of several processes of interest to chemists. Recognize that enthalpy is not the only energy factor that determines spontaneity. Entropy sometimes also plays an important role.

Specific heat. To determine the enthalpy of a process, the mass, temperature change, and the specific heat of the components are needed. The specific heat of a substance is the amount of energy required to raise the temperature of one gram of the substance one degree Celsius. For water, with its many unusual properties (such as the density decrease with freezing), it would appear that its very high specific heat of 1.00 cal. $gram^{-1}\ K^{-1}$ or 4.184 joules $gram^{-1}\ K^{-1}$ (compared with the values for most other substances - metals are 0.1 to 0.8 joules $gram^{-1}\ K^{-1}$) is again another unique and unusual property. However, as you will discover in *Parts b* and *c* of *Prelaboratory Problem 4*, the value is completely consistent with the values for other substances and theory. Today you will use the value for the specific heat of water in an experiment to determine the specific heat of a metal. The specific heat values necessary for enthalpy calculations for other substances will also be provided.

Enthalpies of solution and reaction. The change in enthalpy for a process is a convenient thermodynamic property for laboratory study because it is the change in heat content at constant pressure. As most laboratory experiments are conducted open to the atmosphere, the pressure is constant for the process. If the temperature change for a process can be measured and the mass and specific heat of the system are known; $\Delta H = mC_p\Delta T$ (m = mass, C_p = specific heat of substance, ΔT = temperature change). There are at least two basic complications to this procedure. First the system changes temperature in a container and the container also changes temperature and absorbs or evolves energy. Today you will run the reactions in polystyrene cups. Although it is possible to correct for the "calorimetry constant" of the cup, our experiments indicate the error in the cup corrections are as large as the error itself. We will therefore assume a calorimetry constant of zero and recognize that more accurate measurements would need to take the constant for the cup into account.

Second, changes in the system are not instantaneous. Time is consumed mixing the reactants and it takes time to read thermometers. Thermal energy is lost during the mixing process and this causes some error. You will correct for these errors by plotting temperature versus time and extrapolating to the time of mixing.

Enthalpy of formation. For a chemical reaction, your text will develop the concept of Hess's law. Hess's law is a powerful predictive tool as it enables you with a short table of values for heats of formation of substances to calculate the heats of reaction for a long list of chemical reactions. The heat of reaction ΔH_r is the difference between the sum of the heats of formation of the products and reactants:

$$\Delta H_r = \Sigma[n\Delta H_{f(products)}] - \Sigma[n\Delta H_{f(reactants)}]$$

(n = moles of substance or coefficient in balanced equation)

In today's experiment, you will compare your experimental value for the heat of a neutralization reaction to the value calculated from heats of formation.

Procedure

A. Specific heat of a metal. Obtain a metal "shot" unknown from your instructor. Weigh out 30 g of the metal to at least the nearest 0.01 g (preferably 0.001 g) and transfer it to a clean, dry test tube. Clamp the test tube in a beaker of water mounted over a Bunsen burner with the top of the test tube at least 5 cm above the water level. Heat the beaker to boiling and then maintain it there for at least five minutes and until you are ready for the next part.

Using a pipet or buret deliver 25.00 mL of water into a polystyrene cup. Insert and carefully read a thermometer every 0.25 minutes (15 seconds) for 1.75 minutes. At precisely the 2 minute mark, quickly transfer the heated metal to the cup being sure to stir immediately after transferring. Starting at the 2.25 minute mark, record the thermometer reading every 0.25 minutes for about three minutes. Be sure that the metal is properly recycled. If time is available, perform a second run.

You are now ready to graphically determine the temperature change for the process. Graph the temperature on the vertical axis and the time on the horizontal axis. The data prior to mixing and subsequent to mixing are extrapolated using straight lines **to the time of mixing**. The temperature difference between the extrapolated lines at the time of mixing is the temperature change. Refer to *Figure 12-1* for an example of the analysis. Note however, that the temperature change in *Figure 12-1* is much larger than the change you will obtain here (a few degrees). Use of a computer interfaced temperature probe can considerably facilitate this procedure.

The specific heat of the metal can now be calculated from the following treatment. The sum of the energy changes for the process must be zero or the energy gained by the water plus the energy lost by the heated metal must equal zero.

Let:
m_w = mass of water
C_w = specific heat of water
Δt_w = change of temperature of water

m_u = mass of unknown metal
C_u = specific heat of unknown
Δt_u = change of temperature of unknown

$$m_w C_w \Delta t_w + m_u C_u \Delta t_u = 0$$

$$C_u = -\frac{m_w C_w \Delta t_w}{m_u \Delta t_u}$$

In 1819, it was observed by Dulong and Petit that for most metallic elements, the product of the atomic mass and the specific heat is close to the value 25 J/mol deg. Theoretical equations indicate that the value should be 3 times the gas law constant (8.313 J/mol deg) or 24.9 J/mol deg. We notice that theory and the early experimental results agree quite closely. By assuming the relationship, (atomic mass)(specific heat) = 25 J/mol deg [$(M_m)(C_p) = 25$], you can calculate an approximate atomic mass of your unknown and hopefully identify it.

B. Heats of solution. When an ionic compound dissolves in water, energy is required to break up the crystal lattice (the attraction between ions is called the crystal lattice energy) but energy is released by the formation of bonds between ions and water (energy of hydration). When the energy of hydration is greater than the crystal lattice energy, the dissolving process will be exothermic and causes a temperature increase. When the crystal lattice energy exceeds the energy of hydration, the dissolving process is endothermic and the temperature of the system will drop (You should ask why it dissolves when the process is endothermic - see ***Experiments 25*** and ***26*** on the other part of the energy balance - entropy).

The temperature change upon dissolving of two different ionic compounds will be measured to determine the relative magnitudes of the crystal lattice energy and the heat of hydration. As calculations will not be performed on this part of the experiment, only approximate temperature changes will be determined.

1. Add 3 grams of lithium chloride to a dry test tube. Measure the temperature of 15 mL of water and transfer it to the tube. Thoroughly mix to dissolve the solid and read the temperature.

2. Repeat the above with 3 grams of ammonium chloride.

C. Enthalpy of neutralization. The endothermicity or exothermicity of several acid and base reactions will be determined by measuring the temperature change upon mixing. From the results of the series of mixtures, you will be able to evaluate the effects of the nature and concentration of the acids and bases on the enthalpy changes of the reaction.

For run 1, weigh a clean, dry polystyrene cup to at least the nearest 0.01 g. Add 50.0 mL of 2.00 M HCl to the cup with a graduated cylinder. Add 50.0 mL of 2.02 M NaOH to a beaker. Insert a thermometer into the acid solution and read its temperature every 0.50 minutes for 2.5 minutes. We will assume that both solutions are at the same temperature since they have been sitting in containers in the lab for some time. At exactly the 3 minute mark, pour the sodium hydroxide solution into the acid solution and stir vigorously with the thermometer. Resume taking temperature readings at 0.50 minute intervals at the 3.5 minute mark until a trend is established (about 5 minutes). Weigh the cup and contents.

Repeat the above procedure for runs 2, 3 and 4 substituting the appropriate solutions as indicated in the chart below.

run	first reagent (50 mL)	second reagent (50 mL)
1	2.00 M HCl	2.02 M NaOH
2	2.02 M NaOH	2.00 M HCl
3	2.00 M HNO_3	2.02 M NaOH
4	1.00 M HCl	2.02 M NaOH

Temperature is plotted as a function of time. The data prior to mixing and subsequent to mixing are extrapolated using straight lines. The temperature difference between the extrapolated lines at the time of mixing is the temperature change. Refer to *Figure 12-1* for an example of the analysis. Use of a computer interfaced temperature probe can considerably facilitate this procedure.

Fig. 12-1

Temperature vs Time

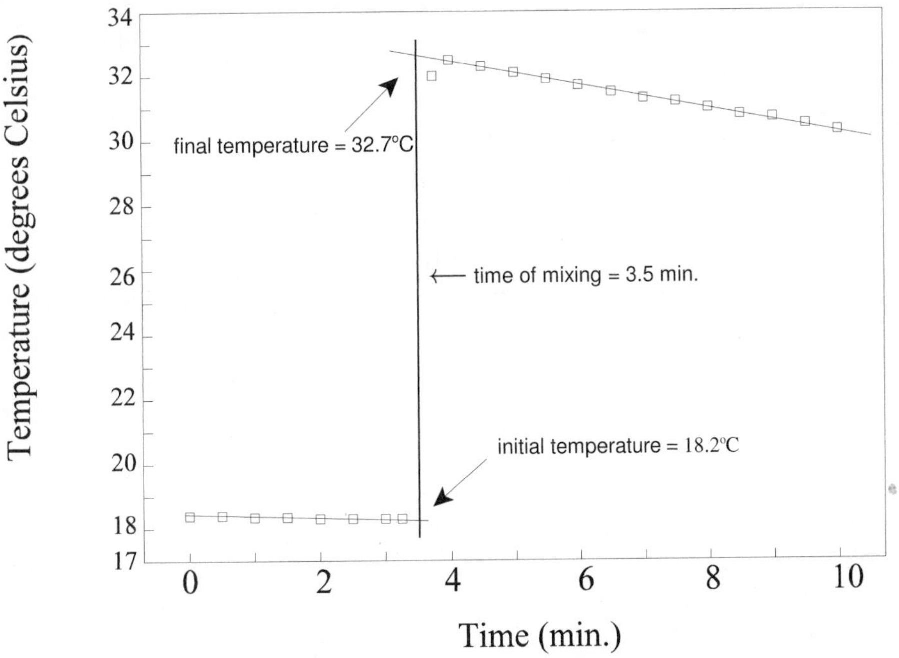

Graph the temperature on the vertical axis and the time on the horizontal axis. Extrapolate the data between 0 and 2.5 minutes using a straight edge to the 3.0 minute mark. Extrapolate the data between 3.5 and 9 minutes (or whenever you stopped) back to the 3.0 minute mark. Determine the temperature change between the two extrapolations at the 3.0 minute mark. Calculate the heat of neutralization and the molar heat of neutralization from the treatment that follows.

Let:

ΔH_n	=	heat of neutralization
$\underline{\Delta H_n}$	=	molar heat of neutralization
Δt	=	temperature change of solution
m_s	=	mass of final solution
C_s	=	specific heat of final solution
a	=	number of moles of HCl

$$\Delta H_n + m_s C_s \Delta t = 0$$

$$\Delta H_n = -m_s C_s \Delta t$$

$$\underline{\Delta H_n} = \Delta H_n / a$$

Hess's law can be used to calculate the molar heat of neutralization from literature values for heats of formation of HCl, NaOH, NaCl and H_2O. These values and the specific heat needed for the calculations are in the table below.

$$\Delta H_r = \Sigma[n\Delta H_{f(products)}] - \Sigma[n\Delta H_{f(reactants)}]$$

solution	concentration (mol/L)	specific heat (J/g deg)	heat of formation (kJ/mol)
hydrochloric acid	1.00		-164.4
sodium hydroxide	1.00		-469.6
sodium chloride	1.00	3.90	-407.1
water			-285.9
nitric acid	1.00		-206.6
sodium nitrate	1.00	3.89	-446.2
0.5 M sodium chloride + 0.5 M sodium hydroxide		3.94*	

*estimated

D. Hess's law calculations using data available on the Internet. The *NIST* (National Institute of Standards and Technology) site on the Internet

http://webbook.nist.gov/chemistry/

can be used to obtain data for Hess's Law calculations. In this exercise, you will find the heats of formation of water (l and g) and carbon dioxide as well as the heats of formation and combustion for methane (g), ethane (g), propane (g), and 2,2,4-trimethylpentane (l). [Note: Commercially 2,2,4-trimethylpentane is called isooctane. The commercial name is not consistent with the naming system used by organic chemists but unfortunately finds common usage. Isooctane is one of the standards for determining the octane rating of gasoline. Its octane rating has arbitrarily been assigned a value of 100.] The heats of formation will be used to calculate the heats of combustion (using Hess's Law) for hydrogen, methane, ethane, propane, and 2,2,4-trimethylpentane (l) and the heat of vaporization of water. The calculated values for the hydrocarbons will be compared to the heats of combustion given at the *NIST* site.

Name______________________________ Date__________ Lab Section__________

Prelaboratory Problems - *Experiment 12* - Thermochemistry

The solutions to the starred problems are in ***Appendix 4***.

1. The temperature change determined graphically is the difference in temperatures at the time of mixing. This is determined by extrapolating the data before and after mixing to the time of mixing.

   a. What does extrapolation mean?

   b.* Determine the temperature change in *Figure 12-1*. __________

2.* The addition of 15.00 g of an unknown metal at 100.0°C to 25.00 g of water at 22.0°C resulted in a final maximum temperature of the system of 27.5°C. Calculate the specific heat of the metal and its approximate atomic mass (see Dulong and Petit method on page 135 and *Problem 3* below). Suggest a name for the unknown metal.

specific heat __________

atomic mass __________

element __________

3. The addition of 21.60 g of an unknown metal at 80.0°C to 20.00 g of water at 21.0°C resulted in a final maximum temperature of the system of 23.0°C. Calculate the specific heat of the metal and its approximate atomic mass (see Dulong and Petit method on page 135. Suggest a name for the unknown metal.

specific heat __________

atomic mass __________

element __________

4. The Dulong and Petit empirical relationship for the product of atomic mass and specific heat, $(M_m)(C_p) = 25$, indicates that the specific heat is inversely proportional to the atomic mass or proportional to the number of moles of particles as with colligative properties.

   a. Using the data in ***Appendix 1*** for at least four metals, test the Dulong and Petit relationship. Is it accurate?

   b. For use of the equation with molecules rather than atoms, the number 25 should be multiplied by the number of atoms per molecule. Is the value 4.184 joules/g deg. for water consistent with this oversimplified approach? Explain your answer (Note that there is a threshold temperature below which this approach will have significant error).

   c. Compared to metals, what two factors cause the specific heat of water to appear to be exceptionally high?

   ________ ________

   d. Does this concept work on other substances such as sodium chloride, potassium chloride, calcium chloride, aluminum chloride, carbon tetrachloride or methanol (CH_3OH)? Explain your answer. (See ***Appendix 1*** for specific heat data)

5. The addition of 50 mL of 2.02 M NaOH to 50 mL of 2.00 M HNO_3 at 22.0°C resulted in a maximum temperature of 36.1°C with a total mass of the solution of 100.0 g. Calculate the molar heat of neutralization and compare it to the result calculated from Hess's law (refer to data on page 138).

experimental ΔH_n ________

calculated ΔH_n ________

Name_________________________________Date_________Lab Section_________

Results and Discussion - *Experiment 12* - Thermochemistry

A. Specific heat of a metal (graph the data below on one of graph sheets that follows).

	Run 1		Run 2	
Time (min.)	Water Temp. (°C)	Mixture Temp. (°C)	Water Temp. (°C)	Mixture Temp. (°C)
0.00	________		________	
0.25	________		________	
0.50	________		________	
0.75	________		________	
1.00	________		________	
1.25	________		________	
1.50	________		________	
1.75	________		________	
2.00	**MIX**		**MIX**	
2.25		________		________
2.50		________		________
2.75		________		________
3.00		________		________
3.25		________		________
3.50		________		________
3.75		________		________
4.00		________		________
4.25		________		________
4.50		________		________
4.75		________		________
5.00		________		________

	Run 1	Run 2
1. Unknown number	______	
2. Mass of metal	______	______
3. Volume of water	______	______
4. Initial temperature of water from graph of page 141 data (see *Fig. 12-1* for method of determining this value and value for #8)	______	______
5. Density of water (from *Handbook of Chemistry and Physics)*	______	______
6. Mass of water calculated from volume and density	______	______
7. Temperature of metal	______	______
8. Final temperature of system from graph of page 141 data	______	______
9. Specific heat of water	______	______
10. Δt_w	______	______
11. Δt_u	______	______
12. Energy gained by water	______	______
13. Specific heat of unknown metal	______	______
14. Average value of specific heat if two runs performed		______
15. Atomic mass of unknown metal (Dulong and Petit method on page 135)		______
16. Probable identity of unknown metal (in addition to atomic mass, use logic and consider other observations). Explain your answer.		______

B. Heat of solution.

1. Temperature change when 3 grams of lithium chloride is added to 15 mL of water (indicate + or −) ______

2. Temperature change when 3 grams of ammonium chloride is added to 15 mL of water (indicate + or −) ______

3. Which is greater for ammonium chloride, the lattice energy or the hydration energy? Explain your answer. ______

C. Enthalpy of neutralization. (Remember to weigh the empty, dry cup before starting. Graph the data below on one of graph sheets that follows.)

Run 1 (2.00 M HCl + 2.02 M NaOH)			**Run 2** (2.02 M NaOH + 2.00 M HCl)		
Time (min.)	HCl(aq) Temp. (°C)	Mixture Temp. (°C)	Time (min.)	NaOH(aq) Temp. (°C)	Mixture Temp. (°C)
0.00	________		0.00	________	
0.50	________		0.50	________	
1.00	________		1.00	________	
1.50	________		1.50	________	
2.00	________		2.00	________	
2.50	________		2.50	________	
3.00	**MIX**			**MIX**	
3.50		________	3.50		________
4.00		________	4.00		________
4.50		________	4.50		________
5.00		________	5.00		________
5.50		________	5.50		________
6.00		________	6.00		________
6.50		________	6.50		________
7.00		________	7.00		________
7.50		________	7.50		________
8.00		________	8.00		________
8.50		________	8.50		________
9.00		________	9.00		________
9.50		________	9.50		________

Be sure to weigh the cup and its contents before disposing of it. On the accompanying pieces of graph paper, plot temperature on the vertical axis and time on the horizontal axis. Following *Figure 12-1*, use a straight edge to draw the best straight line through the data prior to mixing and subsequent to mixing and extrapolate both to the 3.0 minute mark. Determine the temperature difference between the lines at the 3.0 minute mark and enter the data on page 145.

(Remember to weigh the empty, dry cup before starting. Graph the data below on one of graph sheets that follows)

Time (min.)	**Run 3** (2.00 M HNO_3 + 2.02 M NaOH) HNO_3(aq) Temp. (°C)	Mixture Temp. (°C)	Time (min.)	**Run 4** (1.00 M HCl + 2.02 M NaOH) HCl(aq) Temp. (°C)	Mixture Temp. (°C)
0.00	______		0.00	______	
0.50	______		0.50	______	
1.00	______		1.00	______	
1.50	______		1.50	______	
2.00	______		2.00	______	
2.50	______		2.50	______	
3.00	**MIX**			**MIX**	
3.50		______	3.50		______
4.00		______	4.00		______
4.50		______	4.50		______
5.00		______	5.00		______
5.50		______	5.50		______
6.00		______	6.00		______
6.50		______	6.50		______
7.00		______	7.00		______
7.50		______	7.50		______
8.00		______	8.00		______
8.50		______	8.50		______
9.00		______	9.00		______
9.50		______	9.50		______

Be sure to weigh the cup and its contents before disposing of it. On the accompanying pieces of graph paper, plot temperature on the vertical axis and time on the horizontal axis. Following *Figure 12-1*, use a straight edge to draw the best straight line through the data prior to mixing and subsequent to mixing and extrapolate both to the 3.0 minute mark. Determine the temperature difference between the lines at the 3.0 minute mark and enter the data on page 145.

	run 1	run 2	run 3	run 4
1st cup - 50 mL of:	2 M HCl	2 M NaOH	2 M HNO_3	1 M HCl
beaker - 50 mL of:	2 M NaOH	2 M HCl	2 M NaOH	2 M NaOH
1. Mass of empty polystyrene cup	______	______	______	______
2. Mass of cup + mixture	______	______	______	______
3. Mass of mixture	______	______	______	______
4. Initial temp. of 1st reagent (see *Fig. 12-1*, from extrapolation forward to 3.00 min.)	______	______	______	______
5. Final temp. of mixture (see *Fig. 12-1*, from extrapolation back to 3.00 min.)	______	______	______	______
6. Temperature change (Δt)	______	______	______	______
7. Total heat evolved in reaction (ΔH_n)	______	______	______	______
8. Number of moles of H_2O formed	______	______	______	______
9. Molar heat of neutralization ($\underline{\Delta H}_n$)	______	______	______	______
10. Molar heat of neutralization calculated from Hess's law and data on page 138	______	______	______	______
11. Percentage difference between experimental and calculated values	______	______	______	______

12. Write net ionic equations for the reactions in runs 1-4.

#1

#2

#3

#4

13. Comparing runs 1 and 2, does the order of mixing have a significant effect on the temperature change? Explain your answer. _________

14. Comparing runs 1 and 3, does the nature of the acid have a significant effect on the temperature change? If not, why not? _________

15. Compare values for runs 1 and 4 and explain any differences.

D. Hess's law calculations using data available on the Internet.

This exercise should be performed using the Internet site: http://webbook.nist.gov/chemistry/ After writing balanced equations for the combustion reactions listed below, locate a computer with Internet access and enter the *NIST* site above. At this site, click on "Name" and then enter the name of the compound of interest in the box. For carbon dioxide, water, methane, ethane and propane, click on "gas phase thermochemistry" data and scroll down and find the standard heats of formation data [$\Delta_f H°_{gas}$]. and fill in the data in the table below. For water and 2,2,4-trimethylpentane, select "Condensed phase thermochemistry" and record the standard heats of formation [$\Delta H_f°_{liquid}$]. For methane, ethane, propane and 2,2,4-trimethylpentane, you should also select "Reaction thermochemistry data" and record the top value [$\Delta H_c°$]. To make sure you are in the right place, the value for methane is -890.7 kJ/mol.

1. Write the reactions for the combustion of hydrogen, methane, ethane, propane and 2,2,4-trimethylpentane.

a. $H_2(g)$ + $O_2(g)$ =

b. $CH_4(g)$ + $O_2(g)$ =

c. $C_2H_6(g)$ + $O_2(g)$ =

d. $C_3H_8(g)$ + $O_2(g)$ =

e. $C_8H_{18}(l)$ + $O_2(g)$ =

2. Table of data and Hess's law calculation results

substance	ΔH_f° gas (kJ/mol)	ΔH_f° liquid (kJ/mol)	ΔH_c° (calculated) (kJ/mol)	ΔH_c° (from *NIST*) (kJ/mol)
carbon dioxide		na	na	na
water			na	na
methane		na		
ethane		na		
propane		na		
2,2,4-trimethylpentane				(from liquid)

3. Which of the potential fuels, hydrogen, methane, ethane, propane, 2,2,4-trimethylpentane is the most efficient? Explain your answer.

4. From the values for the heats of formation of liquid and gas phase water, calculate the heat for the phase change of liquid water to gas.

____

5. The value for the heat of vaporization of water is -40.7 kJ/mole. Can you account for any difference between the value calculated in #4 and the value of -40.7 kJ/mole?

6. Suggest any methods you can think of to improve any part(s) of this experiment.

7. Some of the *Learning Objectives* of this experiment are listed on the first page of this experiment. Did you achieve the *Learning Objectives*? Explain your answer.

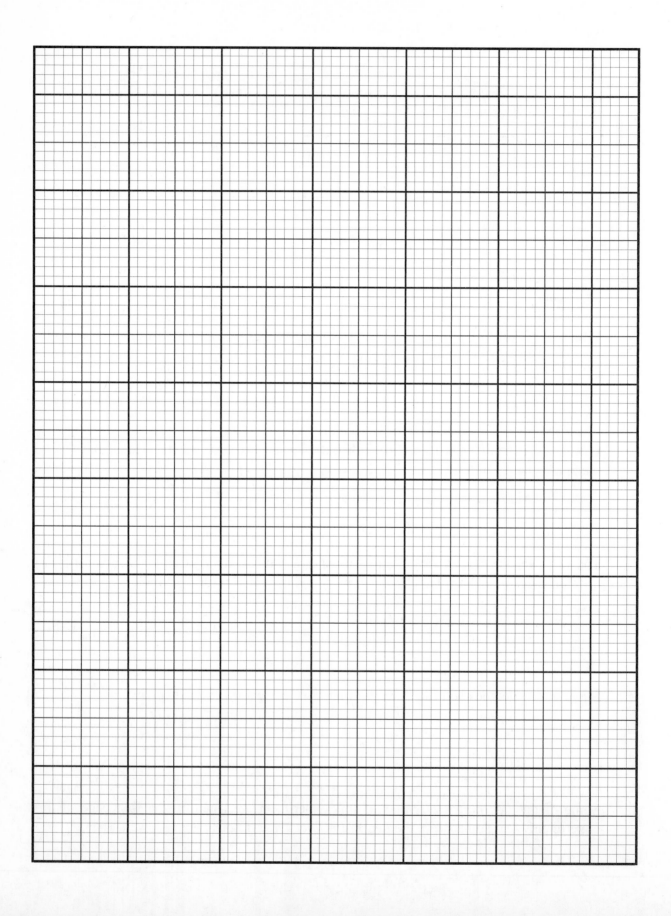

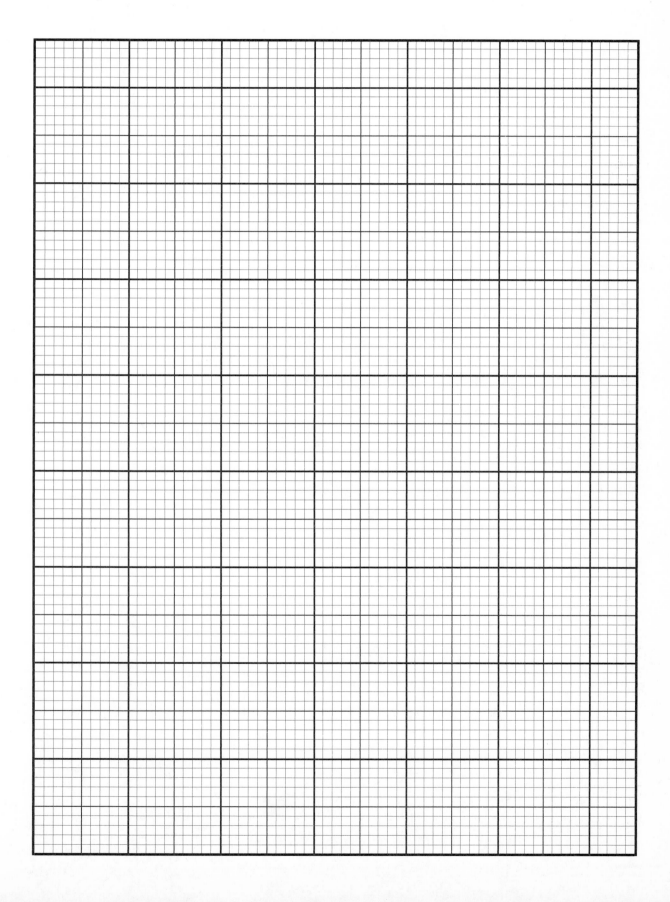

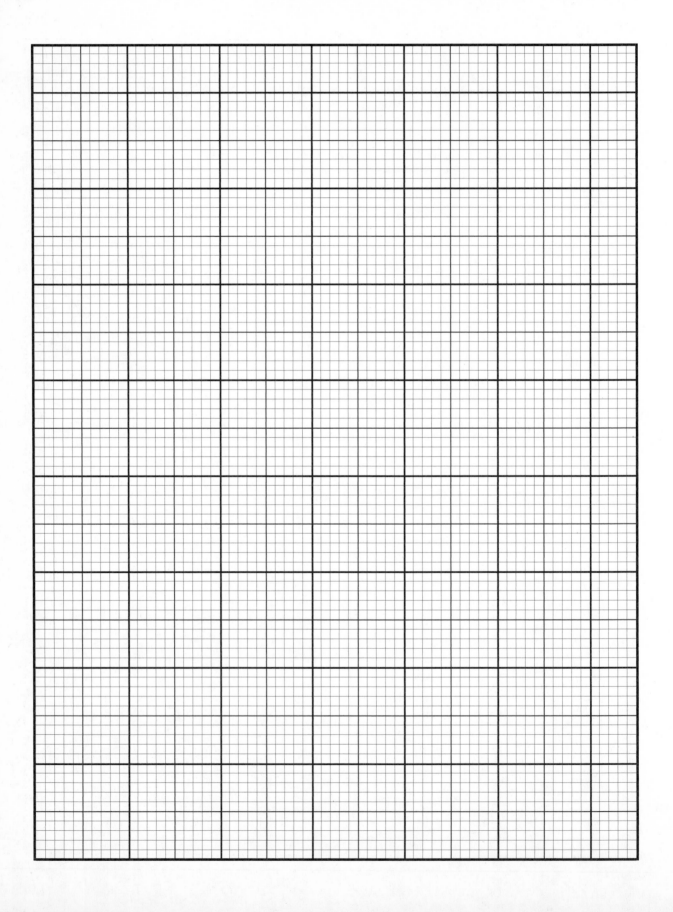

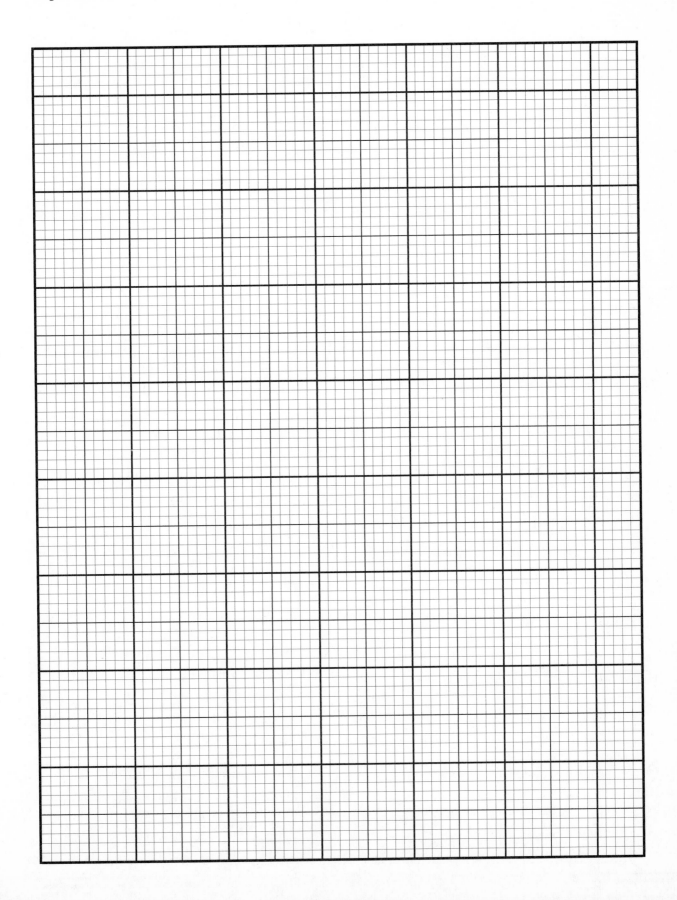

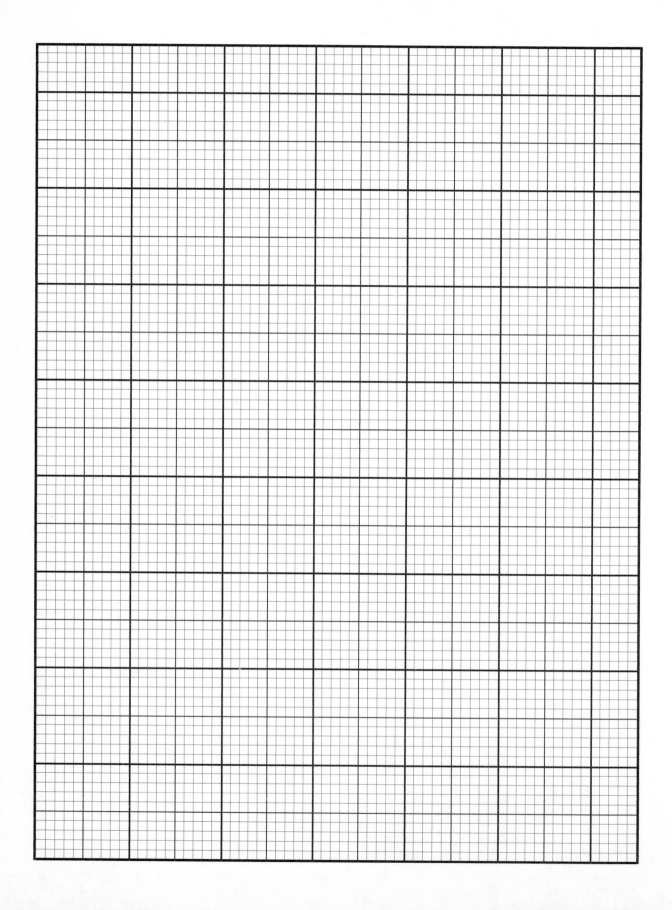

Experiment 13

Dimitri Mendeleev
1834 - 1907

PROPERTIES OF ELEMENTS AND COMPOUNDS: AN INTERNET STUDY

Learning Objectives

Upon completion of this experiment, students will have experienced:
1. The use a personal computer to navigate the Internet.
2. An investigation of properties of the elements.
3. A search for properties including toxicities of compounds.

Text Topics

Periodic properties of the elements (for correlation to some textbooks, see page ix).

Notes to Students and Instructor

To do this experiment, you need access to a computer that has an Internet browser and has a connection to the Internet.

Discussion

The library is as important to a chemist as the laboratory. Chemists depend on journals, handbooks and textbooks as the sources of data, concepts and ideas that serve as the foundation of current and future research. The Internet is also a very valuable and rapidly expanding source of many different types of information useful to chemists. This exercise will introduce you to a few of the valuable chemistry sites on the Internet. To make this experience more valuable, searches will focus on topics that you are currently studying in the lecture portion of your chemistry course including some of the periodic properties of the elements.

As with any secondary source of information including handbooks, it should be recognized that there is always the possibility that the Internet information contains recording errors. In addition, most journals are refereed (to be accepted for publication, papers are subject to peer review before publication) but currently, there are no restrictions to publication on the Internet. Therefore, some of the information and concepts presented might not be reliable or correct. Again, with any secondary source of information that is going to serve as the basis of future work, it is imperative that primary sources be read and used as the references in written reports. Despite these limitations, secondary sources such as the Internet can save vast amounts of time and much can be learned and gained from using the Internet as a resource.

There are many different methods to search for information on the www. Search engines and indices such as ***Google*** and ***Yahoo*** can be used to search for subjects, keywords and people. Tools of this type are probably the place to start for general searches but it should be recognized that even the best search engines scan only a portion of the total sites available on the web. Some sites such as *Ixquick* provide access to several search engines from one page.

Google	http://www.google.com
Yahoo	http://www.yahoo.com
Ixquick	http://ixquick.com/

For more specific information such as searches in the field of chemistry, there are many directories that have been uploaded that attempt to organize the topics into categories. Because the number of sites on the Internet increases everyday and addresses (URL's) also change, most of these directories are not completely up to date but they are often still a good place to start. For a site that contains a link to one of the chemistry directories (that in turn contains a listing of many other directories) and provides another learning experience on using the www, see:

http://virtual.yosemite.cc.ca.us/smurov/

For the purposes of this exercise, only a few sites will need to be accessed directly by you so the URL's will be provided and it will not be necessary to use a search engine or chemistry directory. It is highly recommended, however, that you check out the sites given above, try some searches on the search engines and jump to and browse through some of the links provided by any of the sites you visit. Because the sites listed below are some of the most useful sites for chemists on the web, it is recommended that you **establish a bookmark at each site** the first time you visit the site. Future returns to the site can then be made very quickly using your bookmark.

Procedure

Locate a computer with an Internet browser and a connection to the Internet. Establish a connection to the Internet and load your Internet browser. For information on periodicity and periodic properties of the elements, there are several useful sites. Many of these sites contain periodic tables that allow you to jump to information on each element simply by pointing and clicking on the element in the table. Of these sites, ***WebElements*** contains the most information, is the most versatile and has significant graphing capability. It will be very much worth your while to spend some time at the site acquainting yourself with the scope of the information available. For *Part A* of this exercise, you will use ***WebElements***. For **Part B**, you can use ***WebElements*** for most of the questions or a couple of simpler periodic table programs (*Chris Heilman's* or *Environmental Chemistry*).

A. Properties of Elements. This brief search will focus on some unusual properties of two very important but less prominent elements, gallium and selenium. Using your Internet browser, enter the URL below into the appropriate location.

http://www.webelements.com/

After reaching the site, establish a bookmark. Scroll down to the periodic table and click on gallium. Now scroll until you come to *physical properties*. Click on *Thermal properties* and find and record the melting and boiling points of gallium in question 1. Now click on *History* and answer question 2 in section A. The answers to questions 3 and 4 should be in *Uses* and *Key data; description* respectively. After finishing with gallium, click the "Back" button until you return to the periodic table and click on selenium. You will probably need to search the *Uses* section to answer the questions.

B. Extreme Properties of Elements. Use one or more of the URL's below.

http://www.webelements.com/
http://chemlab.pc.maricopa.edu/periodic/periodic.html (*Chris Heilman's* site)
http://environmentalchemistry.com/yogi/periodic/

If you use ***WebElements***, click on any element such as carbon and then on *Thermal Properties* under *physical properties*. You will find some grey boxes that will enable you to search for melting and boiling point ranges. If ***WebElements*** doesn't have the needed information, try one of the other two sites. For the second site (*Chris Heilman's*), scroll down to the Search option and enter each of the appropriate parameters (click on the down arrow at the right end of the query blank to view and select the desired parameter) and perform the search. For the *Environmental Chemistry* site, scroll down to "Elements Sorted by". Record the answers on your answer sheet.

property	range
density	20 to 100 g/cm^3
melting point	3000 to 10,000 K
boiling point	0 to 90 K
electrical conductivity	0.37×10^6 to 1×10^7 /cm Ω

For an optional extension to this element search on material selection, go to **Part E** in the answer section of this experiment.

C. Periodic Properties of the Elements. The three sites listed below can be used to obtain the information needed for the first part of this section. The instructions below are for *Chris Heilman's* site. (If you prefer to use ***WebElements***, the navigation techniques are up to you.) The *Graph* function enables you to produce on screen bar graphs of a limited number of parameters versus your selection of atomic number range. To facilitate your answering the questions about the behavior of the properties in the table below for the elements listed in the second column, it is suggested that you choose the range of atomic numbers given in the third column. These selections have been made because they are the easiest trends to see given the limitations of the graphing abilities of the site.

http://www.webelements.com/
http://chemlab.pc.maricopa.edu/periodic/periodic.html (*Chris Heilman's* site)
http://www.schoolscience.co.uk/periodictable.html

	property	*vs*	*elements*	*suggested atomic numbers*
1.	1st ionization energy		period 2, 3	3 - 18
2.	1st ionization energy		inert gases	2 - 54

To select the desired parameter, click on the down arrow on the right of the *Graph* query box. With the graph on the screen, answer the questions in Part C of the ***Results and Discussion*** section.

The next part of this section has been designed for use with ***WebElements*** but other sites can also provide the needed information For instance, isotopic information is available at:

http://encyclopedia.thefreedictionary.com/Isotope%20table%20(divided)
http://www.sisweb.com/mstools.htm
http://www.noble.org/PlantBio/MS/isotope_table_main.html
http://atom.kaeri.re.kr/ton/nuc5.html

Before proceeding into these questions, it will be helpful and save you time if you read all the questions. Each time you select an element, you will be able to find the information required for several different questions. Information will be needed on all the elements from sodium (#11) across the period to argon (#18). Access ***WebElements*** and click on sodium in the periodic table. Scroll down until you find *Naturally occurring isotopes* on the left side and click on that heading. On this web page, find the number of naturally occurring isotopes for sodium and enter the number in the table on page 158. To advance to another element, rather than returning to the periodic table, you can click on the next element in a chart of neighboring elements on the right side. For magnesium and chlorine, you will also need to record the mass (to 4 significant figures) and percentage of each naturally occurring isotope. Finally, for each element, on the right side, scroll down until you find the formulas of compounds of the element with oxygen. From the oxygen compounds listed, calculate the oxidation number of the element in the compound (assuming oxygen is -2) and record the highest oxidation number found for the element of interest on page 158 (e.g., the oxidation number of arsenic in As_2O_5 can be calculated from $2x + 5(-2) = 0$ since oxygen is assumed to be -2 and the sum of the oxidation numbers must be the charge on the species which in this case is 0). Disregard peroxides (such as Na_2O_2, MgO_2 and unusual compounds including $NaO_{2,}$

D. Properties of Compounds. Other than the gases in air, it is unusual to find substances in elemental form in nature. The vast majority of the substances found in nature are compounds. It is very useful to be familiar with the many resources available for quickly finding properties of compounds. The *Handbook of Chemistry and Physics* is one of the most useful resources. The Internet also contains a substantial amount of information about compounds including some properties such as toxicity data that are not available in the *Handbook*. There are many different sites on the web that contain data or links to sites that have data. The best of these are given below.

ChemFinder Web Server -	http://chemfinder.cambridgesoft.com/
ChemExper Chem. Directory -	http://www.chemexper.com/
Vermont Safety Info. Resources, Inc. -	http://hazard.com/msds
Cornell Univ. -	http://msds.ehs.cornell.edu/msdssrch.asp

Bookmarks for these sites will be especially useful. Most of the information you will obtain is intended to acquaint you with toxicity data in the form of oral LD_{50} values for rats. LD_{50} represents the amount of a substance in mg/kg of body weight (be careful - sometimes it is given in g/kg) that will kill half of the population of the animals. Values for rats will be collected here because oral LD_{50} data for rats is by far the most abundant. Recognize that extrapolation to humans is not necessarily appropriate.

Name________________________________ Date__________ Lab Section__________

Results and Discussion - *Experiment 13* - PROPERTIES OF ELEMENTS AND COMPOUNDS: AN INTERNET STUDY

A. Properties of Elements

1. Physical Data

element	melting point (°C)	boiling point (°C)
gallium	________	________
selenium	________	________

2. In what year and how was gallium discovered? ________

3. What are the some of the uses of gallium and why is it suitable for these uses?

4. One of the unusual properties of gallium is also a property of water. What is this property and what precautions must be taken because of it?

5. What are the some of the uses of selenium and why is it suitable for these uses?

B. Extreme Properties of Elements.

a. Fill in the following table:

property	*range*	*elements*
density	>20 g/cm^3	________________
melting point	>3000 K	________________
boiling point	≤ 90 K	________________
electrical conductivity	$\geq 0.37\ 10^6$/cm Ω	________________

b. For electrical conductivity, is there any correlation with the periodic chart? Give the outer electron structure (e.g., Cr = $4s^1 3d^5$) of the best conducting elements and comment on a possible relationship between the electron structure and the conductivity.

C. Periodic Properties of the Elements. Give and explain the trend for the elements below.

property	*vs*	*elements*
1. 1st ionization energy		period 2, 3

There are two small discrepancies in the trend. Where do these occur and why?

2. 1st ionization energy		noble gases

3. Fill in the information below (information is only needed when blanks have been included)

element #	# of isotopes	mass	percentage	oxide formulas	maximum oxid.
Na	_____			__________________	_____
Mg	_____	________	________	__________________	_____
		________	________		
		________	________		
Al	_____			__________________	_____
Si	_____			__________________	_____
P	_____			__________________	_____
S	_____			__________________	_____
Cl	_____	________	________	__________________	_____
		________	________		
Ar	_____				

a. Using the data above, complete the graph for elements 11 - 18 below.

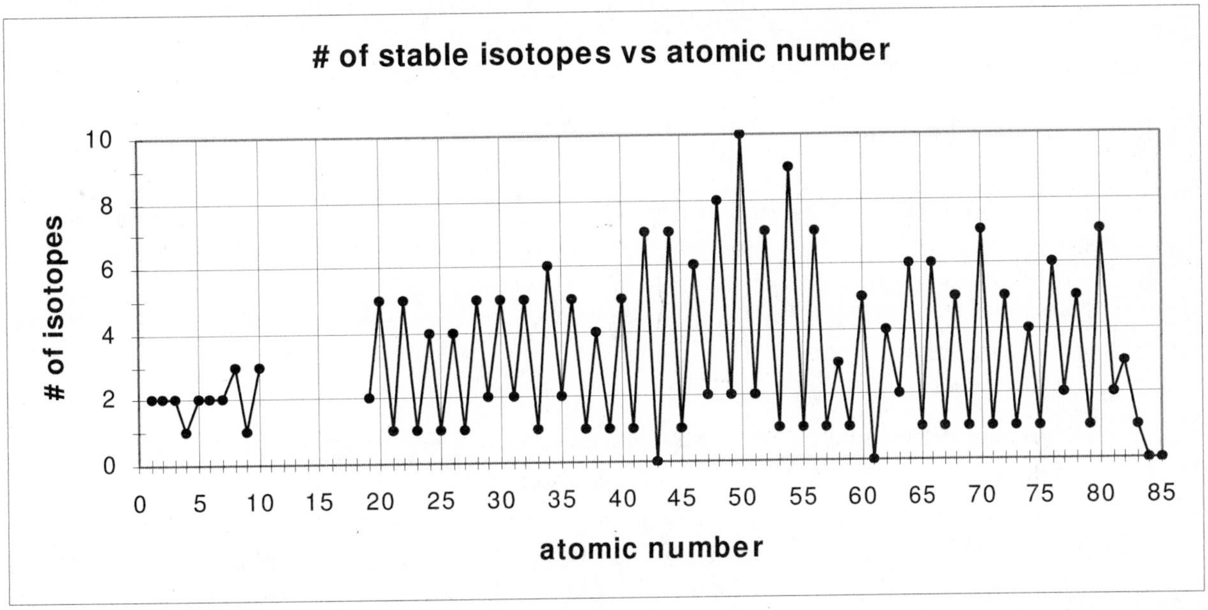

b. From the graph above, describe the periodicity that exists for the number of isotopes for each element as a function of the atomic number. Suggest a reason for the periodicity.

c. Some people might say that if the atomic number is subtracted from the atomic mass that appears in the periodic table, the result is the number of neutrons in the nucleus. Is this a factual statement? If not, how would you rephrase the statement to make it factual?

d. Based solely on information in a periodic table, comment on the expected nuclear stability of phosphorous nucleii with 16 neutrons and with 13 neutrons. Explain your answer.

e. Based on the mass and percentage data that you have recorded above, calculate the atomic masses that should appear for magnesium and chlorine in the periodic table. Show your calculations.

Mg ____________ Cl ____________

f. Is there any periodicity in the maximum oxidation number for elements 11 through 17? If so, explain your answer.

D. Properties of Compounds.

1. Record the melting range of vanillin. Does it agree with the value given ***Appendix 1***?

____

2. For the compounds listed, record the oral LD_{50} values for rats and calculate the amount that would kill half the population of 70 kg humans assuming that the extrapolation is valid.

compound	oral LD_{50} values for rats	extrapolated amount for 70 kg human
vanillin	_________	_________
potassium cyanide	_________	_________
glucose	_________	_________
strychnine	_________	_________
acetaminophen	_________	_________

3. Would you consider vanillin to be a safe food additive? Explain your answer.

4. Strychnine is used as a rat poison. Should it be used carefully? Explain your answer.

5. Acetaminophen is the active ingredient in "non-aspirin" pain relievers. Extra strength tablets contain 500 mg of acetaminophen. How many tablets would be toxic to half the population of 70 kg humans (assuming the rat LD_{50} values apply to humans)?

6. Hetch Hetchy reservoir in the Sierra Nevada Mountains near Yosemite was formed when the O'Shaughnessy Dam was completed in 1923 on the Tuolomne River to provide a water source and storage for San Francisco.[1] The reservoir holds 360,000 acre feet of water. An acre ft. = 1.23×10^6 L. Terrorist threats are in the news today on a daily basis and one of the many concerns is that our water supplies could be contaminated with a very toxic substance.

   a. To determine whether this is feasible, calculate the amount of potassium cyanide that would have to be dumped into Hetch Hetchy to kill half of the people who drink 230 grams (about 8 oz.) of the water. Assume the density of the reservoir contents is 1.0 g/mL.

   b. Do you think it would be possible to dump the amount calculated above into the reservoir without being detected? Explain your answer.

   c. Answer the previous question assuming you have a toxin that is ten times more toxic than potassium cyanide. Explain your answer.

7. You have probably heard it said that people with high blood pressure should decrease their consumption of foods that are high in sodium ion content. One alternative to the use of table salt is the use of potassium chloride. It is possible that at least a part of the cause of blood pressure increase is due to chloride. Explain why it is difficult to perform a controlled study to distinguish between the blood pressure effects of sodium and chloride and perform an Internet search to try to ascertain the best and most recent available evidence on this issue. Is substitution of KCl for NaCl a wise therapy? Examples of possible Internet sites that might be useful are:

   http://www.americanheart.org/presenter.jhtml?identifier=1795
   http://content.nejm.org/cgi/content/abstract/322/9/569
   http://content.nejm.org/cgi/content/abstract/317/17/1043

---

[1]The construction of the dam was strongly opposed by environmentalists including the famous naturalist John Muir, the first president of the Sierra Club. John Muir described Hetch Hetchy as "a wonderfully exact counterpart" of Yosemite Valley, and therefore "one of nature's rarest and most precious mountain temples." Despite John Muir's efforts, the dam was built but even today, environmental groups are working to have the dam removed and the valley restored to its natural beauty.

E. Material Selection (optional). Consider for a moment that you have been asked to select a material to use as electrical wire. Assume for this exercise that you are restricted to the use of a pure element. First, you need to think about the pertinent properties involved in this selection. For electrical conductivity, the element should: have very high electrical conductivity (or minimal resistivity), be very ductile so it can be drawn into wires, have corrosion resistance and have a low cost. Next, the relative importance of each property needs to be considered and then a search should be performed for elements that satisfy the criteria. Taking all properties into account, you would probably conclude that copper and aluminum would be the best candidates. For the applications below, list the criteria you would use to make a selection and, recognizing that the information available in *WebElements* (or another periodic table site such as *Chris Heilman's* or *Environmental Chemistry* - see page 155) is rather limited, do your best to select at least one element for the application. Perhaps an even better search site for properties is:

http://www.matweb.com/

At the above site, click on "Physical Properties - Metric" and then select metal. Next choose properties and specify values for the properties. This program will find alloys as well as elements and it is ok to list alloys instead of elements below.

	criteria	elements
1. pots and pans	____________________	__________
2. a friendship ring	____________________	__________
3. the head of a hammer	____________________	__________

4. Suggest any methods you can think of to improve any part(s) of this experiment.

5. Some of the *Learning Objectives* of this experiment are listed on the first page of this experiment. Did you achieve the *Learning Objectives*? Explain your answer.

Experiment 14

SPECTROSCOPY OF COBALT(II) ION

Learning Objectives

Upon completion of this experiment, students will have experienced:

1. Preparation of a standard solution.
2. Quantitative dilutions.
3. Use of a spectrometer to determine an absorption spectrum and a Beer's law plot.

Text Topics

Quantitative dilutions, spectroscopy, Beer's law (for text correlations, see page ix).

Notes to Students and Instructor

This experiment should take about two hours.

Discussion

Can you imagine the world without color? Have you thought about how you are able to perceive color? The fact that colors are distinguishable suggests that information may be obtained from color determinations. Chemists use spectra for the analysis of composition, structure and concentration. Studies of spectra also led to the development of our current theory of the electronic structure of atoms and molecules. The observation that energetically excited atoms emit light of specific wavelengths that are unique for each element led Bohr, Schrodinger, and others to formulate our present day quantum theory of electronic orbitals.

Organic chemists routinely use infrared and nuclear magnetic resonance spectroscopy to provide important puzzle pieces for the determination of the structures of compounds. Analytical chemists utilize the relationship that the amount of light of a specific wavelength absorbed is proportional to the concentration of the absorbing species (Beer's Law: $A = \epsilon bc$ where A is the absorbtion, ϵ a proportionality constant that is determined by the nature of the absorbing species, b is the path length of light through the sample and c is the molarity of the absorbing species).

In our investigation today, visible light will be used to determine the value of the product ϵb for cobalt(II) solutions. Then Beer's law is used to determine the molarity of the cobalt(II) ion in a solution of unknown concentration. In round tubes of the same diameter, b is constant, but its value can only be approximated so it is better to simply determine the product of the two constants, ϵ and b than to try to determine them individually.

The visible and ultraviolet regions are only a small portion of the electromagnetic radiation spectrum but they encompass the region of energy necessary to promote electrons from ground state orbitals to higher energy orbitals. Considering the wave characteristics of light, its energy as you might have intuitively expected is directly proportional to its frequency, $E = h\nu$. Since the speed of light is the product of its wavelength and frequency, $c = \lambda\nu$, the energy is inversely proportional to the wavelength, $E = hc/\lambda$. It is important to recognize that the longer the wavelength, the lower the energy. Gamma rays, x-rays and ultraviolet radiation have shorter wavelengths and higher energy than visible light. Infrared, radio and TV have longer wavelengths and lower energy than visible.

Procedure

Preparation of solutions. Prepare 50 mL of a stock solution of 0.150 M cobalt nitrate by dissolving the appropriate amount of $Co(NO_3)_2 \cdot 6H_2O$ in 25 mL of deioinized water in a small beaker. Transfer the solution to a 50 mL volumetric flask. Be sure to rinse the beaker and add the washings to the flask. Dilute to the mark and **thoroughly mix**.

Rinse a 25 or 50 mL buret with the cobalt(II) solution and then fill the buret. Clean, dry and number 0 - 5, six 13 x 100 mm test tubes or better yet colorimetry cuvettes. Deliver with the buret, 1.00 mL into test tube 1, 2.00 mL into test tube 2 and so on. Rinse the buret several times with deionized water and fill it with deionized water. Deliver 5 mL of deionized water into test tube 0, 4.00 mL into test tube 1 and so on. Check by eye to see that there is now a total of 5 mL in each test tube. **Thoroughly mix the contents of each test tube.** Obtain a cobalt nitrate solution of unknown concentration from your instructor.

Absorption spectrum. Before you perform the concentration study, it is first necessary to determine the absorption spectrum of cobalt(II) ion. The spectrum will be determined using only tube 5 and tube 0 for a blank. The other tubes would give identical absorption profiles with proportionally lower absorption values. The spectrum enables you to select the best wavelength for the concentration study. Familiarize yourself with the appropriate spectrometer instructions. For a *Spectronic 20*, set the wavelength to 430 nm (see *Figure 14-1* on page 167). With nothing in the sample compartment, set the left knob so that the transmission reads zero (when the sample compartment is empty, the light is mechanically blocked from reaching the detector therefore the amount of light transmitted is zero). In theory this setting of the left knob should not depend on the wavelength and should not have to be reset. It is wise to recheck it occasionally however. Now insert the water blank (tube 0) and calibrate the instrument by setting the right hand knob so that the instrument reads 100% transmission or 0.00 absorption. Insert tube 5 and record the absorption reading. Remove tube 5, change the wavelength to 460 nm and recalibrate the right hand knob with tube 0. [Note: Every time the wavelength is reset, the right knob of the instrument must be reset with a blank. For readings of different samples at the same wavelength, resetting is not required or recommended.] Insert tube 5 and read the absorption. Repeat the above process at 480 nm, 500 nm, 510 nm, 520 nm, 540 nm, 570 nm and 610 nm. Plot the absorption (y axis) vs the wavelength (x axis) and determine the wavelength of maximum absorption.

Beer's law Plot. Set the wavelength to the optimum wavelength determined immediately above and recalibrate the instrument. Successively insert tubes 1 - 5 and the unknown and record the absorption values. Plot the absorption (y axis) versus the concentration of cobalt(II) ion (x axis) and determine the concentration of the cobalt(II) ion in the unknown.

Name________________________________Date__________Lab Section__________

Prelaboratory Problems - *Experiment 14* - Spectroscopy of Cobalt(II) Ion

The solutions to the starred problems are in ***Appendix 4***.

1.* The microwave oven in your house uses electromagnetic radiation that has a wavelength about 0.01 cm. The wavelength range of visible light is 400 - 700 nm. Which is more energetic, microwave or visible light and is it appropriate to say that a microwave "nukes" the food? Explain your answer.

2. a.* How many grams of $CuSO_4 \cdot 5H_2O$ are needed to prepare 50.0 mL of a 0.30 M $CuSO_4$ solution?

____

   b.* If 2.00 mL of the 0.30 M $CuSO_4$ solution is diluted to 5.00 mL with water, what is the resulting $CuSO_4$ concentration?

____

3. a. How many grams of $NiSO_4 \cdot 6H_2O$ are needed to prepare 200×10^2 mL of a 3.5×10^{-2} M $NiSO_4$ solution?

____

   b. If 15.0 mL of the 3.5×10^{-2} M $NiSO_4$ solution is diluted to 50.0 mL with water, what is the resulting $NiSO_4$ concentration?

____

4. a.* The light transmission ($-\log_{10}T = A = \epsilon bc$) at 600 nm of a 0.25 M $CuSO_4$ solution on a shoulder of its absorption peak is 59% in a 1.00 cm cell. What is the value of ϵ for $CuSO_4$ at 600 nm?

b. The absorption values for 1 mm of pyrex at 320 nm and 280 nm are 0.15 and 1.3 respectively. What are the percent transmissions for 1 mm of Pyrex at these wavelengths?

c. Would cells made out of Pyrex be useful for obtaining ultraviolet spectra below 270 nm? Explain your answer.

5.* In this experiment, you will prepare two graphs. The first will be absorption vs wavelength and the second, absorption vs concentration. Should either (or both) graph result in a straight line? Explain your answer.

6. At a wavelength of 270 nm, a 0.040 M solution of acetone in water has an absorption of 0.64 in a 1.00 cm cell. The absorption of a solution of unknown concentration of acetone in water was 0.48 at the same wavelength and in the same cell. What is the concentration of acetone in the unknown? (Assume that there is zero absorption of light by water at 270 nm.)

7. Can visible light cause excitation of an electron in water? Explain your answer.

Name______________________________ Date__________ Lab Section__________

Results and Discussion - *Experiment 14* - Spectroscopy of Cobalt(II) Ion

Fig. 14-1 Spectrophotometer

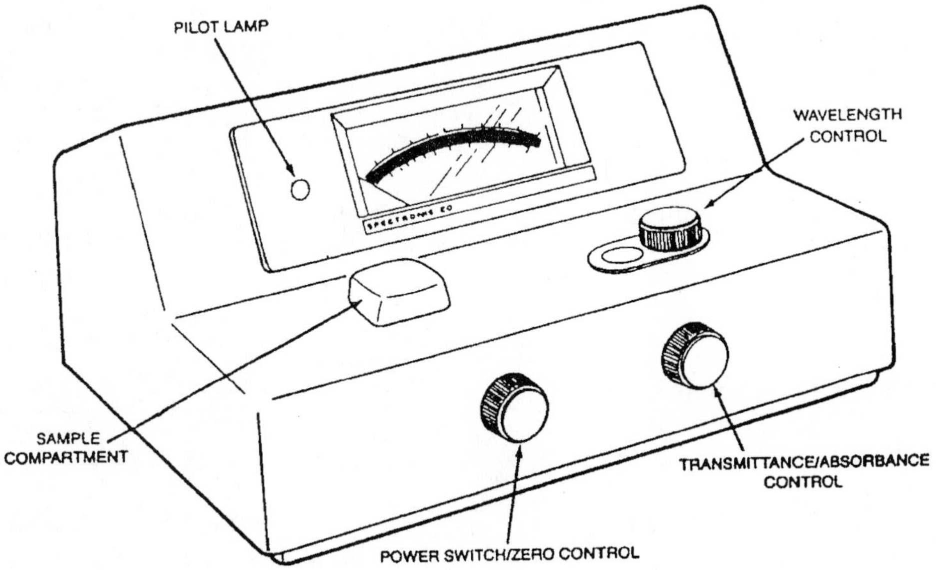

1. Formula mass of $Co(NO_3)_2 \cdot 6H_2O$ __________
2. Mass needed to prepare 50 mL of 0.150 M $Co(NO_3)_2$ __________
3. Mass of beaker + $Co(NO_3)_2 \cdot 6H_2O$ __________
4. Mass of beaker __________
5. Mass of $Co(NO_3)_2 \cdot 6H_2O$ __________
6. Moles of $Co(NO_3)_2 \cdot 6H_2O$ __________
7. Molarity of $Co(NO_3)_2$ __________

8. Absorption spectrum of tube 5

wavelength (nm)	absorption	wavelength (nm)	absorption
430	________	520	________
460	________	540	________
480	________	570	________
500	________	610	________
510	________		

Graph A (y axis) vs λ and determine the optimum wavelength.
Explain your selection. ________

9. Unknown identification number ________

10. Beer's Law Plot

Tube #	Concentration (moles/L)	Absorption	Tube #	Concentration (moles/L)	Absorption
1	____________	________	4	____________	________
2	____________	________	5	____________	________
3	____________	________	**unknown**		________

11. Graph A (y axis) vs c (x axis) and determine the product ϵb from the slope of the line. ________

12. Concentration of cobalt(II) read directly from graph. ________

13. Concentration of cobalt(II) calculated using ϵb. Show calculations. ________

14. Was Beer's Law obeyed? Explain your answer. ________

15. Some of the *Learning Objectives* of this experiment are listed on the first page of this experiment. Did you achieve the *Learning Objectives*? Explain your answer.

Name________________________________Date_________Lab Section_________

Postlaboratory Problems - *Experiment 14* - Spectroscopy of Cobalt(II) Ion

1. List some criteria that you think were used for the selection of cobalt(II) for this experiment.

2. Are there limitations to the concentrations of cobalt(II) ion that can be determined using this technique? Explain your answer.

3. List several additional cations and anions that you think could be analyzed quantitatively by visible spectroscopy.

4. Could visible spectroscopy be used for qualitative analysis of the ions you have listed in *#3*? If so, explain how.

5. In ***Experiment*** *7*, copper(II) ion was quantitatively analyzed by a precipitation technique. Could visible spectroscopy have been used instead? If so, what criteria would you use to choose between the two techniques?

6. The method used to prepare the diluted concentrations of cobalt(II) solutions was not a technically correct method for preparing diluted solutions. The method was used here because of ease and to minimize the need for expensive glassware. What was not technically correct about the method and how should the dilutions have been performed?

7. Suggest any way you can think of to improve any part(s) of this experiment?

Experiment 15

LEWIS STRUCTURES AND MOLECULAR MODELS

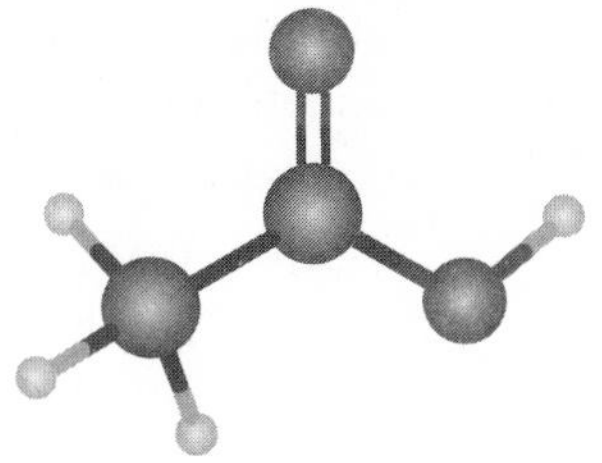

acetic acid

Learning Objectives

Upon completion of this experiment, students will have experienced:

1. Drawing Lewis structures of simple molecules and polyatomic ions.
2. Construction of models of simple molecules and polyatomic ions.

Text Topics

Lewis structures of molecules and polyatomic ions, (see page ix).

Notes to Students and Instructor

Students should work on Lewis structures before coming to the laboratory.

Discussion

Although it has recently become possible to image molecules and even atoms using a scanning tunneling microscope, most of our information about molecular structure comes from X-ray diffraction and interpretation of physical, chemical and spectroscopic properties of substances. This information often enables us to piece together a 3-dimensional picture or model of the molecule. On paper, one of the best methods we have of representing this model is by drawing a Lewis structure of the molecule or ion. The ability to draw Lewis structures for covalently bonded compounds and polyatomic ions is essential for the understanding of polarity, resonance structures, chemical reactivity and isomerism. Molecular models are useful tools to help you visualize the structures especially when the ion or molecule is not planar. It is not the intention here to teach you all aspects of the drawing of Lewis structures or the construction of molecular models but to guide you through some of the more fundamental aspects and provide a few clues where difficulty is often encountered.

Lewis structures. Remember the absolute rule that the Lewis structure must show the correct number of electrons. For a molecule, the sum of the valence electrons is the correct number. For an ion, it is the sum of the valence electrons and the negative of the charge on the ion. For example, formaldehyde, CH_2O, must show $4 + 2 + 6 = 12$ electrons and nitrite, NO_2^-, must show $5 + 6 + 6 + 1 = 18$ electrons. Then, whenever possible try to complete each atom's outer shell so that it has an octet (or duet for hydrogen) of electrons.

One general sequence to follow when constructing Lewis structures is:

1. Determine the correct number of electrons that must be showing (sum of the valence electrons and the negative of the charge on the ion).
2. Arrange the atoms in the correct sequence or sequences if there is more than one logical sequence. For polyatomic ions, the atom listed first (e.g. S in SO_4^{2-}) is usually the central atom (an atom attached to two or more atoms). For more information on choosing the correct sequence, see the discussion below.
3. Connect each of the atoms with a line or one bond.
4. Distribute the remaining electrons in pairs (subtract two each from the total for each bond already inserted) around the external atoms (the ones attached to the central atom) attempting to complete each atom's octet. Remember that the octet (duet for hydrogen) should never be exceeded for atoms through the second period and should only be exceeded for atoms beyond the second period when the atom is a central atom. If electrons still remain, attempt to fill the octets of central atoms or even exceed the octets for atoms beyond the second period.
5. If central atoms are short of achieving an octet, move electron pairs from external atoms to form multiple bonds between the external atom and the central atom in an attempt to provide all atoms with an octet (or, for atoms beyond the second period the octet might be exceeded).
6. If the total number of electrons is odd, it will be impossible to satisfy the octet rule for all atoms and in fact the species will be a free radical (it will have an unpaired electron) and probably will be a relatively reactive species.
7. Visually inspect the resulting structure and decide if it is consistent with other structures you have seen (generalities are difficult but the following are usually true):
   a. hydrogen has one bond
   b. carbon has four bonds (except for carbon monoxide, cyanide ion and unstable species)
   c. halogens have one bond unless the halogen is a central atom
   d. oxygen usually has two bonds except for polyatomic ions
   e. nitrogen usually has three bonds unless the nitrogen is the central atom

If the structure is not intuitively satisfying or if you cannot decide between different bonding sequences, determine the formal charge on each atom. You are probably familiar with the concept of oxidation number. Basically, the oxidation number method assumes that bonds are 100% ionic and assigns all the electrons in bonds to the more electronegative partner. For sodium chloride, this results in oxidation numbers of +1 for sodium and -1 for chloride. While this method gives a good model of the compound when the bonding is ionic, it should be considered nothing more than a bookkeeping method for covalent bonds. In other words, it is a useful method for determining if a reaction is a redox reaction and if so, what is oxidized and what is reduced. For covalent bonds, oxidation numbers give a very distorted view of the charges in the molecule. For instance, for HCl, the oxidation number method results in a +1 for hydrogen and -1 for chloride. In actuality, the hydrogen chlorine bond is best described as polar covalent with a partial positive charge on the hydrogen and a partial negative charge on the chlorine.

The formal charge method assumes that the bonds are 100% covalent with bonded electrons equally shared by the two partners. Thus the formal charge method gives an indication of the locations of charges in covalently bonded compounds. Although formal charges ignore differences in electronegativities, the method still provides useful information. In general, everything else being equal (e.g., octet rule is satisfied) **the structure with the minimum number of formal charges will be favored.** The formal charge is calculated from the following formula:

formal charge = valence electrons - bonds - nonbonded electrons

After you have determined the formal charges, choose the structure with the minimum number of formal charges. For some structures with central atoms that are beyond the second period (commonly phosphorous, sulfur, chlorine, bromine, iodine), it is sometimes preferable to move electrons from external atoms to form multiple bonds with the central atom. Apparently minimizing formal charge is energetically better than maintaining an octet for these atoms.

8. If it is possible to draw more than one reasonable structure by moving electrons only, then all of these structures should be drawn and connected by double headed arrows. These structures are resonance structures and the actual structure is a hybrid of all of the resonance structures. It is very important to realize that the actual structure is not going back and forth between the resonance structures but is a hybrid of the structures. For instance, if formal charges are different from one structure to the other, the formal charges on each atom are probably best represented by an average of the values in each structure.

9. Consider the geometry of the structure by applying VSEPR theory or a hybridization model to the resulting Lewis structure. Both models, when appropriately used will predict with very few exceptions the same shape. The table below summarizes the theories.

groups[1] of electrons around central atom	electronic shape[2]	bond angles	hybridization
2	linear	180°	sp
3	planar	120°	sp^2
4	tetrahedral	109.47°	sp^3
5	trigonal bipyramid	90°, 120°, 180°	dsp^3
6	octrahedral	90°, 180°	d^2sp^3

[1]The number of groups of electrons is equal to the sum of the number of neighbor atoms and nonbonded electron pairs.
[2]Be sure to distinguish electronic shape from molecular shape. If one or more of the groups of electrons are nonbonded pairs, the molecule needs to be described by the relative positions of the atoms; not by the shapes of the electronic orbitals.

For those of you who still want further help on drawing Lewis structures, one technique that can help is to determine the value of the expression $\pi_b = (6n + 2 - \#)/2$ where n is the number of atoms (not counting hydrogens) and # = the number of available electrons. If π_b is a positive whole number, it usually represents the number of extra or π bonds in the species. If π_b contains a fraction, the species will have an unpaired electron (a free radical) and not all atoms will have an octet. If π_b is negative, the species has as atom that exceeds the octet. For most examples in this course, π_b will be either zero or a positive number. When it is zero, all atoms are connected by single bonds. When the value is above zero, the value usually represents the number of π bonds that need to be included and this can save considerable time.

To illustrate the use of the preceding guidelines, the procedure will be applied to a couple of examples. Consider the molecule formaldehyde, CH_2O. First, determine the correct number of electrons that should show in the final structure which is 4 + 2 + 6 = 12. Now we must consider the sequence of bonding. Three possible choices are (other possibilities can be eliminated because they require two bonds to hydrogen which except for some unusual boron compounds should be absolutely avoided):

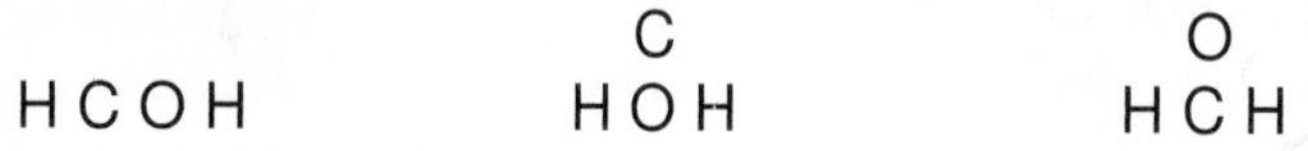

As symmetrical choices are often favored, the first structure could be eliminated because it has lower symmetry than the other two. We will leave it in for this discussion and continue by inserting the three single bonds (6 e^-) and adding the remaining 6 e^- to the external atoms (C and O respectively for the second and third structures) or 6 e^- to the C and O of the first structure. To satisfy the octet rule in each structure, one nonbonded electron pair is moved from the external atom in the second and third structure to form a double bond with the central atom. For the first structure, the second nonbonded electron pair on either the C or O is moved between the two atoms to form a double bond.

Alternatively, the equation $\pi_b = (6n + 2 - \#)/2$ could be used to provide useful information before the electrons are inserted. For formaldehyde, π_b has the value 1 → [(6)(2) + 2 - 12]/2 (remember not to count hydrogens). This means that formaldehyde has 1 π bond. Why do this? Once you gain experience, you will simply write down the number of elements and fill in the total number of electrons while satisfying the duet and octet rules. But to begin with, this technique enables you to ascertain the right number of bonds and avoid some random guessing.

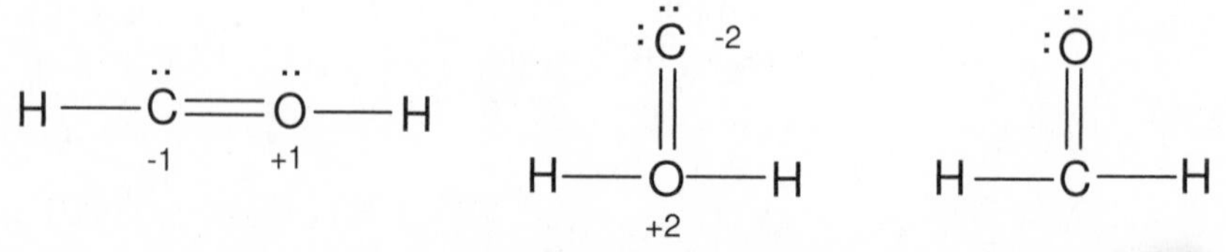

The three structures are technically correct Lewis structures for CH_2O but only one correctly represents formaldehyde. In addition to considering symmetry, it is possible to choose between structures by evaluating the formal charge on each atom in a structure. The minimization of formal charges for CH_2O enables us to strongly favor the third and correct structure.

To draw Lewis structures of ions, the procedure is similar but a modification is needed. In determining the correct number of electrons, the number of valence electrons should be added to the negative of the charge. If the ion has a negative charge, it has extra electrons that it has acquired from its partners. The following example will illustrate the technique for nitrite, NO_2^-. The correct number of electrons is 5 + 12 + 1 = 18. Use of the symmetry guideline leads to the sequence ONO rather than OON and in addition, a formal charge analysis on the completed Lewis structures (try it!) also favors ONO. Calculation of π_b yields a value of 1 indicating the presence of 1 π bond. By following the guidelines on page 174 and/or using the value of π_b, you should end up with the structures below. In this case, you should draw two structures that differ by the position of electrons only and are resonance structures. The double headed arrow below is the convention used to indicate that the two Lewis structures are resonance structures.

Remember that when resonance structures can be drawn, none of the Lewis structures correctly depicts the structure but one must try to imagine a hybrid (or enhanced average) as a better model for the species.

Now lets look at the geometry of formaldehyde and nitrite. For the formaldehyde molecule drawn earlier, there are 3 groups of electrons around the central carbon. Notice the double bond counts as 1 group of electrons and not 2! Focus on the shape around each atom and do not look at the oxygen when determining the shape around carbon. The 3 groups result in a prediction that the atoms around carbon are 120° apart in a plane.

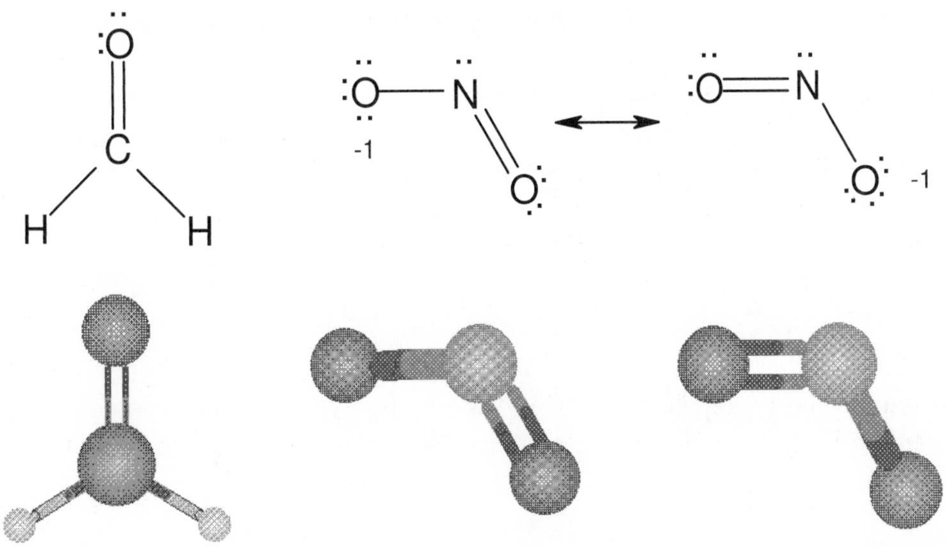

For nitrite, there are also 3 groups of electrons around the central nitrogen as the nonbonded electron pair counts as a group. Again VSEPR theory predicts 120° bond angles and a bent ion.

When 4 groups of electrons surround the central atom, a tetrahedral structure results which can be represented on paper as a projection. While projections are commonly used, CH_4 is often written in planar form with bond angles that appear to be 90°. You should recognize when you see the planar drawing that the bond angles are actually 109.5° and that the molecule is tetrahedral.

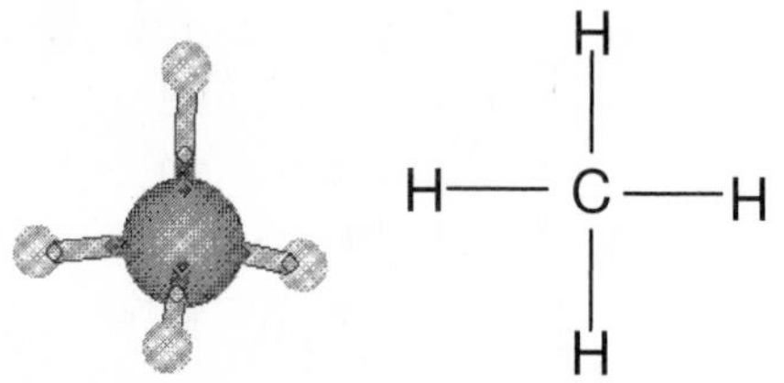

Molecular polarity. Once the Lewis structure has been drawn and the geometry resolved, it is possible to determine if the molecule is polar. A knowledge of polarity is extremely useful for predicting relative boiling points, solubility, and chemical reactivity. For a molecule to be polar, it first must have polar covalent bonds and then it must have a geometry that does not result in the cancellation of bond dipoles. Except for carbon - hydrogen bonds, almost all bonds between different nonmetals are polar covalent. Thus if a molecule contains two different nonmetals and lacks the symmetry necessary for cancellation of bond dipoles, the molecule will be polar. For the formaldehyde (CH_2O) structure above, the carbon - hydrogen bonds are nonpolar but the carbon - oxygen bond is polar with a partial positive charge on the carbon and a partial negative charge on the oxygen. This means that formaldehyde should be a polar molecule. For further discussion of molecular polarity, see ***Experiment 16***.

Molecular models. Most molecular model kits will help you visualize the structures, especially for molecules that are three dimensional. However, the most common ball and stick models compromise correctness for ease when multiple bonds are present. For instance, for formaldehyde, the carbon used will have its holes in a tetrahedral arrangement and the H-C-H bond angle will come out 109.5° and the H-C-O bond angle 125.3° instead of correct bond angles of 120°. Also, for most models kits, the double bond is made up of two identical bent bonds but the hybridization model used in most organic chemistry texts characterize the double bond as two different kinds of bonds; a σ bond and a π bond.

Procedure

Recognizing the discrepancies discussed above, draw Lewis structures and construct models for the molecules and polyatomic ions listed in the ***Results and Discussion*** section.

Prelaboratory Exercises - *Experiment 15* - Lewis Structures and Molecular Models

Try to do the Lewis structures on the following pages before coming to laboratory.

Name__________________________________Date__________Lab Section_________

Results and Discussion - *Experiment 15* - Lewis Structures and Molecular Models

1. For each molecule in the chart below, draw a Lewis structure, construct a model, determine the bond angle(s), molecular polarity (P = polar, N = nonpolar) and the hybridization of the central atom.

Molecule	Lewis Structure	Bond Angle	Polarity	Hybridization
F_2			____	
N_2			____	
ICl			____	
CO_2		____	____	____ C
H_2O		____	____	____ O

Molecule	Lewis Structure	Bond Angle	Polarity	Hybridization
NH_3		_____	_____	_____ N
CH_4		_____	_____	_____ C
C_2H_6		_____ H-C-H _____ H-C-C	_____	_____ C
C_2H_4		_____ H-C-H _____ H-C-C	_____	_____ C
C_2H_2		_____	_____	_____ C
HCN		_____	_____	_____ C

2. For the molecules below, draw the two reasonable, structurally different, possible Lewis structures. Calculate and indicate values of nonzero formal charges. Based on formal charges, circle the preferred structure, construct a model of it and answer the questions on bond angles and hybridization.

Molecule	Lewis Structures	Bond Angle	Hybridization
CH_4O		_____ H-C-H	_____ C
		_____ H-C-O	_____ O
		_____ H-O-C	
N_2O		_____	

3. For each of the molecules below, draw the two reasonable resonance structures and indicate the nonzero formal charges that are present (if any) in each of the structures. For each, construct a model of one of the resonance structures. (Hint: sulfur and nitrogen are the central atoms respectively and the hydrogen in nitric acid is bonded to an oxygen.)

Molecule	Lewis Structures	Bond Angle	Hybridization
SO_2		_____	_____ S
HNO_3		_____ H-O-N	_____ N
		_____ O-N-O	

4. For each polyatomic ion in the chart below, draw all the reasonable resonance structures, indicate nonzero formal charges and construct a model of one of the resonance structures of each ion. Determine the indicated bond angle(s).

Polyatomic Ion	Lewis Structures	Bond Angle
OH^-		
CN^-		
ClO_2^-		____
ClO_3^-		____
CO_3^{2-}		____

Polyatomic Ion	Lewis Structures	Bond Angle
SO_3^{2-}		____
SCN^-		____
HCO_2^-		____ H-C-O ____ O-C-O
NO_2^+		____

Problems 5-10 attempt to demonstrate the importance of Lewis structures. In addition to providing a view of the shape and polarity of the molecule, Lewis structures sometimes provide insight into chemical reactivity. When the Lewis structure indicates some unusual or undesirable characteristic such as high formal charges or unusual oxidation numbers, strained bond angles, unpaired electrons or lack of an octet, there is a strong possibility that the molecule will exhibit extraordinary behavior. Basically this exercise is designed to show you that you can apply your knowledge of chemistry to new situations and think like a chemist.

5. Ozone (O_3) is needed in the stratosphere to absorb (and filter out) potentially damaging ultraviolet light. However, in the lower atmosphere it is a dangerous pollutant as it is a very reactive form of oxygen and as a result, very toxic and destructive. Draw the two reasonable resonance structures (Hint: it is not a ring) and indicate the bond angle and the nonzero formal charges. Suggest a reason for its high reactivity.

6. The combustion of gasoline in a car cylinder generates gas at a high temperature that pushes the cylinder down and powers the car. Unfortunately, nitrogen and oxygen combine at the high temperature in the cylinder to produce some undesired nitrogen monoxide. After emission from the exhaust, the nitrogen monoxide is oxidized by oxygen to nitrogen dioxide. NO_2 is one of the brown colored gases present in L. A. smog and is a very dangerous pollutant because of its very high reactivity. NO_2 establishes an equilibrium with its dimer N_2O_4. Draw Lewis structures of nitrogen dioxide and its dimer, dinitrogen tetroxide.

$$2\ NO_2(g) \rightleftarrows N_2O_4(g)$$

Suggest a reason for the high reactivity of NO_2 and its dimerization reaction.

7. Hydrogen peroxide (H_2O_2) is a reactive molecule, often used as an antiseptic and sometimes used for bleaching. Draw a Lewis structure for hydrogen peroxide. Calculate the oxidation number of oxygen in hydrogen peroxide and suggest a reason for its reactivity.

8. Use of an atomic orbital approach to bonding without adding modifications for electron promotion and hybridization leads to the naive conclusion that carbon atoms should combine with hydrogen atoms to give CH_2. Draw a Lewis structure for this result. CH_2 (usually called methylene or carbene) actually can be made as a transient species and plays a very important role in synthetic organic chemistry. It has been described as one of the most indiscriminate reagents in organic chemistry. For example, it reacts with ethylene (C_2H_4) to give cyclopropane (C_3H_6). Show this reaction using Lewis structures for reactants and products and suggest a reason for the very high reactivity of methylene.

9. Bromomethane (CH_3Br) reacts with hydroxide ion to give methanol (CH_4O - see *Problem* 2) and bromide ion. Show this reaction using Lewis structures and use an explanation involving polarity to give a reason for the site of attack of hydroxide on bromomethane.

10. Sodium borohydride ($NaBH_4$) and lithium aluminum hydride ($LiAlH_4$) are very useful reducing reagents in organic chemistry. They are commonly used to reduce carbonyl compounds (aldehydes and ketones) to alcohols [e.g., acetone (CH_3COCH_3) to isopropyl alcohol ($CH_3CHOHCH_3$)].

    a. Draw the Lewis structures of acetone and isopropyl alcohol.

    b. Draw the Lewis structures of the borohydride and aluminum hydride ions. Calculate the oxidation numbers (Hint: consider electronegativities in chart on page 189) of boron, aluminum and hydrogen and suggest a reason for the reducing capability of the two ions.

11. Suggest any ways you can think of to improve any part of this experiment.

12. Some of the *Learning Objectives* of this experiment are listed on the first page of this experiment. Did you achieve the *Learning Objectives*? Explain your answer.

Experiment 16

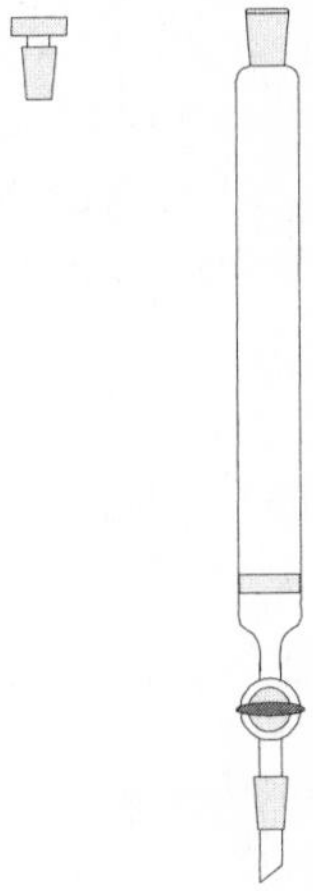

MOLECULAR POLARITY AND CHROMATOGRAPHY

Learning Objectives

Upon completion of this experiment, students will have experienced:

1. Application of molecular polarity concepts.
2. The techniques of extraction, paper chromatography, and fabric dyeing.

Text Topics

Electronegativity, molecular polarity, intermolecular forces, extraction, chromatography (for correlation to some textbooks, see page ix).

Notes to Students and Instructor

This experiment can be completed in a reasonable amount of time with careful planning. The chromatograms should be started first and the miscibility and fabric dyeing performed while the chromatograms progress.

Discussion

In the liquid state, the molecules of a substance are adjacent to each other and attracted together by intermolecular attractions. For covalently bonded substances, the attractions are usually classified into three groupings: dispersion forces (or London forces), dipole-dipole attractions and a sub category of the latter, hydrogen bonds. In the gaseous state, molecules behave as though they are independent of each other and do not experience significant intermolecular attractions except at high pressure or very low temperature. To boil a liquid requires that sufficient energy be supplied to overcome the intermolecular attractions. The boiling point should and does generally correlate with the strength of intermolecular attractions.

Very briefly, London forces are temporarily induced polarizations that result from the approach of two molecules. The magnitude of these forces depends on the size and shape of the molecules. As polarization of electron clouds are of prime importance in this effect, it would be expected that the strength of the attractions should depend partially on the number of electrons. As the number of electrons correlates with the molecular mass, the boiling point would be expected to correlate with molecular mass if other parameters do not change.

Consider the boiling points of the hydrocarbons in *Table 16-1*. These compounds consist of carbon-carbon and carbon-hydrogen bonds only. The former are obviously non-polar covalent bonds and the latter are about as non-polar as possible for bonds between two different elements. Thus the boiling points exhibit a steady increase due to increasing London forces. The hydrocarbons in *Table 16-1* do not have significant dipoles or charge separations and are good models for comparison with other compounds. [Note that another consideration is that the average molecular velocity decreases as the molecular mass increases (at a given temperature). This factor also contributes to the tendency of boiling points to increase with molecular mass.]

Table 16-1

alkane	formula	boiling point (°C)
methane	CH_4	-161.7
ethane	C_2H_6	-88.6
propane	C_3H_8	-42.1
butane	C_4H_{10}	-0.5
pentane	C_5H_{12}	36.1
hexane	C_6H_{14}	68.7
heptane	C_7H_{16}	98.4
octane	C_8H_{18}	125.7

Table 16-2

CH_4	NH_3	H_2O	HF
-161.7	-33.4	100	19.5
SiH_4	PH_3	H_2S	HCl
-111.8	-87.4	-60.7	-85.0
GeH_4	AsH_3	H_2Se	HBr
-88.5	-55	-41.5	-67.0
SnH_4	SbH_3	H_2Te	HI
-52	-17.1	-2.2	-35.4

In this paragraph and in *Prelaboratory Problem 4*, the data in *Table 16-2* and *Fig. 16-1* will be considered. Going down the periods in Group 4 from CH_4 to SnH_4, there is a steady increase in boiling points. This increase is consistent with an increase in dispersion forces and lower average molecular velocities as the molecular mass increases. However, *Table 16-2* reveals two anomalies that deserve our attention. First, the compounds of the second period elements, C, N, O and F with H all have similar numbers of electrons thus dispersion forces should be similar. On this basis, the boiling points for CH_4, NH_3, H_2O and HF should be similar. Since the data contradict this prediction, another attraction must come into play. This attraction is called hydrogen bonding. The second anomaly is apparent when we compare the boiling points of NH_3 with that of PH_3, H_2O with that of H_2S, HF with that of HCl. Despite the greater number of electrons of the second compound of each comparison, the first compound has a much higher boiling point. Hydrogen bonding again accounts for the apparent discrepancy. Hydrogens bonded to N, O and F are capable of hydrogen bonding to an N, O or F in another molecule. None of the other elements is electronegative enough to have this capability.

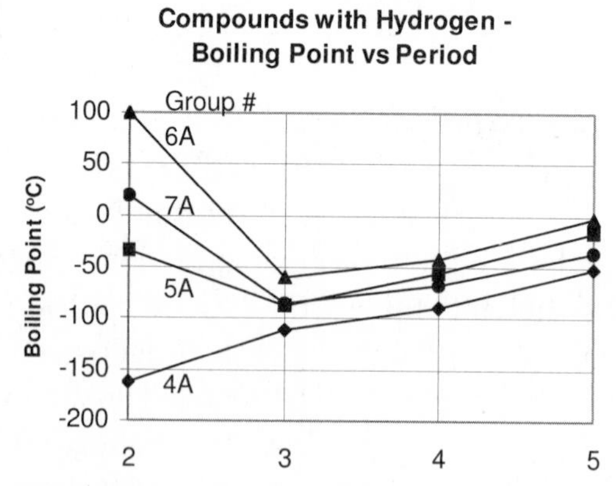

Fig. 16-1

Another consequence of the increasing bond polarity going across the second period is that the non-polar methane (non-polar because of non-polar bonds and its high symmetry) is not soluble in the very polar water. However, methanol (CH_3OH) with polar carbon - oxygen and oxygen - hydrogen bonds and ammonia with polar nitrogen - hydrogen bonds are miscible with water. Methanol and ammonia both form strong hydrogen bonds with water.

Fig. 16-2

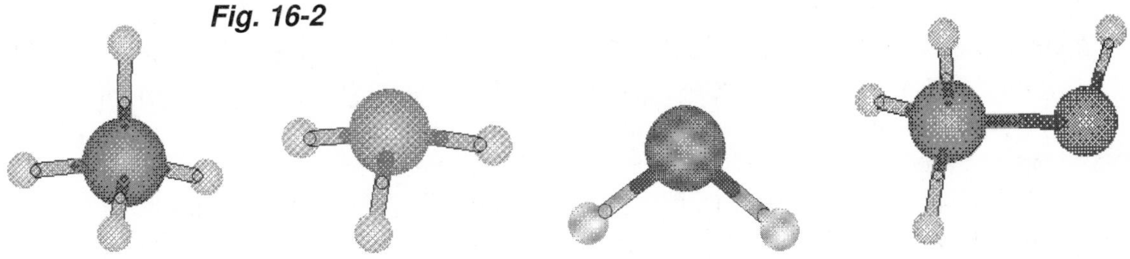

methane (CH_4) ammonia (NH_3) water (H_2O) methanol (CH_3OH)

Intermolecular attractive forces and especially hydrogen bonds are extremely significant in chemistry. They influence boiling points, solubility, chemical reactivity and are responsible for determining the 3-D structures of proteins and DNA. Bond polarity can be predicted by an examination of Pauling's empirical values of electronegativities. One set of values for many elements are in the following chart: [Extracted from A. L. Allred, *J. Inorg. Nucl. Chem.*, 17, 215 (1961).]

Table 16-3

H = 2.20							
Li = 0.98	*Be* = 1.57	*B* = 2.04	*C* = 2.55	*N* = 3.04	*O* = 3.44	*F* = 3.98
Na = 0.93	*Mg* = 1.31	*Al* = 1.61	*Si* = 1.90	*P* = 2.19	*S* = 2.58	*Cl* = 3.16
K = 0.82	*Ca* = 1.00	*Ga* = 1.81	*Ge* = 2.01	*As* = 2.18	*Se* = 2.55	*Br* = 2.96
Rb = 0.82	*Sr* = 0.95	*In* = 1.78	*Sn* = 1.96	*Sb* = 2.05	*Te* = 2.1	*I* = 2.66

Generally, when the electronegativity difference between two bonding partners is very small (such as for the carbon-hydrogen bond), the bond behaves as though it is non-polar. As the electronegativity difference increases, the bond polarity increases to the limiting point where the bond is completely ionic. **It is easier to just remember that bonds between identical elements and the important carbon - hydrogen bond are non-polar while most other nonmetal - nonmetal bonds are polar covalent. Metal to nonmetal and metal to polyatomic ion bonds usually have predominantly ionic character.** For a molecule to be polar, there must be polar bonds and a lack of symmetry in structure such that the bond dipoles do not cancel each other out. CCl_4 has polar bonds but because of its tetrahedral geometry, is a non-polar molecule. Water has polar bonds and because of its bent geometry is highly polar. In general, polar molecules will have higher boiling points than non-polar molecules with similar molecular mass. Molecules with similar polarities will have a greater tendency to dissolve in each other or "like dissolves like."

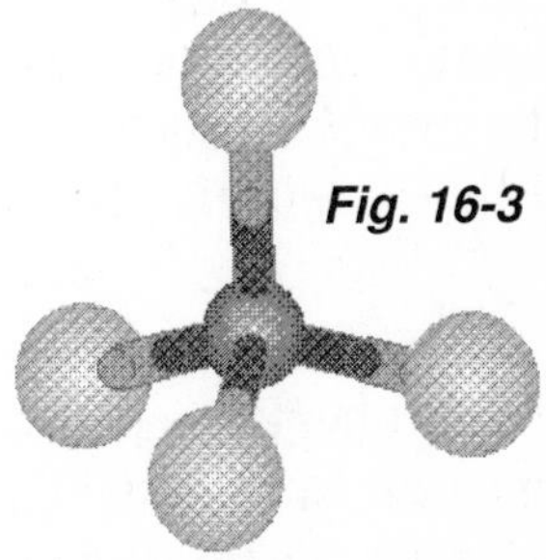

Fig. 16-3

The above considerations will be applied to miscibility tests, extraction experiments, dyeing of fabrics and paper chromatographic separations. You are aware that ethanol (CH_3CH_2OH) and water are miscible in all proportions (consider vodka - 40 to 50% ethanol). This is because both ethanol and water are polar molecules and due to *hydrogen bonding* are strongly attracted to each other. But what happens when the organic compound has no polar bonds such as in kerosene? We will investigate the miscibility of water and kerosene in this experiment.

Extraction. Consider an immiscible pair of liquids such as vinegar (polar - 95% water, 5% acetic acid) and oil (non-polar). Suppose the vinegar has a small amount of a non-polar solute dissolved in it and you shake the vinegar with some oil. The solute will have two options. As its polarity is closer to that of the oil than of vinegar, the solute will probably shift to the oil. We say that the solute has been extracted from the vinegar by the oil. If the solute had been polar, it would have remained in the vinegar.

Paper chromatography. Chromatography can be thought of as a dynamic extraction where the solute continually has an option of two phases. This is accomplished by having one phase move by a stationary phase. Chromatography was discovered by Michael Tswett (a Russian botanist) early in the 20th century. Tswett allowed a mixture of pigments extracted from plants to percolate down through a column of calcium carbonate. Solvent was added from the top as needed to cause continuous movement of the pigments down the column. Pigments that were more strongly attracted to the stationary phase (calcium carbonate) and had less affinity for the solvent (moving phase) moved down the column more slowly than pigments that had greater affinity for the solvent and had weaker attraction for the stationary phase. Tswett observed that the pigments had separated into several differently colored bands as a result of the fact that they moved down the column at different rates. The term chromatography was coined to describe the phenomenon.

The paper chromatography experiments you will perform today utilize the same principles Tswett employed. A piece of paper, spotted with pure compounds and mixtures is placed in a solvent as illustrated in *Figure 16-4*. Assume spots 1, 2, and 3 are pure compounds and spot 4 is an unknown mixture of the compounds. The solvent (moving phase) will move up the paper (stationary phase). When the solvent reaches the spots, the components of each spot have options. They can dissolve in the solvent and progress up the paper or they can stay adsorbed on the paper and resist movement. The preference depends on several factors including the polarities of the compounds, the solvent and the paper. The solvent and the stationary phase are selected so that the components spend some time in each phase and do not move right along with solvent front or stay at the origin. If the solvent and the stationary phase are selected properly, different compounds will move up the paper at different rates and separate.

The spots of compounds that are less strongly adsorbed on the paper will move up the paper faster than the spots of the more strongly adsorbed compounds. When the solvent front nears the top of the paper, the chromatogram is removed from the solvent and the solvent front marked with a pencil. To find out the composition of the fourth spot two factors are considered, color and relative distance moved. To quantify the relative distances, the R_f (ratio to front) value of each spot is calculated.

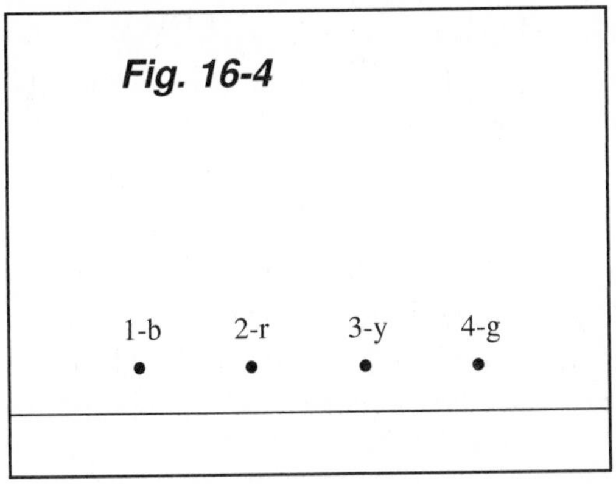

b = blue
r = red
y = yellow
g = green

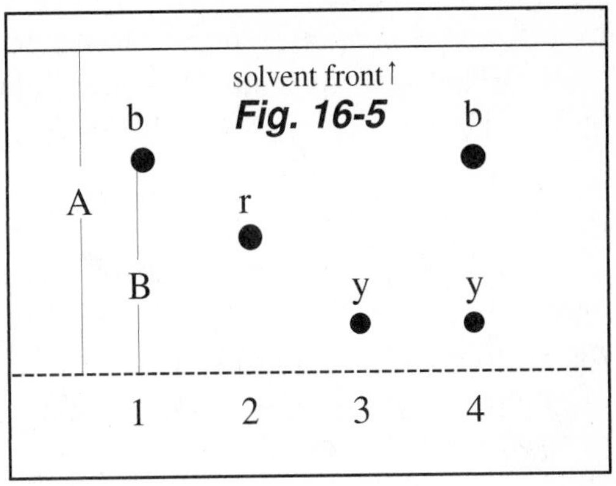

$$R_f = \frac{\text{distance from origin to center of spot}}{\text{distance from origin to solvent front}} = \frac{B}{A} \text{ for spot 1}$$

In the chromatogram above, it can be seen that spot 4 (the green spot) has compounds with the same color and R_f values as pure compounds 1 and 3. This provides strong evidence (but not proof) that spot 4 contains compounds 1 and 3 and does not contain compound 2. Paper chromatography then serves as a separation technique and can also assist with identification if the possible compounds in a mixture are available in pure form.

Fabric dyeing. Many dyes used to color fabrics are polar molecules. Several different types of polymers are used to make fabrics; some have polar groups and others do not. Based on the principle "like dissolves like", polar dyes would be expected to stick to and color polar polymers and be rather ineffective dyes for non-polar fabrics.

Procedure

[Note: It is strongly recommended that you start Part C (paper chromatography) first and do the remaining parts while the chromatograms progress.]

A. Miscibility. Transfer the following liquids into test tubes and attempt to mix. Determine whether the mixtures are miscible or immiscible.

1. 5 mL water + 2 mL ethanol (CH_3CH_2OH)
2. 5 mL water + 2 mL kerosene (hydrocarbon with 10 to 16 carbons)

B. Extraction.

1. Consider the polarities of water and elemental iodine. Should iodine be very soluble in water? Visually inspect the stock solution of aqueous iodine. Can you tell if you were right? Look up the solubility of iodine in water in the *Handbook of Chemistry and Physics*. Do you think iodine should be very soluble in kerosene? Will the iodine prefer water or kerosene if you give the iodine the option? Decant 5 mL of the aqueous iodine solution into a test tube, add 2 mL of kerosene and shake vigorously. Record your observations.

2. Methylene blue is an ionic compound (and therefore polar) but it is a very large molecule and therefore has limited solubility in water. Will the methylene blue prefer water or kerosene? Pour 5 mL of the aqueous methylene blue solution into a test tube, add 2 mL of kerosene, stopper the tube and shake vigorously. Record your observations and conclusions.

C. Paper chromatography. The same general procedure will be followed for your two chromatograms. Following *Figures 16-6* and *16-8*, draw straight pencil lines two centimeters from the bottoms of both pieces of chromatography paper and place pencil dots at equal intervals along the lines. Spot the larger paper with the felt-tip pens, as shown in the diagrams. The spots should be about this size ●.

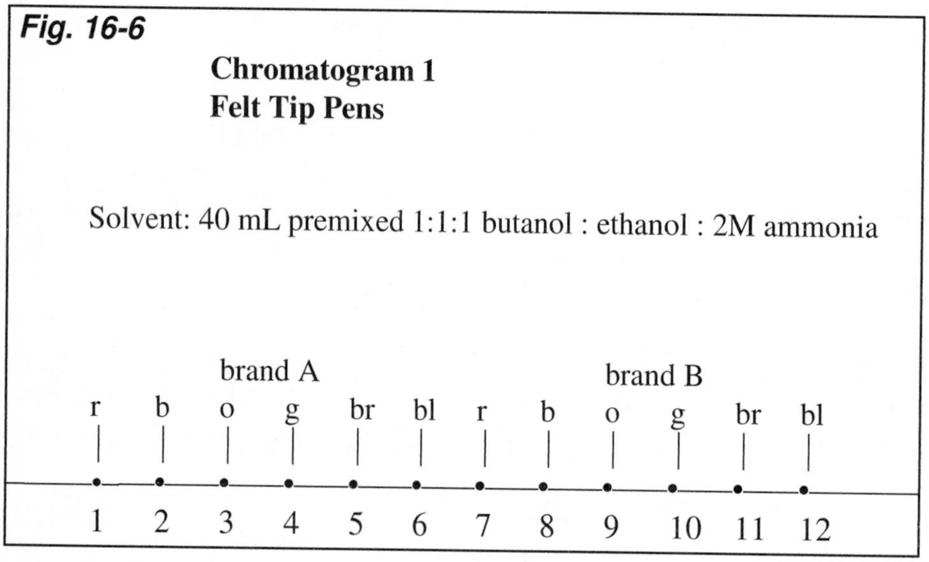

Fig. 16-6

r-red
b-blue
o-optional
g-green
br-brown
bl-black

After the spotting (*Figure 16-6*) is complete, roll the paper into a cylinder and staple it so that there is a small gap between the two ends (*Figure 16-7*). The ends of the paper should **not** overlap. Add 40 mL of the butanol, ethanol, ammonia solvent to a 600 mL beaker. Gently put the paper cylinder (spotted edge down) into the beaker and cover the beaker with plastic wrap. Do not move or turn the beaker again until you remove the paper. At this point, you should begin your second chromatogram but keep your eye on the first one. When the solvent front reaches about 2 cm below the top of the paper, remove the chromatogram. Mark the solvent front with a pencil and dry the paper in the hood, with a hot air blower.

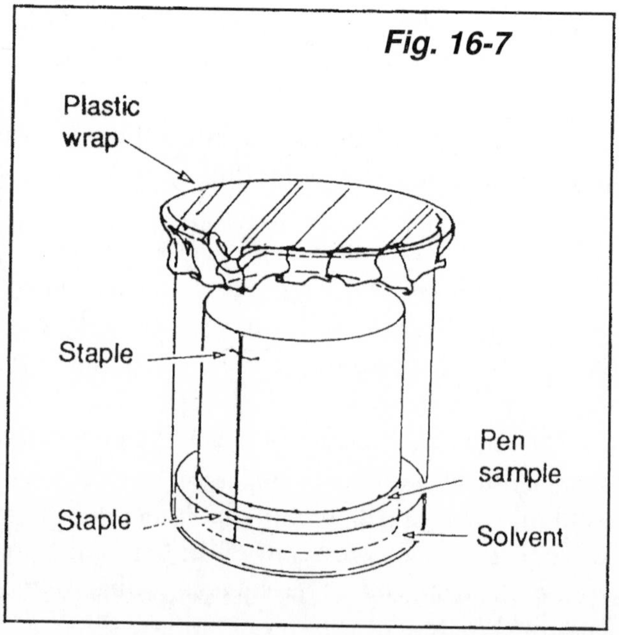

Fig. 16-7

Following *Figure 16-8*, spot the smaller piece of paper with the metal ion solutions and with your unknown. To do this, draw some of the solution up into a capillary tube and apply it to the paper on the appropriate pencil dot. Again, the spots should be about this size, ●. Before spotting your chromatogram, practice your spotting techniques on a scrap piece of chromatography paper. Put 7 mL of 6 M hydrochloric acid and 25 mL of acetone into your 400 mL beaker and stir the mixture until the liquids are ***thoroughly mixed***. Staple the paper into the form of a cylinder, as you did with the larger paper; gently place it into the 400 ml beaker and cover the beaker with the plastic wrap. When the solvent front reaches about 1 cm below the top of the paper, remove the paper from the beaker and mark the solvent front with a pencil.

1 - iron(III)
2 - copper(II)
3 - cobalt(II)
4 - manganese(II)
5 - mixture of 1,2,3,4
6 - unknown

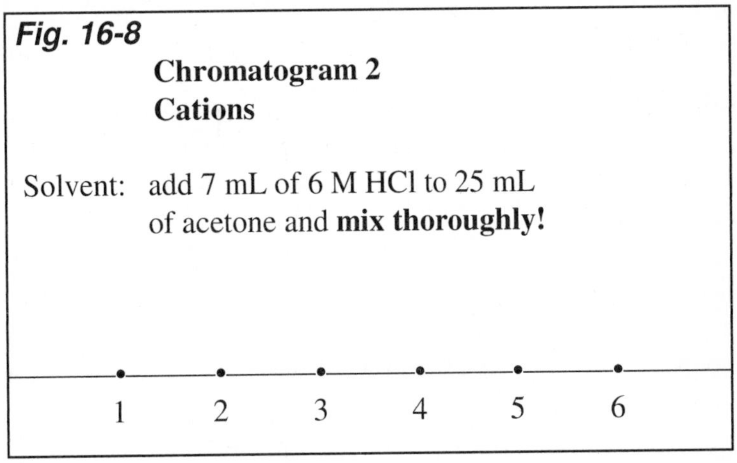

Chromatogram 2
Cations
Solvent: add 7 mL of 6 M HCl to 25 mL of acetone and mix thoroughly!
1
2
3
4
5
6

Fig. 16-8

Dry the metal ion chromatogram carefully with a hot air blower, outline and record colors of any spots that are visible, place it over a Petri dish **in the hood** containing concentrated ammonia ***[Caution: avoid touching the ammonia or breathing the vapors]*** and cover with a watch glass for a few minutes until the spot labeled Mn^{2+} has appeared. Be sure to expose the entire chromatogram to the ammonia vapor. Outline any new spots that appear and record any changes in color in the matrix on page 200. Heat the chromatogram a second time with the hot air blower and again record any changes in spot color.

On both chromatograms, outline each of the spots with a pencil and measure the distance from the origin to the center of each spot. The distance from the origin to the center of the spot, divided by the distance form the origin to the solvent front (see *Figure 16-5*), is the R_f of the spot. Record the R_fs of all spots on your data sheet. Turn in your chromatograms with your report.

D. Fabric Dyeing. A company called *TESTFABRICS, INC.* (200 Blackford Ave., P.O. Box 420, Middlesex, New Jersey 08846) produces a product called "Multifiber Fabric 43" that you will use for this experiment (Alternatively *TESTFABRICS, INC.* also has a less expensive fabric with six different fabrics). Each 2" x 4" piece of this fabric contains 13 different fibers in ⅓" wide bands. The list and order of the fibers is given in the ***Results and Discussion*** section. Transfer 20 mL of an aqueous dye solution containing 0.05% eosin to a 50 mL beaker. Place the testfabric in the solution and heat it to boiling for about 5 minutes. Use tongs to remove the fabric from the bath, allow it to cool, wash it thoroughly with running water and set it aside for drying. Record the color of each fiber. Based on the knowledge that eosin is a polar dye, determine the relative polarities of each fiber.

Note: Other dye solutions (such as methyl orange) also work but most use dyes that are suspected carcinogens. For instructions on the preparation of other dyes for use in this experiment, see for example, Williamson, K. L., *Macroscale and Microscale Organic Experiments*, Heath, 614-620 (1989). Also *Testfabrics, Inc.*, (phone # 201 469-6446) sells identification stains that give many more colors and can be used to identify fabrics.

Chromatogram Summary

Description	Chromatogram 1 Felt-Tip Pens	Chromatogram 2 Cations
paper size	11 x 19.5 cm	9.5 x 14 cm
beaker size	600 mL	400 mL
solvent	40 mL premixed 1:1:1 1-butanol, ethanol, 2 M ammonia	prepare immediately before use - 7 mL of 6 M HCl + 25 mL acetone, mix thoroughly
distance between spots	1.5 cm	2.0 cm
distance of spots from bottom	2.0 cm	2.0 cm
spots	Brand A[1]	Cations
1	red	Fe^{3+}
2	blue	Cu^{2+}
3	optional	Co^{2+}
4	green	Mn^{2+}
5	brown	Fe^{3+},Cu^{2+},Co^{2+},Mn^{2+}
6	black	unknown
	Brand B[2]	
7	red	
8	blue	
9	optional	
10	green	
11	brown	
12	black	
visualization	dry (hot air blower)	a. dry with hot air blower b. place over Petri dish of concentrated NH_3 c. dry again
analysis	Report color, spot distance and R_f value for each pigment in every pen.	After each step (a,b,c), outline each spot and report its color. Determine all R_f values and identify cations in unknown.

[1]Flair or Paper Mate (same company) suggested
[2]Instructor's option

Name__________________________________Date__________Lab Section_________

Prelaboratory Problems - *Experiment 16* - Molecular Polarity and Chromatography

The solutions to the starred problems are in ***Appendix 4***.

1. Which of the following would you expect to have a higher boiling point? Explain your answers.

   a.* CH_4 or HF

   b. CH_3OCH_3 or CH_3CH_2OH

   c. CH_4 or C_2H_6

   d.* CH_4 or SiH_4

2. Should the first compound listed be more soluble in the first solvent or the second? Explain your answer.

   a.* NaCl in water or kerosene

   b. HCl in water or kerosene

   c. Wax in water or kerosene

3. Calculate the R_f values for A, B and C in *Figure 16-9*. What are the components of mixture M?

R_f = ____ (A)

R_f = ____ (B)

R_f = ____ (C)

R_f = ____, ____ (M)

Components of M = ____________

____

Fig. 16-9

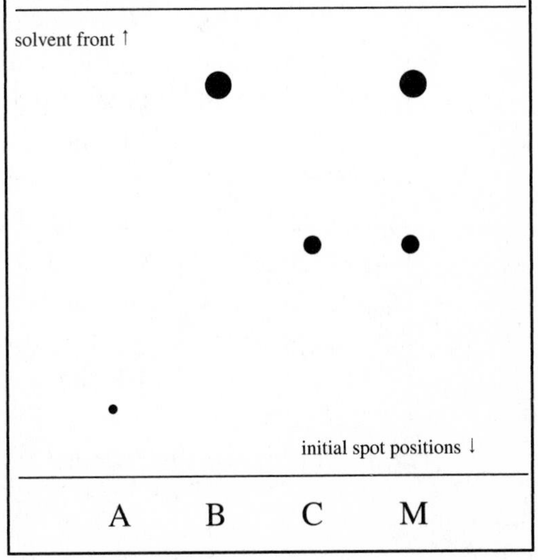

4. Please refer to *Tables 16-2, 16-3* and *Figures 16-1* and *16-10* for these questions.

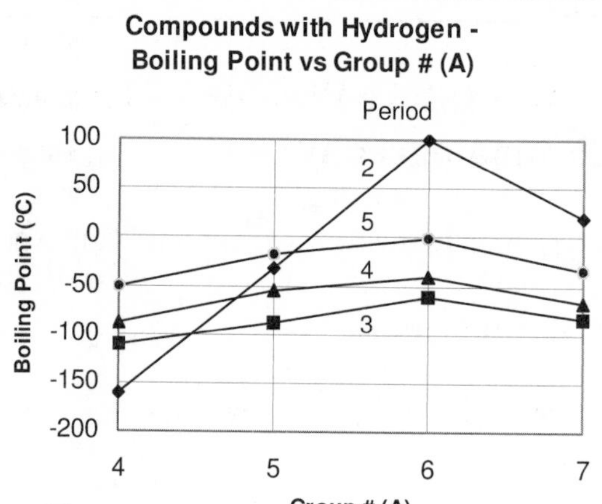

Fig. 16-10

a. Briefly summarize the evidence that hydrogens intramolecularly bonded to N, O or F can form intermolecular hydrogen bonds to N, O or F.

b. Is there any evidence that chlorine forms hydrogen bonds? Explain your answer.

c. General guidelines for determining the nature of a bond from electronegativity scales are that the bond can be considered: ionic if the electronegativity difference between the bonded elements is greater than 1.7, polar covalent if the electronegativity difference is between 0.5 and 1.7 and non-polar covalent if the electronegativity difference is between 0 and 0.5. According to this guideline, bonds from carbon to sulfur and iodine should be non-polar covalent but you will find in organic chemistry that these bonds react as though they are polar covalent. That is why this text prefers the bold statement on page 189 for the determination of the nature of bond. Additionally, use of the bold statement has the advantage that it does not require access to an electronegativity table. Another apparent problem generated by the use of electronegativities is that chlorine is at least as electronegative as nitrogen. Explain how your answers to 4-a and 4-b above seem to be in conflict with the electronegativities of N and Cl.

Name________________________________Date_________Lab Section_________

Results and Discussion - *Experiment 16* - Molecular Polarity and Chromatography

A. Miscibility

mixture	Miscibility (M = miscible, I = immiscible)
1. water + ethanol	__________
2. water + kerosene	__________

B. Extraction

1. Should iodine be very soluble in water? Explain your answer. _________

2. Solubility of iodine in water from *Handbook of Chemistry and Physics*. edition______page______ _________

3. Should iodine be very soluble in kerosene? Explain your answer. _________

4. Observations when aqueous iodine and kerosene are mixed.

   Explanations and conclusions.

5. Observations when aqueous methylene blue and kerosene are mixed.

   Explanations and conclusions.

C. Paper chromatography

1. Chromatogram 1 - Ink Pigments

Brand A _________ Brand B_________ Origin to solvent front_________

	spot #		*pigments* color	dist.	R_f	color	dist.	R_f	color	dist.	R_f	color	dist.	R_f
A	1	red												
	2	blue												
	3	______												
	4	green												
	5	brown												
	6	black												
B	7	red												
	8	blue												
	9	______												
	10	green												
	11	brown												
	12	black												

a. Evidence that two compounds are the same is provided when colors of spots from different sources match and R_f values are within experimental error of each other. Based on color and R_f value, some of the Brand A pigments are used in several different colored Brand A felt-tip pens. Give the color and R_f values of the pigments in the red, blue, optional and green pens that are also apparently used to make black ink.

black		red		black		blue	
R_f	color	R_f	color	R_f	color	R_f	color
___	______	___	______	___	______	___	______
___	______	___	______	___	______	___	______

black		______		black		green	
R_f	color	R_f	color	R_f	color	R_f	color
___	______	___	______	___	______	___	______
___	______	___	______	___	______	___	______

b. Which pigments, if any, are present in the brown and black Brand A pens and not present in any of the other Brand A pens? Give the color and R_f of any such pigments.

black pen ____________________ brown pen ____________________

c. Give the color and R_f values of any Brand A pigments that are the same as those used by Brand B.

Brand A pen color	R_f	common color	Brand B pen color	R_f
________	___	____________	________	___
________	___	____________	________	___
________	___	____________	________	___
________	___	____________	________	___
________	___	____________	________	___
________	___	____________	________	___
________	___	____________	________	___
________	___	____________	________	___

2. Chromatogram 2. Metal Ions

Unknown #____________ Origin to solvent front ____________

solution #	ion	Color: after drying	Color: after ammonia	Color: after 2nd dry	dist (cm.)	R_f	ion
1							
2							
3							
4							
5							
6 unknown	spot 1						
	spot 2						
	spot 3						
	spot 4						

D. Fabric Dyeing. Record the color of each fiber. Based on the color of each fiber, rate the relative polarity of the fiber on a scale of 1 to 5 with 1 being very polar and 5 non-polar.

Fiber	Color	Relative polarity
Acetate (bright Celanese staple)[1]	________	____
SEF (Monsanto Modacrylic)	________	____
Arnel (bright filament)	________	____
Cotton (bleached)[2]	________	____
Cresian 61	________	____
Dacron 54 Polyester[4]	________	____
Dacron 64 Polyester	________	____
Nylon 66[3]	________	____
Orlon 75 Polyester[5]	________	____
Silk	________	____
Polypropylene	________	____
Viscose	________	____
Worsted Wool[6]	________	____

[1-6]The fabrics given with superscripts are listed in the order that they appear in the six fabric sample.

Name________________________________Date__________Lab Section_________

Postlaboratory Problems - *Experiment 16* - Molecular Polarity and Chromatography

1. Solubility tests are often used to help distinguish and identify ionic compounds (see for example, ***Experiment 9***). Solubility is sometimes also useful for the identification of organic compounds. After referring to ***Appendix 1***, suggest a method for distinguishing between ethanol and cyclohexane.

2. Extraction is a technique that is commonly used to separate organic compounds from inorganic compounds as well as organic acids, bases and neutral compounds.

   a. Benzoic acid is soluble in ether. Suggest a step by step procedure for the separation of sodium chloride and benzoic acid.

   b. Could sodium chloride and potassium nitrate be separated using extraction? Explain your answer.

3. The most difficult part of the chromatography experiment was done by the scientists who selected the solvents for the separations. While a consideration of polarities is certainly used for selection of the stationary and moving phases, much of the work must be done by trial and error. If you were given the assignment of analyzing a mixture of organic non-polar pesticides using thin layer chromatography (paper chromatography is a specific type of thin layer chromatography - with thin layer chromatography there are additional choices for the stationary phase besides cellulose), would you start with a solvent with high polarity such as water? Explain your answer.

4. In addition to using paper chromatography as an identification technique, it is possible to cut out spots from the developed chromatogram and to extract the separated compounds into appropriate solvents. Thus the technique can serve as a separation and purification technique. Comment on limitations this method would have based on sample size.

5. Suggest any ways you can think of to improve any part of this experiment.

6. Some of the *Learning Objectives* of this experiment are listed on the first page of this experiment. Did you achieve the *Learning Objectives*? Explain your answer.

Experiment 17

GAS LAW STUDIES

Sir Robert Boyle
1627 - 1691

Learning Objectives

Upon completion of this experiment, students will have experienced:

1. Experimental testing of Boyle's law.
2. Graphical analysis of data.
3. Application of the ideal gas law, Dalton's law of partial pressures, stoichiometry of reactions involving gases.

Text Topics

Gas laws, stoichiometry (for correlation to some textbooks, see page ix).

Notes to Students and Instructor

The first part of this experiment (a test of Boyle's law) should be performed as a demonstration by the instructor with the students performing their own analysis of the data collected from the demonstration.

Discussion

Fig. 17-1

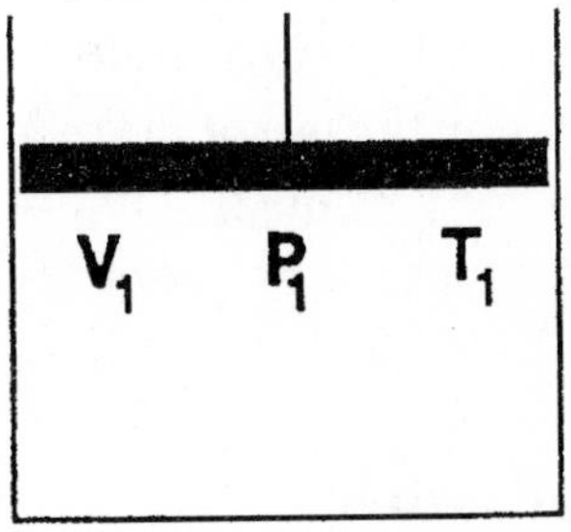

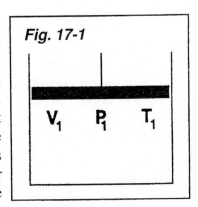

Intuition is a useful attribute. When you apply intuition to the behavior of gases, you can arrive at the same relationships discovered 200 years ago by Robert Boyle, Jacques Charles, Amedeo Avogadro, Joseph Gay-Lussac and Guillaume Amontons. Consider a gas confined in a cylinder with a volume, V_1, at temperature, T_1, and pressure, P_1. What should happen to the volume of the gas if the temperature is held constant but mass is added to the piston (increasing the pressure on the gas)? The answer is that the volume will decrease or as the pressure increases, the volume decreases. The simplest algebraic equation that can be written that predicts this behavior is an inverse proportionality, $V = m/P$ or $PV = m$ where m is a constant. Other relationships such as $V = m/P^2$ are also consistent with a volume decrease as the pressure increases. The question is which equation not only agrees with the direction of the changes but quantitatively agrees with the behavior. It turns out as Boyle discovered that the simple inverse proportionality correctly predicts the behavior of gases and we call $PV = m$, Boyle's law. Today's goal is to experimentally test Boyle's law.

Now return your thoughts to the cylinder and ask what should happen to the volume if the temperature is increased and the pressure held constant. To maintain constant pressure the volume has to increase and drive the piston up. The simplest and again the correct algebraic expression of this relationship that is consistent with this observation is the direct proportionality or Charles' law, $V = mT$ or $V/T = m$. Inspection of this equation reveals that a temperature of zero predicts a volume of zero. Now you know that the volume of a gas (e.g., the interior of a balloon) will not go to zero at 0°C (put a balloon in your freezer). A temperature scale where the volume *theoretically* would be zero needs to be devised. The Kelvin scale conforms to this criteria where 0 K = -273.15°C. **Remember to use Kelvin for all gas law calculations.**

Amontons and Gay-Lussac explored the relationship when the piston is held in fixed position and the pressure is increased. As you would intuitively expect, the temperature has to increase to keep the piston in the same position. Again the direct proportionality, $P = mT$ or $P/T = m$ accurately predicts experimental behavior and is called either Amontons' or Gay-Lussac's law.

Since volume is proportional to temperature and inversely proportional to pressure, these two relationships can be combined into one equation , $V = mT/P$ or $PV/T = m$ which is called the combined gas law.

Finally consider the effect on the volume of adding more gas to the cylinder while maintaining constant temperature and pressure. Wouldn't you expect the volume to double as the number of moles of gas is doubled? The relationship is again the simplest one with the volume proportional to the number of moles, $V = mn$ (n equals the number of moles and m is a constant). Combining all these relationships yields $V = nRT/P$ where R is the proportionality constant and is appropriately called the gas constant. In the second part of this experiment, you will determine the value of R and compare it to the accepted literature value of 0.08206 L atm/mol K (alternatively depending on units desired 1.987 cal/mol K or 8.313 joules/mol K).

The equation is usually written in the form $PV = nRT$ and called the ideal or perfect gas law. All of the previously discussed gas laws can be easily derived from this one equation. For example, if n and T are held constant, PV = constant which is Boyle's law. Note that many elementary texts focus on the use of the relationship that 1 mole of gas at STP (0°C and 1 atm) occupies 22.4 L. The unit conversion 22.4 L/mol has very limited use as it applies only at STP whereas $PV = nRT$ applies whenever gas behavior approximates that of an ideal gas.

Procedure

A. An experimental test of Boyle's law. Using the apparatus in *Figure 17-2*, the "volume" of a gas will be determined at several different pressures. ***(Extreme Caution: The apparatus contains a large and potentially dangerous amount of mercury. The apparatus should be kept in a dishpan so that a leak or a spill can be contained.)*** According to Boyle's law (assuming constant temperature and number of moles) the product of the pressure and volume, PV, should be constant. This prediction will be checked mathematically with the data obtained today.

Graphing is an alternative method of analyzing data to test for agreement with an equation. If the theoretical equation is in the form $y = mx + b$, a plot of y vs x should yield a straight line. For example, Charles' law, $V = mT$, is in the form $y = mx + b$ with $b = 0$. If Charles' law is true, a plot of V vs T should yield a straight line with an intercept of zero [see ***Prelaboratory Problem 7***]. If a plot of y vs x does not yield a straight line, then the mathematical relationship between x and y is not $y = mx + b$ and different, reasonable equations should be tested.

For volume as a function of pressure, if the relationship is in the form $V = rP^q$, then the relationship can be determined using a log-log plot. If logarithms are taken of both sides of the equation, the relationship:

$$\log V = \log r + q \log P$$

results. A plot of log V vs log P should yield a straight line as the equation is in the form of a straight line with $\log V = y$, $\log r = b$, $q = m$ and $\log P = x$. You will apply this technique to the data obtained today to see if the applicable equation is in the form $V = rP^q$. If the log-log plot is linear, then you can determine the value of q from the slope. [If Boyle's law is correct, what should the value of q be?]

An alternative treatment that is more hit and miss is to assume a relationship and graph the data using a method that will verify or refute the relationship. Assuming that the relationship between V and P is $V = m/P$, a plot of V vs P should yield a curve (in this case a hyperbola) and not a straight line as it is in the form $y = m/x$ and not in the form $y = mx + b$. As it is fairly easy to verify visually if data points fall on a straight line, the simplest way to verify graphically that data do conform to an equation is to transform the equation into a form where a straight line graph will result. For $V = m/P$, this is done by letting $x = 1/P$ and plotting V vs x ($V = mx$). In other words, plot the volume versus the reciprocal of each pressure measurement. If Boyle's law is correct, a straight line should result with a zero intercept as $b = 0$ for this equation.

To obtain experimental data for a test of Boyle's law, a Boyle's law apparatus is needed (see *Figure 17-2*). This apparatus can be assembled using a 50 mL pinchcock buret, a meter stick, a ringstand, a ring, a leveling bulb, several feet of flexible rubber tubing, a rubber stopper, and about 250 mL of mercury.

The gas to be studied will be the air confined in the buret between the rubber stopper and the top of the mercury. The buret is cylindrical and the volume of a cylinder equals $\pi r^2 h$. Since π is a constant and r^2 is constant for this apparatus, the volume of the air column in the buret is proportional to its height ($V = \pi r^2 h$ or V = constant x height). Therefore, in both the mathematical and graphical analyses, the volume can be replaced by the height of the air column. For the mathematical analysis, Boyle' law, $PV = m$ becomes $P\pi r^2 h = m$ or $P \cdot h = m'$ (a new constant) and the product $P \cdot h$ should still equal a constant. For the graphical analysis, $V = m/P$ becomes $\pi r^2 h = m/P$ or $h = m'/P$. Plots of log h versus log P and h versus 1/P should yield straight lines if Boyle's law is correct.

If the mercury in the reservoir is at the same level as the mercury in the buret, the pressure of the confined gas should be the atmospheric pressure outside the apparatus. If the reservoir is raised up such that the mercury level is higher in the reservoir than in the buret, the pressure inside will be atmospheric pressure plus an additional amount due to the pull of gravity on the additional column of mercury. The interior pressure can be calculated by adding atmospheric pressure in millimeters of mercury to the length of the additional column of mercury measured in millimeters. If the reservoir is lowered such that the mercury level in the reservoir is lower in the reservoir than in the buret, the pressure inside will be atmospheric pressure minus the difference in the heights of the two mercury measurements.

Record the barometric pressure and the height of the top of the air column. Now a series of measurements of the mercury levels in the buret and the reservoir will be taken and calculations can be performed to check Boyle's law. The reservoir should be placed so that at least three of the readings are above atmospheric pressure, one at atmospheric pressure and at least three are below atmospheric pressure. The graphical analysis can be done on paper or by using a computer with graphical analysis software.

Fig. 17-2

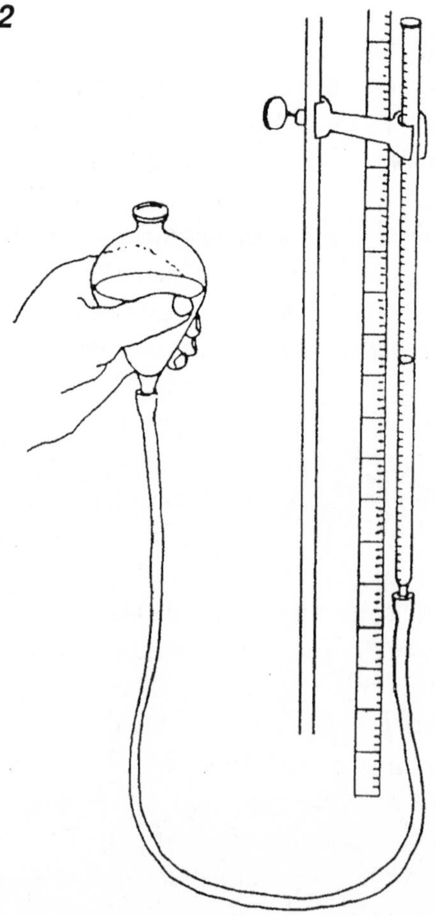

B. Determination of the value of the gas constant, R. The determination of the gas constant R from the ideal gas law (R = PV/nT) requires the values of P, V, n, and T. P, V and T will be measured and the value of n determined from a stoichiometric calculation. Hydrogen will be prepared by a single replacement reaction:

$$Zn(s) + 2\,HCl(aq) = ZnCl_2(aq) + H_2(g)$$

By utilizing a weighed amount of zinc, the number of moles of hydrogen that will be produced can be calcluated.

Assemble the apparatus pictured in *Figure 17-3*. ***Be sure to keep all flames far away from this apparatus once you start to generate hydrogen.*** Fill the 500 mL Florence flask (or Erlenmeyer flask) up to the neck with water and add 25 mL of 6 M hydrochloric acid to the 250 mL Erlenmeyer flask. While the Erlenmeyer flask is unstoppered, apply compressed air (or use a rubber bulb) to the tube inserted through the stopper. This will cause water to flow from the Florence flask into the beaker. As soon as the flow begins, clamp the tube through which the water is flowing while disconnecting the compressed air source. Now that the delivery hose is filled with water, unstopper the 500 mL Florence flask, return the water you collected in the beaker to the Florence flask and restopper it.

Weigh a little less than a gram of zinc metal to the nearest 0.001 g. Holding the zinc sample with one hand and the stopper for the 250 mL Erlenmeyer in the other, drop the sample into the HCl, quickly stopper the flask, and remove the clamp. After the zinc has completely reacted, allow a few minutes for the gas in the flasks to cool to room temperature. Equalize the pressure between the Florence flask and the beaker by raising the beaker or flask (whichever liquid level is lower) so that the level of the water you have collected in the beaker is the same as the level of water in the flask. **Do not unstopper the flasks until after you have measured the water!** Clamp the delivery hose when the water levels are equal. Measure the amount of water you collected using a 500 mL graduated cylinder and also measure the temperature of the water. Read the barometer to determine the current atmospheric pressure.

Because you collected the hydrogen over water, part of the gas in the Florence flask was water vapor. Consequently, to obtain the actual pressure of hydrogen that was produced, you must subtract the vapor pressure of the water from the pressure at which the gas was collected (atmospheric). The number of moles of hydrogen can be calculated from the number of moles of zinc used and the stoichiometry of the reaction. P, V, and T can be determined from measurements made during the experiment.

Fig. 17-3

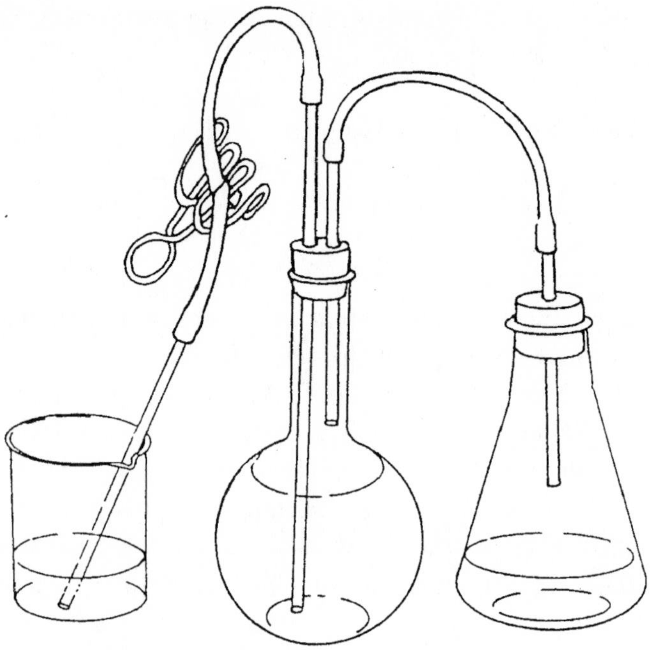

Vapor Pressure of Water as a Function of Temperature

Temperature (°C)	*Vapor Pressure (mm)*
10	9.2
11	9.8
12	10.5
13	11.2
14	12.0
15	12.8
16	13.6
17	14.5
18	15.5
19	16.5
20	17.5
21	18.6
22	19.8
23	21.1
24	22.4
25	23.8
26	25.2
27	26.7
28	28.3
29	30.0
30	31.8

Name____________________________________Date__________Lab Section__________

Prelaboratory Problems - *Experiment 17* - Gas Law Studies

The solutions to the starred problems are in ***Appendix 4***.

1. Use the ideal gas law to derive the combined gas law.

2. Use the ideal gas law (with the gas constant R = 0.08206 L atm/mol K) to calculate the volume of 1 mole of gas at:

   a.* STP

   b. 25°C, 1 atm

3. a. Assuming no elasticity forces in rubber, what volume should a 1.0 L balloon at 25°C and 1.00 atmosphere pressure be if immersed in liquid nitrogen (-196°C, 1 atm)?

   V = ____________

   b. When this experiment is actually performed, the volume of the balloon approaches 0.0 L. How do you account for the discrepancy?

4. A gas is collected over water at 18°C at a pressure 755 mm_{Hg}. What is the partial pressure of the gas?

   P = ____________

5.* 0.525 grams of magnesium is reacted with an excess amount of aqueous hydrochloric acid at 27°C and 752 mm_{Hg}. Calculate the number of moles of hydrogen gas produced and its volume under the above conditions.

   n = ____________

   V = ____________

6. When a 0.134 g popcorn kernel pops, it loses about 13% of its mass. Assuming that the volume of a typical kernel is 9.5×10^{-2} mL and that the mass loss is due to water lost, use the ideal gas law to calculate the pressure of the water vapor in the kernel immediately before it pops at 100°C. Does the resulting calculated pressure account for the popping of the popcorn?

P = ____________

7. The volume of a gas at several temperatures is recorded below. On a piece of graph paper, plot the volume versus the temperature. Is Charles' law apparently obeyed? By extrapolating the graph to a volume of 0.00 L, determine the value of absolute zero in °C according to this data.

Temperature (°C)	Volume (L)	Temperature (°C)	Volume (L)
100	2.98	20	2.33
80	2.87	0	2.21
60	2.68	-20	2.07
40	2.53	-40	1.88

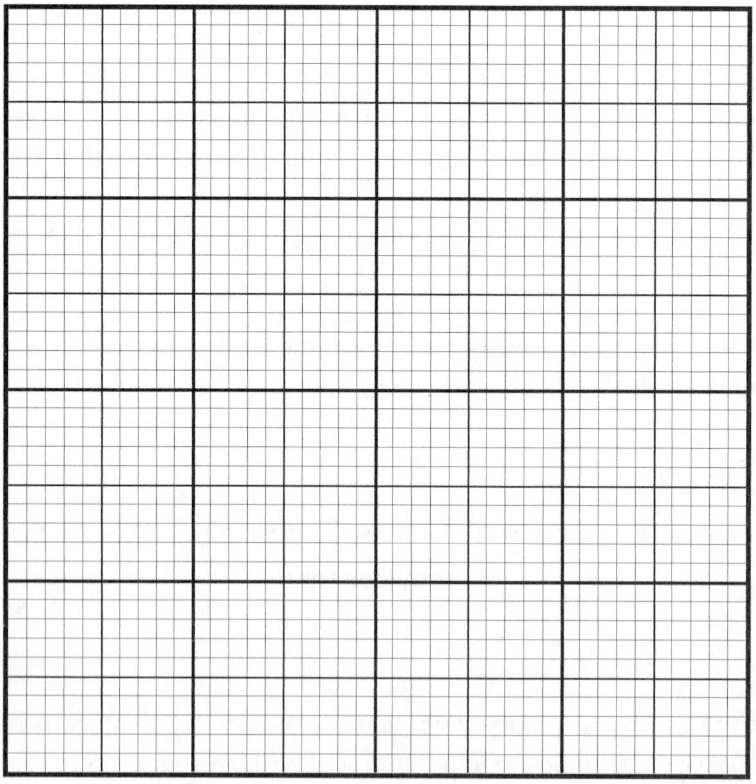

T (extrapolated to zero volume = ____________°C

Name________________________________Date__________Lab Section__________

Results and Discussion - *Experiment 17* - Gas Law Studies

A. An experimental test of Boyle's law.

1. Atmospheric pressure __________
2. Top of air column __________

Trial ⇓	buret level (mm)	reservoir level (mm)	air column height (mm)	log (h)	pressure difference (mm)	pressure (mm)	log (P)	1/P (mm^{-1})	P x h (mm^2)
1									
2									
3									
4									
5									
6									
7									

3. On the first piece of graph paper provided (or using a computer), plot h vs P.

4. On the second piece of graph paper provided (or using a computer), plot log h vs log P. If a straight line results, determine the slope of the line. As a result of the value of the slope, write an equation for the relationship between h and P.

__________ Slope   __________ Equation

5. On the third piece of graph paper provided (or using a computer), plot h vs 1/P.

6. Does your data support Boyle's law? Give at least three reasons one way or the other.

7. It should be possible to use the graph of h vs 1/P to predict the height of the air column that results from a given pressure. What values of h should result at 675 mm and at 1.20 atm.

675 mm h = __________   1.20 atm h = __________

B. Determination of R.

1. Mass of zinc ________
2. Moles of zinc ________
3. Moles of hydrogen ________
4. Volume of water collected (mL) ________
5. Volume of water collected (Liters) ________
6. Temperature of water collected (°C) ________
7. Temperature of water collected (K) ________
8. Atmospheric pressure (mm) ________
9. Vapor pressure of water (mm, from table, p. 214) ________
10. Corrected pressure of hydrogen (mm) ________
11. Corrected pressure of hydrogen (atm) ________
12. Experimentally calculated value of R ________
13. Literature value of R ________
14. Percent error ________
15. Suggest reasons for the percent error (#14)

16. Suggest any ways you can think of to improve any part(s) of this experiment.

17. Some of the *Learning Objectives* of this experiment are listed on the first page of this experiment. Did you achieve the *Learning Objectives*? Explain your answer.

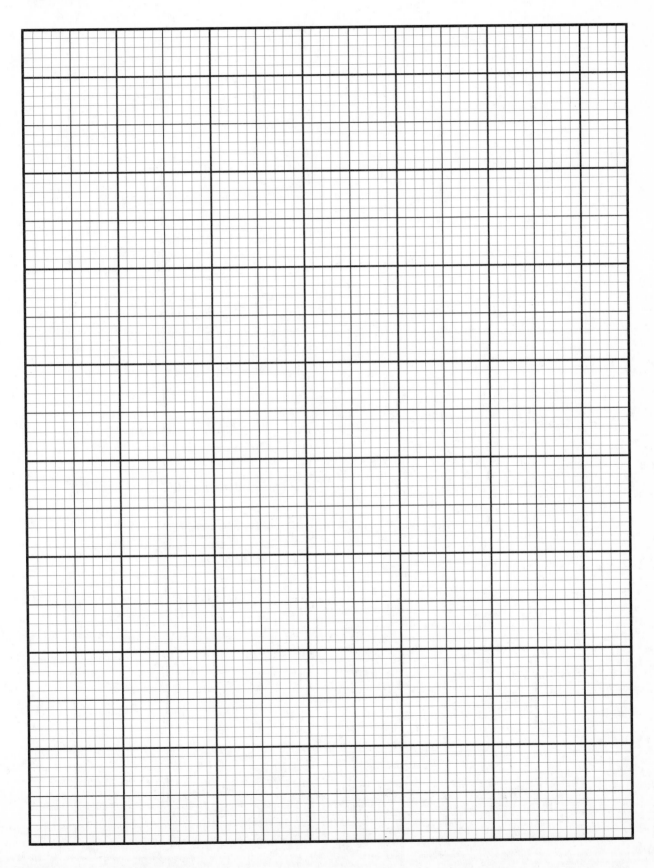

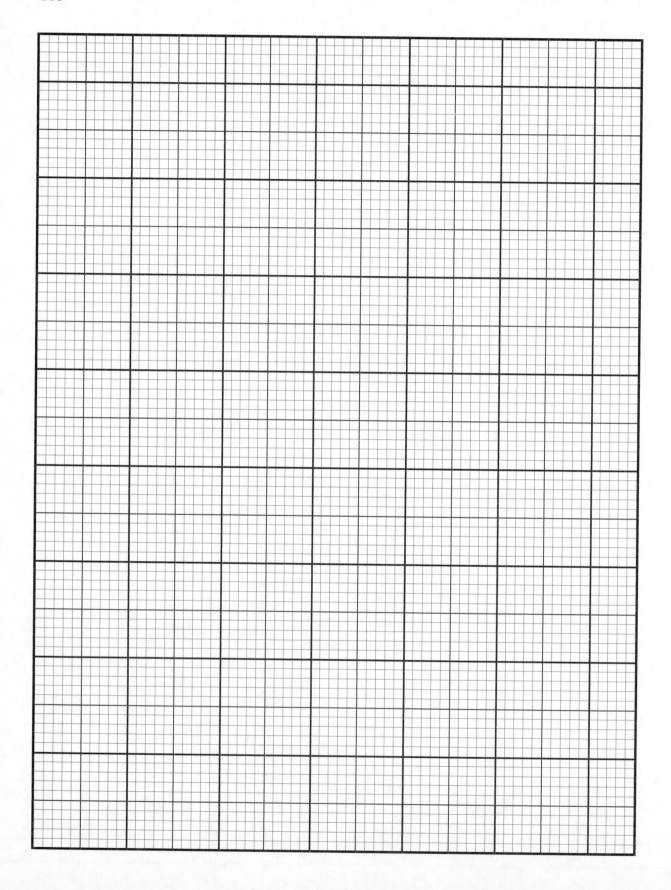

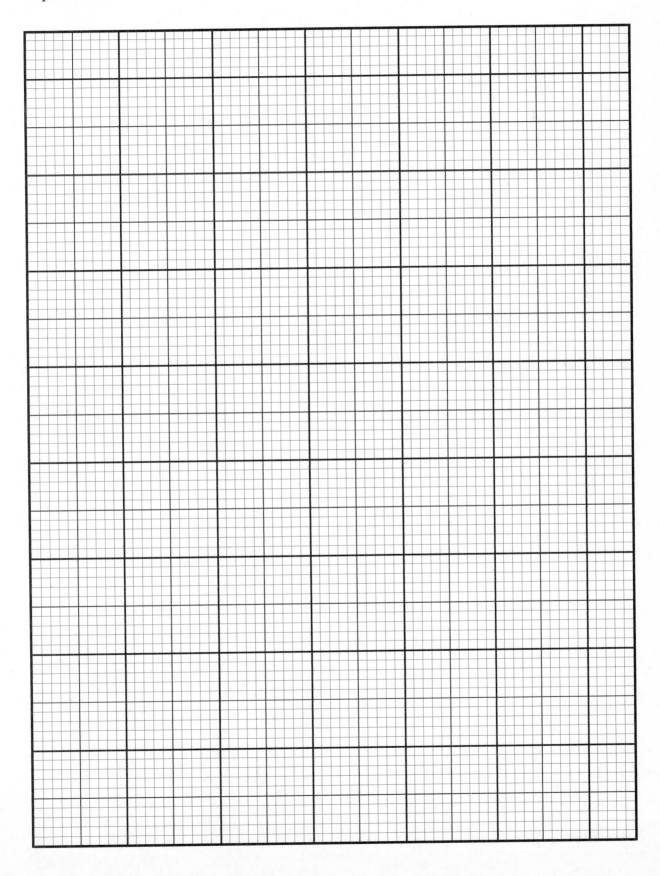

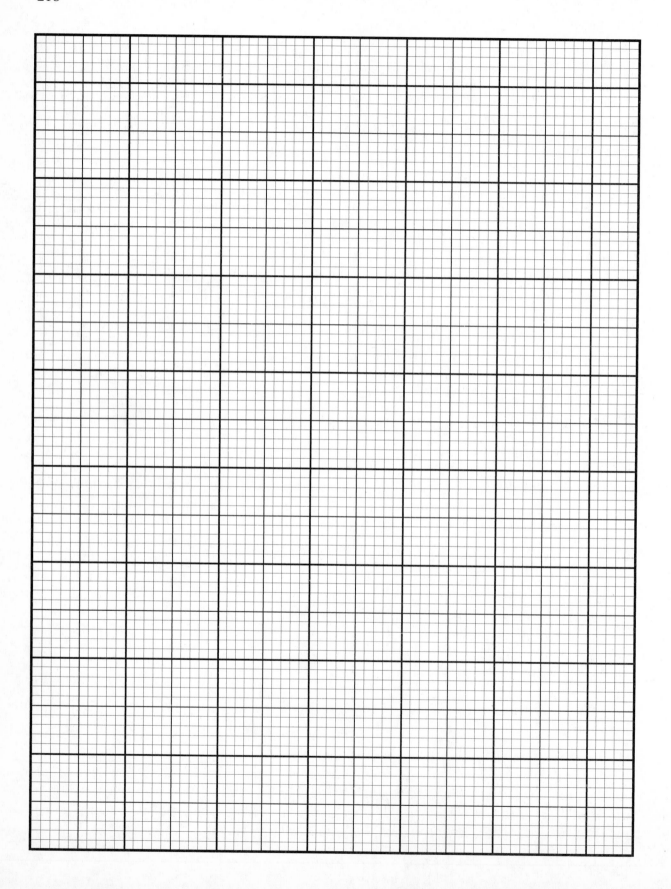

Experiment 18

COOLING CURVES AND CRYSTAL STRUCTURES

Learning Objectives

Upon completion of this experiment, students will have experienced:

1. The use of cooling curves to determine melting points.
2. A study of the effects of impurities on melting points.
3. The determination of the atomic radius of a metal from its density.

Text Topics

Colligative properties, melting points and the effects of impurities on melting ranges, crystal structure, atomic radius (for correlation to some textbooks, see page ix).

Notes to Students and Instructor

The length of this experiment can be varied from 2 to 3 hours depending on the number of cooling curve trials on each sample. Lauric acid is suggested for use here for several reasons: it has a low tendency to supercool, it is not too expensive, and its melting point (around 43°C) means that 0 to 50°C thermometers can be used. If digital thermometers are available, myristic, palmitic and stearic acids are alternative choices and all four acids have the same freezing point depression constant. A computer interfaced temperature probe considerably facilitates this experiment.

Discussion

Cooling Curves. The capillary method of measuring melting ranges that was used to determine the success of your purification of vanillin (***Experiments 2 and 3***) is quick, accurate and uses small amounts of sample. Using the capillary method, you heated the sample and recorded the melting temperature range. Today you will do the reverse; heat the sample above the melting range and observe it while it cools and solidifies. Rather than simply recording the solidification range which is not easily observed, you will obtain data and graph a curve by recording the temperature of the cooling sample as a function of time. While this method is not quick or routine, it does provide information not readily available from the capillary method.

Ideally the temperature of the liquid sample should decrease at a steady rate until it reaches its freezing temperature, hold for some time while all of the sample solidifies and then cooling should resume at a steady rate as is illustrated in *Figure 18-1*. In practice, however, you may obtain curves that look more like *Figures 18-2 and 18-3*. Point ***A*** is due to supercooling of the liquid and can be avoided or at least minimized by continuous stirring of the sample during cooling. The slight slope at ***B*** is due to the presence of impurities. Recall from ***Experiment 3*** that impurities or additives depress melting points (why is salt put on roads in the winter and used with ice to make ice cream?). The changing slope near ***C*** or the slowing of the cooling rate is due to the approach to room temperature.

Fig. 18-1

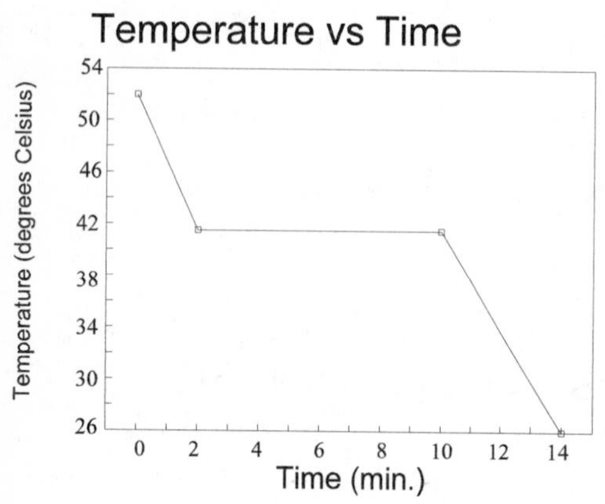

Fig. 18-2

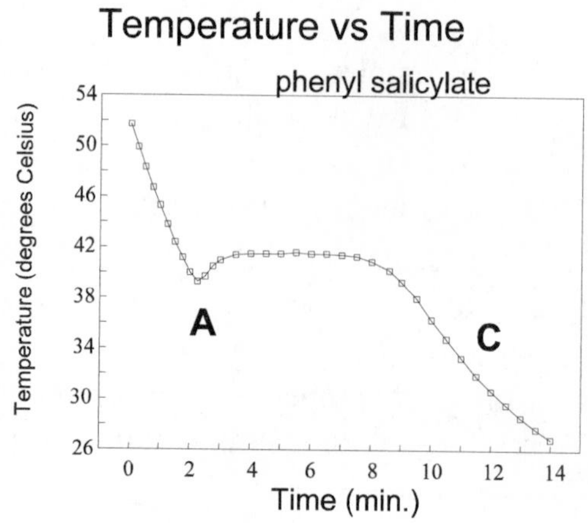

The melting point of the sample is obtained by drawing with a straight edge the best straight line through the steadily declining points of the liquid and another line through the slightly sloped "plateau". The intersection of the lines is taken as the melting point.

Fig. 18-3

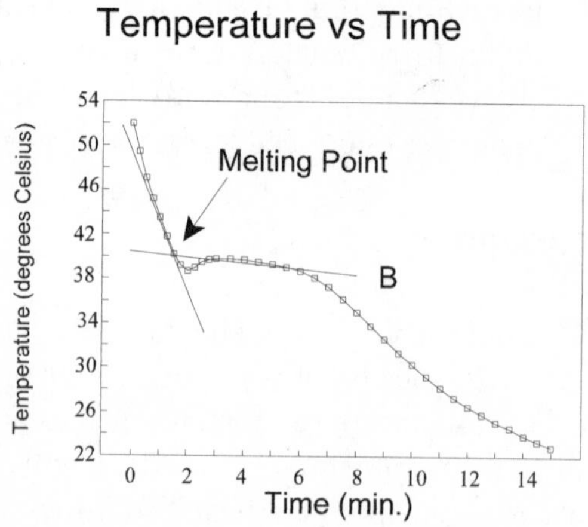

One of the advantages of the cooling curve method is that the amount of depression caused by impurities can be accurately determined. The freezing point depression is a useful parameter as it is a colligative property. This means that the amount of the depression is related simply to the number of moles of particles added and not to the nature of the particles. In other words one mole of acetone will depress the melting point of water the same amount as 1 mole of ethanol. As 1 mole of sodium chloride provides two moles of particles, it will depress the freezing point twice as much as one mole of acetone. Because it is a colligative property, it is possible to demonstrate that the freezing point depression is directly proportional to the number of moles of solute dissolved in a given amount of solvent. Because of the way samples are prepared and the nature of the relationship, concentrations are expressed in molality (moles of solute/kg of solvent).

$\Delta t_f = k_f m$ (Δt = temperature change in °C, k_f = freezing point depression constant, m = molality of solute)

The freezing point depression constant, k_f is dependent on the nature of the solvent and must be experimentally determined for each solvent. If the constant is known and the amount of the freezing point depression is measured, the molality of the additive can be calculated. If the mass of the solute and solvents are measured, the molecular mass of the solute can be calculated. Thus freezing point depression measurements are a method for determining the molecular mass of an unknown. Or for a known solute it is possible to find out the extent of its ionization in the solvent because the depression of the freezing point is determined by the concentration of dissolved particles.

Crystal Structure and atomic radius. In crystalline solids, the atoms, ions or molecules are arranged in a definite patterns. This treatment will focus on two of the simplest repeating arrays, the body centered cube and the face centered cube. Many of the metals contain these crystal structures. These two arrays are illustrated in *Figures 18-4* and *18-5*.

Fig. 18-4

Body Centered Cube

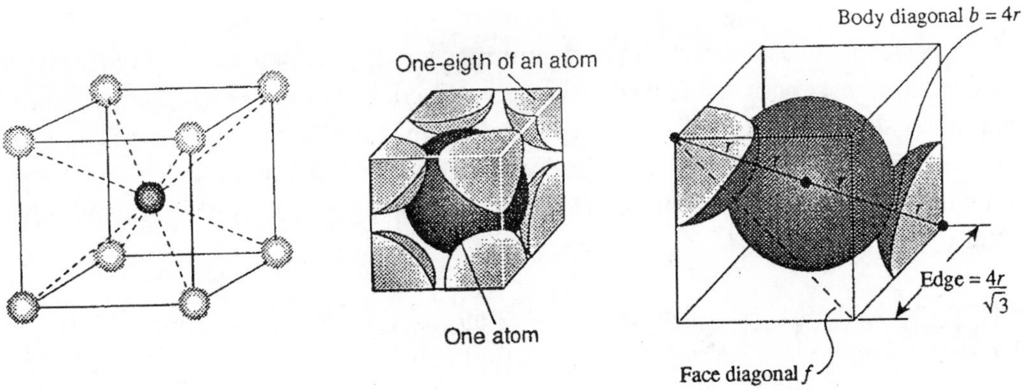

Fig. 18-5 **Face Centered Cube**

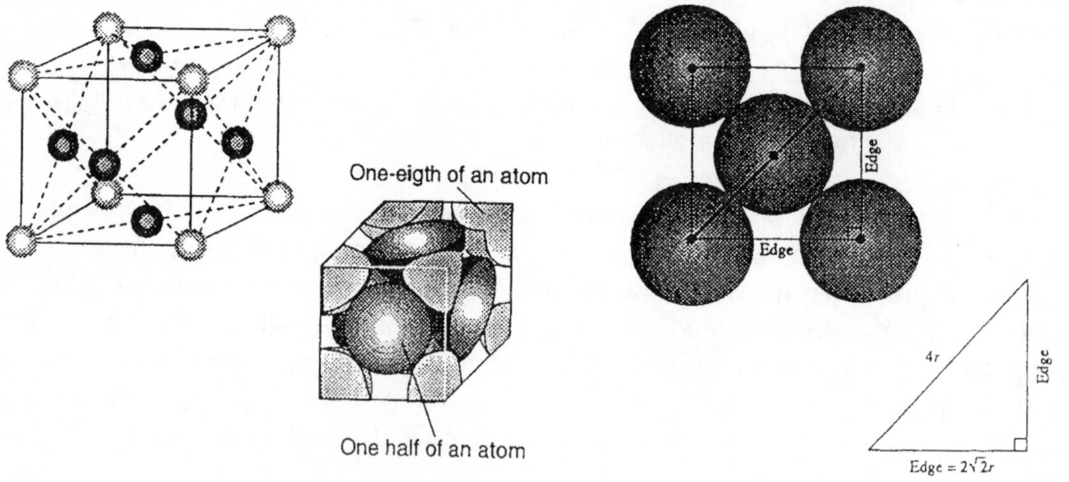

As can be seen from the drawings, the body centered cube has ⅛ of an atom at each corner and one in the center for a total of two atoms per unit cell. The cube diagonal is four times the radius of an atom and can be shown to be $3^{1/2}$ times the edge length of the cube.

The face centered cube has ⅛ of an atom in each corner and ½ in each of the six faces for a total of 4 atoms per unit cell. A face diagonal is four times the radius of an atom and is $2^{1/2}$ times the edge length of the cube.

If you performed ***Experiment 3***, you determined the density of a metal cylinder. If the crystal structure of an element is known and its density determined, it is possible with proper mathematical manipulation to calculate the atomic radius of the element. The calculation involves the use of the atomic mass, Avogadro's number, the number of atoms per unit cell and the density. For example, chromium with a density of 7.19 g/cm³ crystallizes as a **body centered cubic** crystal. The volume V of its unit cell is:

$$V = \left(\frac{1\ \text{cm}^3}{7.19\ \text{g}}\right)\left(\frac{52.0\ \text{g}}{1\ \text{mole}}\right)\left(\frac{1\ \text{mole}}{6.022\text{x}10^{23}\ \text{atoms}}\right)\left(\frac{2\ \text{atoms}}{\text{unit cell}}\right) = 2.40\text{x}10^{-23}\ \text{cm}^3$$

The edge of the unit cell (a) is the cube root of the volume [$V^{1/3}$]

$$a = V^{1/3} = (2.40\text{x}10^{-23}\ \text{cm}^3)^{1/3} = (24.0\text{x}10^{-24}\ \text{cm}^3)^{1/3} = 2.89\text{x}10^{-8}\ \text{cm}$$

$$\text{atomic radius} = r = a \text{ x } 3^{1/2}/4 = (2.89\text{x}10^{-8}\ \text{cm})(0.433) = 1.25\text{x}10^{-8}\ \text{cm}$$

Procedure

A. Cooling Curves of lauric acid. For reasons stated in the ***Notes to Students and Instructors***, lauric acid has been selected as the solvent for this experiment. There are many other possible solvents and some have significantly larger values of k_f. The larger the value of k_f, the larger the temperature depressions and this tends to decrease measurement errors. However, we have found supercooling to be a problem with many of the solvents that melt in the desired temperature range (40 - 50°C). With small modifications, the procedure below can be applied to other solvents but we hope you have more success than we did.

Fig. 18-6

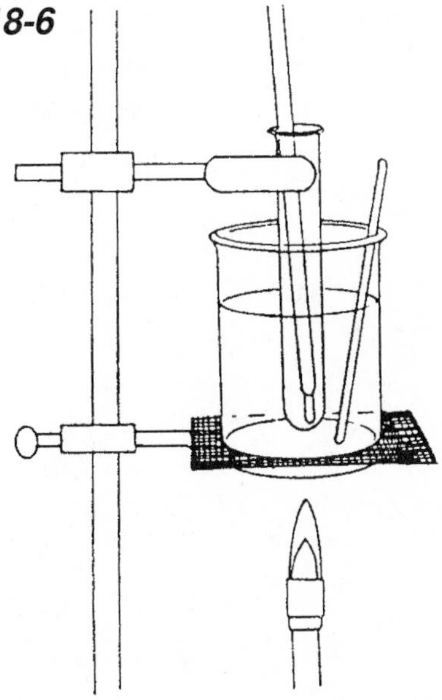

Weigh a clean and dry 18x150 mm test tube to 0.001 gram. Transfer about 4 g of lauric acid to the tube and reweigh the tube to the nearest 0.001 g. Clamp the test tube on a ring stand in a 400 mL beaker containing about 200 mL of water mounted over a wire gauze and a Bunsen burner. Insert a thermometer or temperature probe (preferably with a 0.1°C readout) into the test tube. Heat the water to about 50°C **(be sure if the thermometer has a range of 0 - 50°C that you do not exceed 50°C).** Raise the tube out of and away from the water bath and take temperature readings every 0.50 minutes for about the next 15 minutes. It is important to stir the sample with the thermometer between readings. The temperature will decrease at first, but will remain steady as the sample freezes. After the sample has solidified completely, the temperature will start dropping again. After it drops a few degrees below the plateau, the readings can be terminated.

Weigh out about 0.40 g of benzoic acid (weighed to the nearest 0.001 g) and add it to the test tube with the lauric acid. Insert the tube back into the water and repeat the procedure above. **Be sure to stir after melting.** Graphically determine the melting point depression caused by the benzoic acid and calculate the molecular mass of benzoic acid (k_f for lauric acid = 4.40°C kg/mol).

If the instructor desires, more than one trial of each of the above runs can be made or the molecular mass of an unknown can be determined.

B. Crystal structure and the atomic radius of a metal. Follow *Procedure C* in ***Experiment 3*** to determine the density of an unknown metallic element provided by your instructor. Use a method of your choice to determine the volume of the metal and weigh the metal to the nearest milligram. Compare your experimental density to values for aluminum, iron, copper and lead from the *Handbook of Chemistry and Physics* or *Appendix 1* and identify your metal. Using the information from the table below, your experimental density, the atomic mass and Avogadro's number, calculate the atomic radius of the metal. For example calculations, refer to the calculation on page 222 and to Problem 4 in the ***Prelaboratory Problems*** on page 226.

element	crystal structure
aluminum	face centered cubic
copper	face centered cubic
iron	body centered cubic
lead	face centered cubic

Name________________________________Date__________Lab Section_________

Prelaboratory Problems - *Experiment 18* - Cooling Curves and Crystal Structures

The solutions to the starred problems are in ***Appendix 4***.

1. Melting point depression is termed a colligative property.

   a. What is the meaning of the word colligative?

   b. Which, if any of the following are colligative properties?

   density _____ boiling point elevation _____ color _____

2.* Naphthalene ($C_{10}H_8$) melts at 80.5°C. Addition of 0.38 grams of biphenyl ($C_{12}H_{10}$) to 5.00 g of naphthalene caused a melting point depression to 77.1°C.

   a. What is the freezing point depression constant of naphthalene?

   b. 0.150 g of anthracene ($C_{14}H_{10}$ - melting point = 216°C) is added to 3.00 g of naphthalene. At what temperature should the mixture start melting?

3. The freezing point depression constant of water is a rather small 1.86°C kg/mol.

   a. Addition of 0.100 g of an unknown compound to 2.50 g of water results in a decrease in the melting point of 2.3°C. What is the molecular mass of the unknown?

   molecular mass ________

   b. Addition of 0.454 g of zinc chloride to 3.33 g of water lowers the melting point by 5.2°C. Calculate the expected melting point for this mixture and account for any difference.

   calculated melting point ________

   c. What are some of the problems with the use of water as a solvent for determining the molecular mass of an unknown using the freezing point depression method?

4.* Nickel has a density of 8.90 g/cm^3 and crystallizes as a face centered cube. Calculate the atomic radius of nickel.

Name________________________________ Date__________ Lab Section__________

Results and Discussion - *Experiment 18* - Cooling Curves and Crystal Structures

A. Cooling Curves of Lauric Acid

1. Mass of test tube __________

2. Mass of test tube + lauric acid __________

3. Mass of lauric acid __________

Time (min.)	Temperature (°C)	Time (min.)	Temperature (°C)	Time (min.)	Temperature (°C)
0.00	________	6.00	________	12.00	________
0.50	________	6.50	________	12.50	________
1.00	________	7.00	________	13.00	________
1.50	________	7.50	________	13.50	________
2.00	________	8.00	________	14.00	________
2.50	________	8.50	________	14.50	________
3.00	________	9.00	________	15.00	________
3.50	________	9.50	________	15.50	________
4.00	________	10.00	________	16.00	________
4.50	________	10.50	________	16.50	________
5.00	________	11.00	________	17.00	________
5.50	________	11.50	________	17.50	________

4. Plot the temperature (y axis) versus time (x axis) and determine the melting point of lauric acid. __________

5. Mass of test tube + lauric acid + benzoic acid ________

6. Mass of benzoic acid ________

Time (min.)	Temperature (°C)	Time (min.)	Temperature (°C)	Time (min.)	Temperature (°C)
0.00	________	6.00	________	12.00	________
0.50	________	6.50	________	12.50	________
1.00	________	7.00	________	13.00	________
1.50	________	7.50	________	13.50	________
2.00	________	8.00	________	14.00	________
2.50	________	8.50	________	14.50	________
3.00	________	9.00	________	15.00	________
3.50	________	9.50	________	15.50	________
4.00	________	10.00	________	16.00	________
4.50	________	10.50	________	16.50	________
5.00	________	11.00	________	17.00	________
5.50	________	11.50	________	17.50	________

7. Plot the temperature (y axis) versus time and determine the melting point of lauric acid + benzoic acid. ________

8. Freezing point depression (Δt_f) ________

9. Molality of solution (according to the *International Critical Tables*, k_f for lauric acid is 4.40°C kg/mol) ________

10. Moles of solute in original sample ________

11. Molecular mass of benzoic acid from freezing point depression ________

12. Molecular mass of benzoic acid from formula ($C_7H_6O_2$) ________

13. Percent error ________

14. Suggest explanations for the percent error.

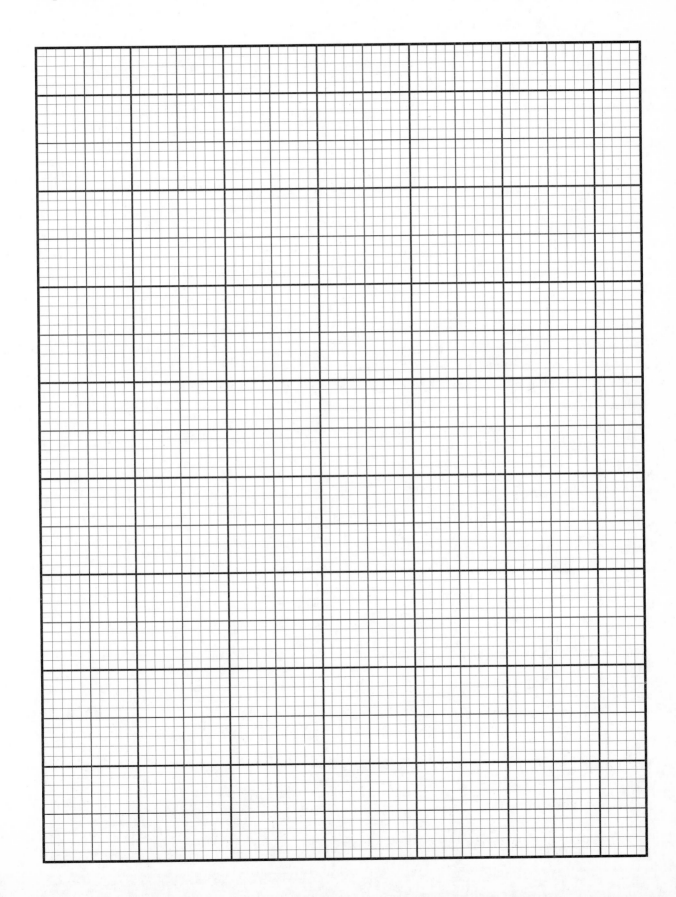

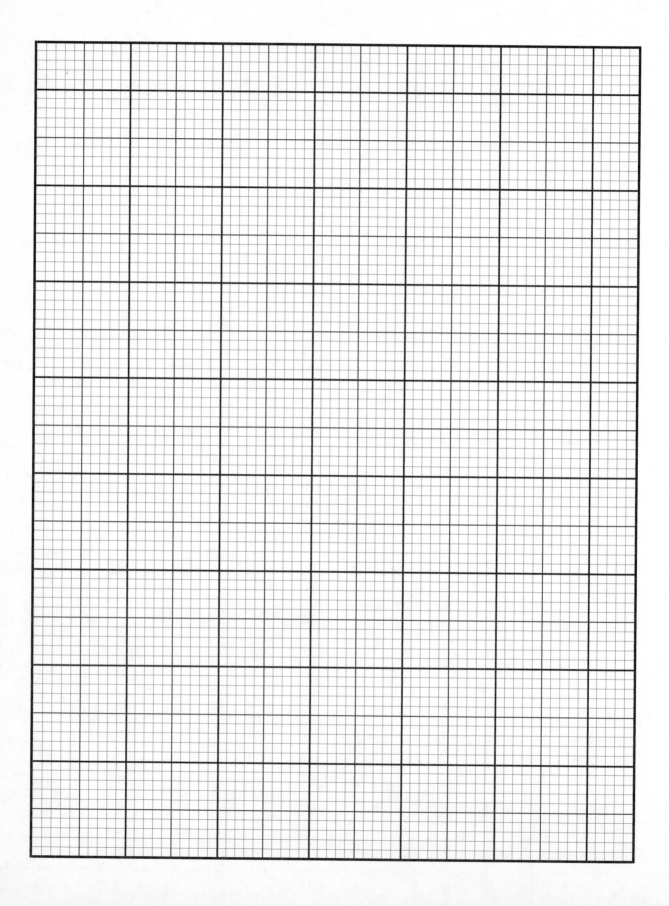

B. Crystal Structure and Atomic Radius of a Metal.

1. Density determination and identification of metal

a. Identification number of metal cylinder ________

b. Mass of metal cylinder ________

c. Describe the method used to determine the volume of cylinder and give the data.

d. Volume of metal cylinder ________

e. Density of metal cylinder ________

Handbook of Chemistry and Physics values for the density of:

f. aluminum ________

g. copper ________

h. iron ________

i. lead ________

j. Identity of metal ________

2. Atomic radius of metal

a. Volume of unit cell (show calculations) ________

b. Edge length of cell ________

c. Atomic radius ________

3. Some of the *Learning Objectives* of this experiment are listed on the first page of this experiment. Did you achieve the *Learning Objectives*? Explain your answer.

Name________________________________Date__________Lab Section__________

Postlaboratory Problems - *Experiment 18* - Cooling Curves and Crystal Structures

1. How do you account for the trend of melting points in the series of acids to the right?

acid	formula	m.p. (°C)
octanoic (caprylic)	$C_8H_{16}O_2$	17.5
decanoic (capric)	$C_{10}H_{20}O_2$	31.5
dodecanoic (lauric)	$C_{12}H_{24}O_2$	44
tetradecanoic (myristic)	$C_{14}H_{28}O_2$	55
hexadecanoic (palmitic)	$C_{16}H_{32}O_2$	63
octadecanoic (stearic)	$C_{18}H_{36}O_2$	71
eicosanoic (arachidic)	$C_{20}H_{40}O_2$	77

2. List criteria that you think were used for selection of lauric acid as the solvent for today's experiment.

3. Lange's *Handbook of Chemistry* lists k_f values for acetic acid, benzene, camphor, strontium chloride and water as 3.9, 5.12, 37.7, 107 and 1.86 °C/m respectively. Why do you think each of the above solvents was eliminated from consideration for this experiment?

acetic acid

benzene

camphor

strontium chloride

water

4. Considering the other methods you have used that could be used to determine molecular mass such as titration and quantitative precipitation, when would the melting point depression method be used? Explain in detail.

5. Suggest any ways you can think of to improve any part(s) of this experiment.

Experiment 19

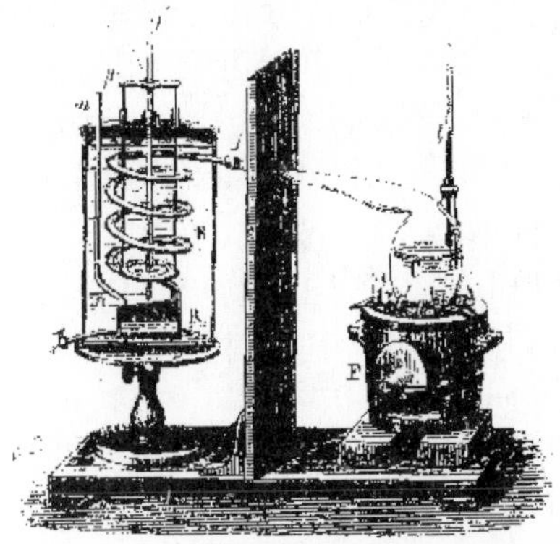
1860 distillation apparatus

WATER PURIFICATION AND ANALYSIS

Learning Objectives

Upon completion of this experiment, students will have experienced:
1. The use of distillation to purify water.
2. The use of a complexiometric titration to test for water hardness.
3. Flame, conductivity and precipitation tests to analyze for water purity.

Text Topics

Distillation, complexes and complexiometric titrations, conductivity, solution stoichiometry (for correlation to some textbooks, see page ix).

Notes to Students and Instructor

The distillation should be set up first and started. While the water is distilling, several of the tests can be performed on other water samples including the titration of tap water. This experiment will be even more interesting if you bring a water sample from home for testing.

Discussion

You have undoubtedly heard water described as hard or soft. These seem to be strange terms to apply to a liquid. Can water really be "hard" in the classical sense of the word? It turns out that hardness has a different and special meaning when applied to water. Hardness is a measure of the combined amounts of calcium and magnesium salts present. As soap consists of fatty acid salts (e.g., sodium stearate) and the fatty acid salts of calcium and magnesium are insoluble in water, the effectiveness of soap decreases as hardness increases due to precipitation with magnesium and calcium. In this experiment, water will be distilled to decrease the water hardness. Analytical chemistry techniques will be used to test the effectiveness of the distillation. Simple qualitative tests will also be performed to check for the presence of sodium and chloride ions and conductivity will be determined as a measure of the total ion content of the water samples.

Distillation. The purification of vanillin in ***Experiment 2*** utilized the most common technique for the purification of solids, recrystallization. Distillation is probably the most commonly used technique for the purification of liquids. Simply put, distillation involves energy input to convert a liquid to its vapor and then condensation of the vapor to liquid in a different part of the apparatus. Low boiling impurities will vaporize and condense first and can be collected and set aside while high boiling impurities will remain in the distilling flask unless more heat is provided. Boiling occurs when the vapor pressure of a substance equals the confining pressure. When the confining pressure is atmospheric pressure, the temperature at which boiling occurs is called the normal boiling point. There are many parameters that can be changed to affect the efficiency of separation including the length and packing of the distilling column, the distillation rate, and the confining pressure. Today's distillation will be a simple distillation at atmospheric pressure.

Water softeners and ion exchange. Other methods of decreasing hardness involve replacement of calcium and magnesium ions by other ions. Commercial packaged water softeners such as *Calgon* react with the calcium and form a soluble complex ion that does not precipitate with soap. The process does result in the addition of sodium ions to the solution. *Calgon* also contains sodium carbonate which results in the precipitation of calcium carbonate but again replaces the calcium ions by sodium ions. Home water softeners usually also replace the calcium ion by sodium ion. Other cation exchange resins replace cations with hydrogen ions. Anion exchange resins usually replace anions with chloride or hydroxide ions. If hydrogen and hydroxide resins are used, hydrogen and hydroxide ions are produced which react with each other to yield water. This is an excellent technique for the purification of water. Notice that the overall process for water softeners and ion exchange resins results in the substitution of the calcium and magnesium ions by other ions. If sodium ions are going to be a problem, the method of choice becomes the hydrogen ion exchange resin.

Complexiometric titration. The concentration of ferrocyanide was determined by titration with a zinc sulfate solution of known concentration in ***Experiment 11***. The total calcium and magnesium concentration in a solution can be determined by titration with a solution of known concentration of the ion of ethylenediamine-tetraacetic acid (EDTA). Many metal ions including calcium and magnesium react one to one with EDTA which surrounds and grabs or chelates with the metal ion to form a new ion called a complex. The hardness of different water samples will be determined using this technique. As the concentrations of calcium and magnesium are very low on a molarity scale, it is common practice to convert molarity to parts per million (ppm) when reporting hardness.

Other analyses. The total ion content of an aqueous solution is related to the conductance of the solution. Relative conductances of the water samples will be determined to compare the success of the different purification methods used. A flame test (see ***Experiments 5*** and ***9***) will be used to qualitatively test for the presence of sodium ions, and precipitation tests will be used to test for insoluble silver salts.

Procedure

A. Distillation. [Note: Reread the **Notes to Students and Instructor**] Set up an apparatus similar to the one in *Figure 19-1*. Use flasks of approximately 150 mL in volume for the distilling pot and receiver. Fill the distilling flask about half full with tap water (never more than ⅔) and add a couple of boiling chips to the flask to prevent bumping. Be sure that the water flows uphill in the condenser (why?) and the mercury bulb of the thermometer is slightly lower than the junction of the condenser with the distillation column. Gently heat the distillation flask with a flame. Eventually boiling, condensation, and collection of the liquid will occur. During the distillation, it should be possible to set up and even partially perform some of the hardness titrations and the other water purity tests.

Discard the first milliliter (~20 drops) as it might contain volatile impurities and contaminants from the distillation glassware. Record the temperature in the distilling flask as soon as you have collected a milliliter of the distillate, then collect distillate until there is about 5 mL of liquid left in the distillation flask. Record the temperature in the distilling flask at the point you stop the distillation; then turn off the burner [Note: Never distill to dryness - it can sometimes lead to explosions]. Stopper the distillate and save it for testing.

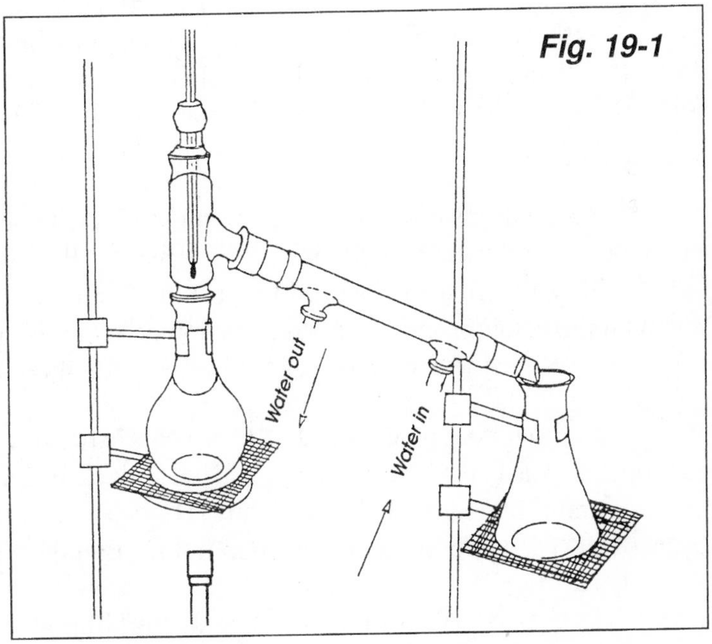

Fig. 19-1

B. Water analysis. Water samples include tap water, 0.1 M NaCl, distilled water, laboratory deionized (or distilled) water and possibly home tap water.

1. <u>Conductivity</u>. While a simple LED conductivity device (like the one in ***Experiment 8***) will suffice for this test, a commercial conductivity apparatus with a probe would be better [or the apparatus suggested by E. Vitz, *J. Chem. Ed.*, <u>64</u>, 628 (1987) with the quantification modification]. If time and the number of these instruments is limited, it is often expedient for the instructor to measure the conductances of the tap water, 0.1 M NaCl, laboratory deionized water and let students measure conductances of their distilled water sample. Some excitement can be added if a contest is held to see which student obtains the lowest conductance for his/her distilled water sample.

2. <u>Chlorides.</u> Test for the formation of precipitates in all the water samples by adding a few drops of 0.1 M $AgNO_3$ to about 0.5 mL of each sample in 13x100 mm test tubes. [Note: Do this part only with the approval of the instructor as silver nitrate is expensive, toxic and a disposal problem].

3. Flame tests. First clean a platinum or nichrome wire by alternately dipping it into 6 M HCl and holding it in a flame until no significant amount of color is observed. Put a few drops of each sample into clean test tubes and then test each water sample in the flame being sure to clean the wire between each test. Do not put the wire into the original water samples as the HCl on the wire will contaminate the samples.

4. Water hardness with a buret. Water hardness will be determined by buret titration on tap water and distilled water (and home tap water if available). Small scale hardness determinations will be determined on all water samples in the next part. Rinse and fill a clean buret with standardized 5.00×10^{-3} M disodium EDTA solution. Open the stopcock and allow the EDTA to displace the air in the buret tip. Close the stopcock and wipe off any liquid adhering to the tip of the buret.

Rinse a 25 mL pipet with tap water and pipet 25 mL of tap water into a 250 mL Erlenmeyer flask. Drain the pipet into the flask and touch the last drop off into the interior wall of the flask. Do not blow the liquid that remains in the pipet tip into the flask. Doing so would give you more than 25.00 mL of liquid.

Add 20 drops of ammonia-ammonium chloride buffer and 2 drops of eriochrome Black T indicator to the flask. Read the buret to the nearest 0.01 mL and begin adding EDTA to the water sample. Slow down to dropwise addition at the first indication that the red color is changing to blue. When all the calcium and magnesium have reacted with EDTA, the red color will be replaced by a blue color. When the color changes to blue, stop the titration, read the buret and record the reading.

Rinse out your pipet with a little of your distilled water and pipet 25.00 mL of it into a clean but not necessarily dry flask (rinsing with deionized water is adequate). Add buffer and indicator as above and titrate with EDTA solution. [Hint: The distilled water should only require a very few drops of EDTA solution before the endpoint is attained.]

Options: Rinse the pipet with your home tap water or an unknown supplied by the instructor and pipet the unknown into a clean flask. Add buffer and indicator and titrate with EDTA solution.

5. Water hardness with a dropper. Water hardness can be determined with smaller samples with a quicker but less accurate method by titrating using an eye dropper or dropper bottle containing EDTA solution and counting the drops required to reach the endpoint. First, using your dropper, determine the mass of 20 drops of 0.00500 M EDTA. As the density of the solution is very close to unity, this value will also be taken as the volume in mL of 20 drops of the solution. Pipet 5.00 mL of the water sample into a large test tube, add 4 drops of the ammonia-ammonium chloride buffer and 1 drop of eriochrome black T indicator and stir. Titrate with the 0.00500 M EDTA using a dropper and count the drops needed to cause a color change. It should be in the range 0-100 drops. The hardness of the water in ppm can be calculated from:

$$\text{hardness (ppm)} \approx (5)(\text{mass of 20 drops})(\text{\# of drops})$$

For typical eye droppers, the mass of 20 drops is close to 1 gram and the equation simplifies to:

$$\text{hardness (ppm)} \approx (5)(\text{\# of drops})$$

This approximate hardness determination should be performed on all the water samples.

Name________________________________Date__________Lab Section_________

Prelaboratory Problems - *Experiment 19* - Water Purification and Analysis

The solutions to the starred problems are in ***Appendix 4***.

1. a. Define boiling point.

   b. Atmospheric pressure in Denver, Colorado is typically about 0.84 atm. Refer to the *Handbook of Chemistry and Physics* to find the boiling point of water at this pressure.

   ____

   c. Will food cooked in boiling water cook slower or faster in Denver? Explain your answer.

   d. What is the function of a pressure cooker?

2. A water sample gives a yellow flame test and yields a precipitate with silver nitrate. What conclusions can you come to from these observations?

3.* Show that for titration with $5.00\text{x}10^{-3}$ M EDTA, the hardness in ppm (assume all hardness is due to $CaCO_3$) can be calculated from:

$$\text{hardness (ppm)} = \frac{(\text{x mL of EDTA}) \text{ x } 500}{\text{y mL of water sample}}$$

4. 25.00 mL of a tap water sample is titrated with 5.00×10^{-3} M EDTA solution. 6.25 mL are required to reach the endpoint. What is the hardness of the water in ppm?

____

5. Use the factor label method to determine what quantity of 5.00×10^{-3} M EDTA would be needed to titrate 25.0 mL of water containing 0.0124 grams of calcium carbonate.

____

6. List the properties an indicator could have to work for this experiment.

Name________________________________Date__________Lab Section_________

Results and Discussion - *Experiment 19* - Water Purification and Analysis

A. Distillation

1. Barometric pressure __________

2. Boiling point of water at measured pressure according to *Handbook of Chemistry and Physics* (edition______ page______) __________

3. Experimental boiling range for collected sample __________

4. What are the bubbles in the water that is boiling as it distills? __________

B. Water softening and ion exchange

1. Brand of water softener ______________

C. Water Analysis

1, 2, 3.

water sample	conductance	$AgNO_3$ results	flame test results
tap water			
0.1 M NaCl			
distilled			
deionized			
home tap			

a. What conclusions can you draw from your conductance measurements? Comment especially on the relative purity of the water samples.

b. Write balanced molecular and net ionic equations for the reaction of silver nitrate with sodium chloride.

c. What conclusions can you draw from your observations on the addition of silver nitrate to each water sample?

d. Did your flame test observations suggest the presence of any cations in any of the water samples? Explain your answer.

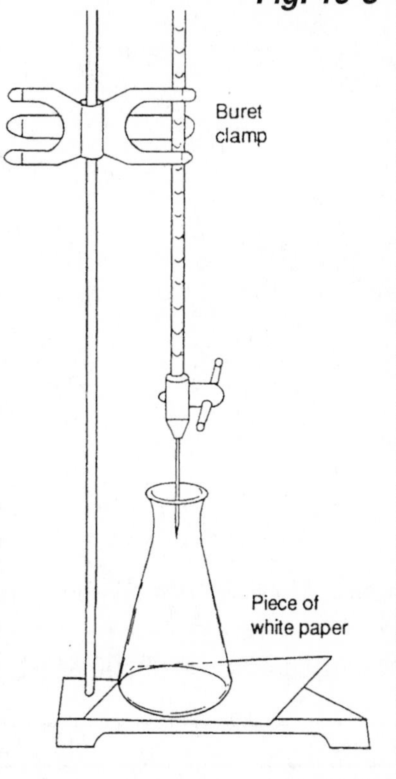

Fig. 19-3

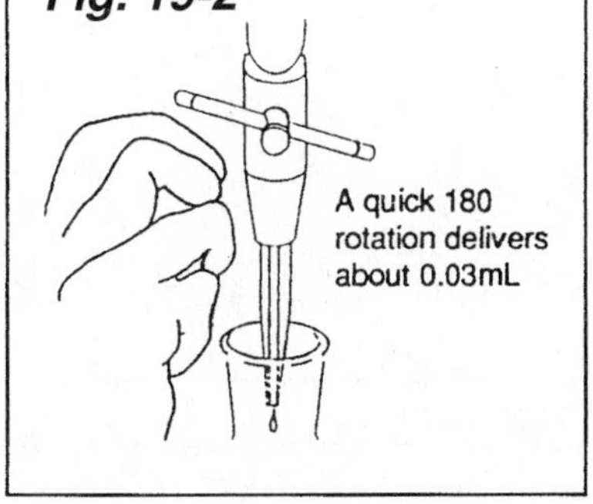

Fig. 19-2

4. Hardness determinations using a buret.

concentration of EDTA ________

	tap water	distilled water	home tap or unk.
final buret reading (mL)			
initial buret reading (mL)			
vol. EDTA (mL)			
vol. water sample (mL)			
$[Ca^{2+} + Mg^{2+}]$[1] (mol/L)			
hardness[2] (ppm)			

[1]Multiply the molarity of EDTA by the volume used in Liters and divide by the volume in Liters of the water sample.
[2]see problem #3 in **Prelaboratory Exercises**

a. Did the distillation have a significant effect on the water hardness? Explain your answer.

b. Water with hardness in the range 0-60 ppm is termed soft, 60-120 ppm medium hard, 120-180 ppm hard and above 180 ppm very hard. Classify the water samples that you titrated.

5. Hardness determinations using a dropper bottle.

mass of 20 drops of 0.00500 M EDTA ________

water sample	# drops EDTA	hardness* (ppm)
tap water		
0.1 M NaCl		
distilled		
deionized		
home tap		

*hardness (ppm) ≈ (5)(mass of 20 drops)(# of drops)

a. How closely did the dropper bottle titration agree with the buret titration? Explain your answer.

6. Some of the *Learning Objectives* of this experiment are listed on the first page of this experiment. Did you achieve the *Learning Objectives*? Explain your answer.

Experiment 20

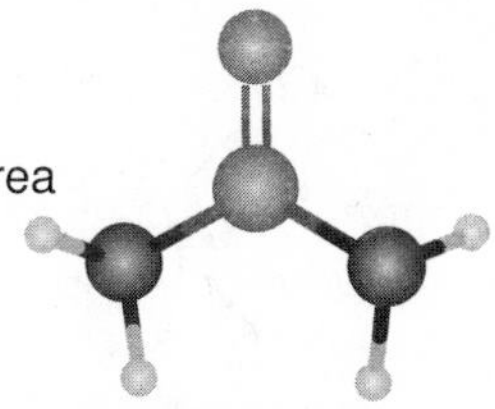

ORGANIC MODELS AND ISOMERISM

Learning Objectives

Upon completion of this experiment, students will have experienced:
1. The construction of molecular models of organic compounds.
2. The concepts of structural, geometric and stereoisomerism.

Text Topics

Structure of organic molecules, σ and π bonding (for text correlation, see page ix).

Notes to Students and Instructor

This exercise should probably not take more than 1½ hours. It is best performed as a class with the instructor leading the discussion as each model is constructed. Students should draw Lewis structures before coming to the laboratory.

Discussion

The world of organic chemistry is fascinating and intriguing. Along with the millions of organic compounds with their multitude of possible reactions, there are patterns and explanations that enable the organic chemist to accurately predict the outcomes of previously untested reactions. To attain the perception necessary for understanding organic chemistry, it is very important that the chemist be able to visualize the 3-dimensional structures of molecules. Knowing the spatial arrangement of the atoms within a molecule enables you to study and predict the effects of polarity, resonance and steric strain on the course and rate of reactions. One of the ultimate goals of organic chemistry is to be able to understand and mimic enzymatic reactions. Enzymes are proteins that have been synthesized by biological systems to perform specific catalytic functions. A substrate must fit into an enzyme just as a key must be an exact match for a lock in order for the enzyme to cause a chemical change in the substrate. Visualization of the fit requires a mind that is adept at thinking in 3-dimensions. Molecular models can considerably facilitate this visualization process.

On a simpler level, the importance of the sequence and spatial arrangement of the atoms is established with the concept of isomerism. Consider the formula $C_2H_2Cl_2$. It is possible to draw three acceptable Lewis structures for this formula that correspond to the compounds 1,1-dichloroethene, *cis*-1,2-dichloroethene and *trans*-1,2-dichloroethene.

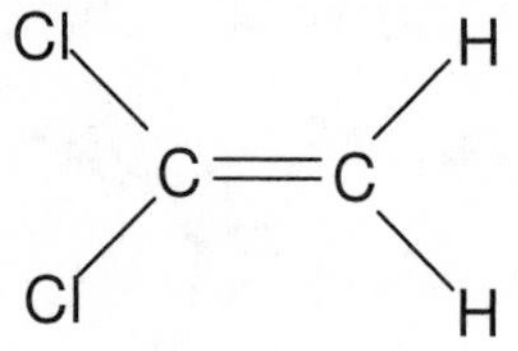

1,1-dichloroethene

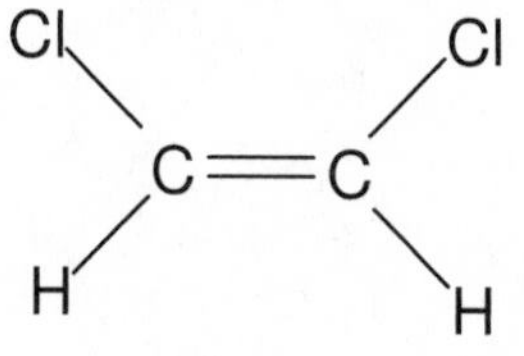

cis-1,2-dichloroethene

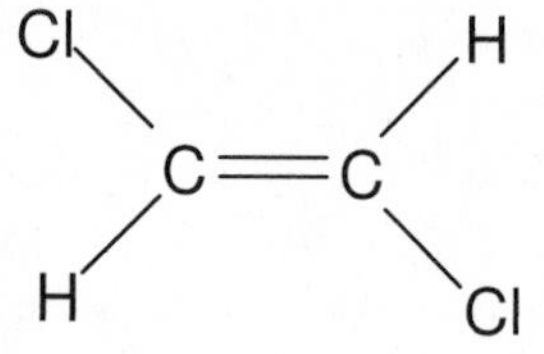

trans-1,2-dichloroethene.

	1,1-dichloroethene	*cis*-1,2-dichloroethene	*trans*-1,2-dichloroethene
b.p.	37°C	60°C	48°C
m.p.	-122°C	-80°C	-50°C

The three possible **isomers** exist and as can be observed from the data above, have different properties. This demonstrates the fact that the spatial arrangement of the atoms has significant effects on the physical and chemical properties of the molecule. The first isomer (1,1-dichloroethene) has a different sequence of bonding than the other two and is a structural isomer of the other two. The second and third isomers have the same sequence of bonding but are spatially different. Because they differ in geometry, they are called geometric isomers. Notice that because of the presence of the double bond, rotation from the *cis* to the *trans* is not possible unless substantial energy input is provided. The absorption of light by rhodopsin in the eye causes a *cis* - *trans* isomerization that leads to the nerve impulse that is sent to the brain and results in vision.

Procedure

Review VSEPR and hybridization in the *Discussion* section in ***Experiment 15***.

A. Construct a model of a methane molecule (CH_4). Observe and record the bond angles.

B. Make a model of ethane (C_2H_6). Observe the free rotation about the carbon - carbon single bond. The different rotational positions (eclipsed, staggered) are called conformers. Record the bond angles.

C. Make two models of propane (C_3H_8). Again notice the rotational possibilities and record the bond angles. Compare the structure to the Lewis structure below. Remember that the Lewis structures of 3-dimensional molecules will not correctly show bond angles and the 90° bond angles typically included in Lewis structures are seldom correct.

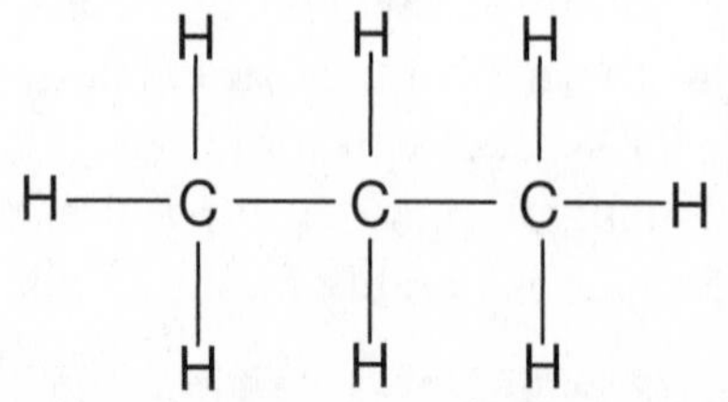

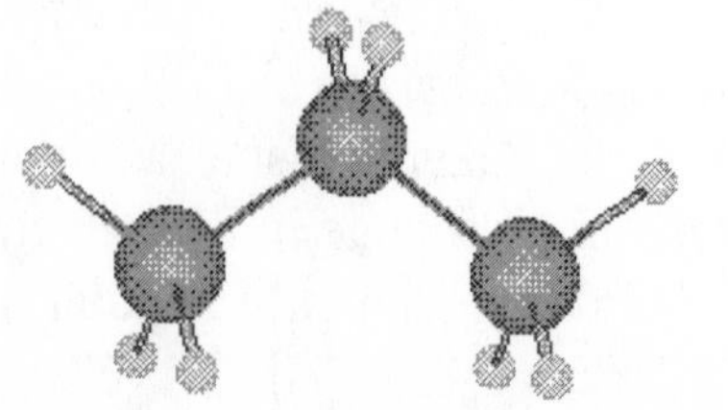

Although the Lewis structure of propane is technically the figure on the left, it is sometimes more convenient to use partially condensed formulas, CH_3–CH_2–CH_3 or even fully condensed formulas, $CH_3CH_2CH_3$.

D. Remove one of the hydrogens from the first carbon on one of the propane models and replace it with a methyl group (- CH_3). This is butane. With the other propane model, replace a hydrogen on the second carbon with a methyl group. This compound is 2-methylpropane (its common name is isobutane). Notice both models have the same formula C_4H_{10}, but different structures and therefore different chemical and physical properties. Look up the boiling and melting points of the two isomers in the *Handbook of Chemistry and Physics* or on the Internet (see page 156). The fully condensed formulas for the two isomers are $CH_3(CH_2)_2CH_3$ (*n*-butane) and $CH_3CH(CH_3)CH_3$ (2-methylpropane).

E. Make all the possible isomers of pentane (C_5H_{12}). On the answer sheet write all the partially condensed formulas for the isomers of pentane.

F. Make all the possible isomers of chlorobutane (C_4H_9Cl). Write the partially condensed formulas for the isomers of chlorobutane.

G. Construct two models of methane.

H. In each of the above models, replace one hydrogen with one chlorine. Verify that both models are identical or superimposable.

I. Replace a second hydrogen on both models with a bromine to make bromochloromethane. Again verify that the models are identical.

J. Replace a third hydrogen on both models with a fluorine to make bromochlorofluoromethane. Are the models still superimposable? Be careful before you answer this. The students who obtain superimposable models should raise their hands (author's prediction: about one-half of the class will raise their hands). Those who raise their hands should take one model and switch the hydrogen and chlorine atoms. Now no one in the class should have superimposable models. You should observe that the models are nonsuperimposable mirror images (enantiomers) of each other. This phenomenon most commonly occurs when at least one carbon in the molecule has four different groups attached to it (in this case H, Br, Cl, F). This creates a stereogenic carbon and in many cases, a chiral molecule. The phenomenon of chirality is quite common in biological systems. For example, amino acids are chiral (except for glycine) and only one of the two enantiomers of each amino acid is biologically active in animals. Although enantiomers have identical physical properties (melting point, boiling point, density, color), they rotate the plane of polarized light in equal but **opposite** directions. Enantiomers also interact differently with other chiral molecules. Since most molecules in your body are chiral, enantiomers of compounds behave differently in your body. Only one of the two enantiomers of each amino acid is of use to your metabolic system. The compounds in spearmint and caraway (*d* and *l*-carvone, see page 251) are enantiomers and yet smell differently to most people because the receptors in our noses are chiral. If vials of *d* and *l*-carvone are available in your laboratory, see if you can distinguish and identify each essence.

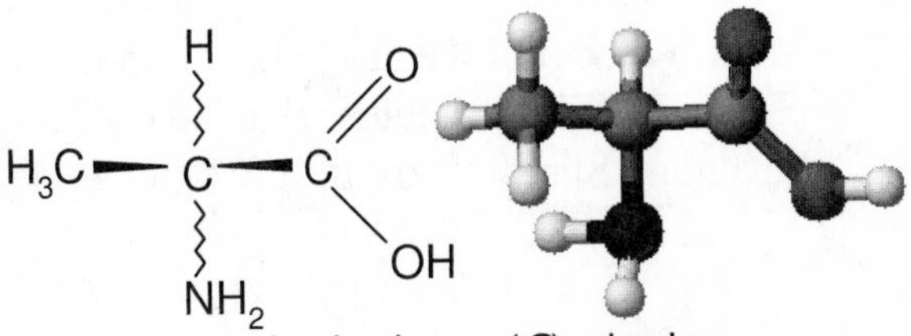

L-alanine = (*S*)-alanine

K. Make a model of ethylene (or ethene, $CH_2 = CH_2$). With most model kits, the model will make it appear that the double bond in ethylene consists of two identical bonds. Other model kits are more consistent with the models in most organic chemistry texts that show that the double bond consists of a σ and a π bond. Additionally, many kits provide only sp^3 hybridized carbons (4 symmetrically spaced holes 109.5° apart). But when carbon is sp^2 hybridized and has one π bond, the bond angles are 120°. Thus the model of ethylene will incorrectly show an H-C-H bond angle of 109.5°. Record the correct bond angles. Is rotation possible about the carbon - carbon double bond?

L. Replace a hydrogen on the ethylene with a methyl group to make propene (C_3H_6).

M. There are four different types of hydrogens in propene thus there are four different ways to replace a hydrogen with a methyl group. Make all four butenes and draw the Lewis structure of each.

N. Make all the isomers of chloropropene (C_3H_5Cl). Write the partially condensed formulas for the isomers of chloropropene.

O. Make a model of ethyne (common name is acetylene - C_2H_2). Record the bond angles.

P. Make a model of cyclohexane (C_6H_{12}). What are the bond angles for a planar regular hexagon? With four groups of electrons about each carbon of cyclohexane, what bond angles do the carbons attempt to achieve? Notice that the model does not assume a planar form but is most stable in either a chair or boat conformation. Cyclohexane molecules spend most of their time in the chair conformation.

Q. Make a model of the simplest alcohol, methanol or methyl alcohol (CH_4O). Note that methanol can be considered to be strongly related to water. Why?

R. Make a model of the simplest carbonyl compound, the aldehyde - formaldehyde (CH_2O).

S. Make a model of the simplest carboxylic acid, formic acid (CH_2O_2).

T. Make a model of the aromatic compound, benzene (C_6H_6). Consider the hybridization of the carbons, the bond angles associated with this hybridization and the angles for a regular planar hexagon. Should there be any bond angle strain in benzene?

Prelaboratory Problems - *Experiment 20* - Organic Models and Isomerism

Students should draw the Lewis structures of the molecules before coming to the laboratory.

Name________________________________Date__________Lab Section__________

Results and Discussion - ***Experiment 20*** **- Organic Models and Isomerism**

Molecule	Lewis structure	Bond Angle	Carbon Hybridization
A. CH_4		____ H-C-H	____
B. C_2H_6		____ H-C-H	____
		____ H-C-C	
C. C_3H_8		____ H-C-H	____
		____ H-C-C	
		____ C-C-C	
D. C_4H_{10}		____ H-C-H	____
		____ H-C-C	
		____ C-C-C	

b.p. ____ ____

m.p. ____ ____

Molecule	Partially Condensed Lewis structure	Bond Angle	Carbon Hybridization
E. C_5H_{12}		____ H-C-H	____
		____ H-C-C	
		____ C-C-C	
F. C_4H_9Cl		____ H-C-H	____
		____ H-C-C	
		____ C-C-C	
		____ H-C-Cl	
		____ C-C-Cl	

G. No recorded answer necessary

H. Alkyl halides such as chloromethane (CH_3Cl) are susceptible to attack on the carbon by nucleophiles such as hydroxide leading to substitution of the hydroxide for the chloride.

1. Write the reaction for this example using Lewis structures and use bond polarities to give a reason for this reaction.

2. What is the meaning of the term "nucleophile"?

I. No recorded answer necessary.

J. Is there a difference in odors of the two samples? Explain your answer.

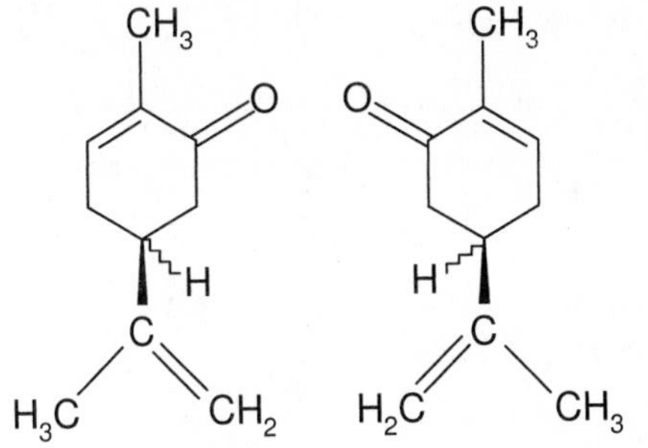

(+)-carvone (caraway seed) (-)-carvone (spearmint)

Molecule	Lewis structure	Bond Angle	Carbon Hybridization
K. C_2H_4		____ H-C-H	____
		____ H-C-C	

Bromine and other electrophilic reagents commonly add to alkenes. For example bromine adds to ethylene (C_2H_4) to give 1,2-dibromoethane (CH_2BrCH_2Br).

1. Write this reaction using Lewis structures and explain why the bromine attacks the π bond of ethylene rather than any of the σ bonds.

2. What is the meaning of the term "electrophilic"?

Molecule	Lewis structure	Bond Angle	Carbon Hybridization
L. C_3H_6		____ H-C-H	____
		____ H-C-C	____
		____ H-C-H	
		____ H-C-C	
		____ C-C-C	

Molecule	Lewis structures
M. C_4H_8	
N. C_3H_5Cl	

Molecule	Lewis structure	Bond Angle	Carbon Hybridization
O. C_2H_2		____ H-C-H	____
		____ H-C-C	
P. C_6H_{12}		____ H-C-H	____
		____ H-C-C	
		____ C-C-C	

Molecule	Lewis structure	Bond Angle	Carbon Hybridization
Q. CH_4O		____ H-C-H	____
		____ H-C-O	
		____ C-O-H	

Explain the relationship of methanol to water.

R. CH_2O		____ H-C-H	____
		____ H-C-O	
S. CH_2O_2		____ H-C-O	____
		____ O-C-O	
		____ C-O-H	

Molecule	Lewis structure	Bond Angle	Carbon Hybridization
T. C_6H_6		____ H-C-C	____
		____ C-C-C	

1. Should there be any bond angle strain in benzene? Explain your answer. ________

2. While 3-membered ring systems such as cyclopropane (C_3H_6) and ethylene oxide (C_2H_4O) do exist, they tend to be more reactive than their open chain analogs, propane (CH_3-CH_2-CH_3) and dimethyl ether (CH_3-O-CH_3). Explain this observation.

3. Some of the *Learning Objectives* of this experiment are listed on the first page of this experiment. Did you achieve the *Learning Objectives*? Explain your answer.

Experiment 21

ACIDS AND BASES: REACTIONS AND STANDARDIZATION

Learning Objectives

Upon completion of this experiment, students will have experienced:
1. Observation of common reactions of acids and bases.
2. The preparation and standardization of a sodium hydroxide solution.
3. The solving of a 4 unknown bottle acid and base system.

Text Topics

Acids and bases, indicators, titrations, standardizations (see page ix).

Notes to Students and Instructor

A solution of sodium hydroxide will be prepared and standardized by titration of the primary standard, potassium hydrogen phthalate. The standardized sodium hydroxide will be used as the titrant in ***Experiments 22 - 25***. Time can be conserved if the instructor prepares the aqueous red cabbage solution.

Discussion

The word "acid" strikes fear in the minds of some people. While acids and bases need to be treated with care and respect, they are extremely important to us. To name a few examples of acids, HCl in our stomachs assists digestion, orange juice and vinegar are part of many meals and sulfuric acid, the most produced synthetic chemical, is used to catalyze a multitude of reactions. Bases such as sodium hydrogen carbonate in baking soda, sodium hydroxide in drain cleaners and ammonia in many household cleaners are also used for many applications. Like most chemicals, acids and bases are dangerous if abused. Solutions containing high concentrations of acids or bases will break down protein and therefore can cause severe burns if not washed off of the skin promptly after contact.

Why is it that compounds with hydrogen as the cation or hydroxide as the anion deserve to have the special names, acids and bases respectively? Sodium and chloride compounds do not get special names. One of the reasons is because hydrogen and hydroxide ions are good catalysts for many types of reactions including the decomposition of proteins such as skin.

This experiment is the first of four experiments that focus on the properties and reactions of acids and bases. Today, you will explore and review some reactions of acids and bases and acid-base indicators, prepare and standardize a sodium hydroxide solution and determine the contents of four unlabeled bottles of acids and bases.

Procedure

A. Reactions and indicators.

1. Measure the temperature of 2 mL of 3 M HCl in a test tube. Add 2 mL of 3 M NaOH to the tube, stir and record the maximum temperature attained.

2. Add 1 mL of 3 M HCl to 1 mL of 1 M sodium carbonate solution in a **large** test tube. Insert a lighted splint into the top portion of the test tube and report your observations.

3. Transfer 3 mL of 3 M NaOH to a test tube, add a small wad of aluminum foil and observe. After the reaction becomes vigorous, hold a lighted splint over the mouth of the test tube. *[Caution: Be sure the tube is not pointing at anyone during the course of this experiment.]*

4. Three solutions will be available.

Solution A: 4 g soluble starch and 2 g of $Na_2S_2O_5$ (or 2.2 g $NaHSO_3$) in 1 liter of water. Prepare by dissolving 4 g of soluble starch in 1 liter of boiling water and adding 2 g of $Na_2S_2O_5$ (or 2.2 g $NaHSO_3$) after cooling.

Solution B: 2 g KIO_3 and 1 mL of 3 M H_2SO_4 in 1 L of water

Solution C: 2 g KIO_3 and 2 mL of 3 M H_2SO_4 in 1 L of water

a. Transfer 10 ml of solution A with a graduated cylinder to a flask. Quickly add 10 mL of solution B to the flask, **swirl** and determine the time elapsed before a change occurs.

b. Repeat a but use 10 mL of solution C in place of solution B.

5. Set up 12 test tubes and transfer 2 mL of 0.01 M HCl to the first four, 2 mL of a pH 7 buffer to the middle four and 2 mL of 0.01 M NaOH to the last four. Add 5 drops of phenolphthalein indicator to the first, fifth and ninth tubes, 5 drops of methyl orange indicator to the second, sixth and tenth tubes, 5 drops of bromothymol blue to the third, seventh and eleventh tubes and concentrated boiled aqueous red cabbage solution (cover some dark purple shredded red cabbage leaves with water in a small beaker and boil until the water is dark purple) to the fourth, eighth and twelfth tubes. Report your observations.

B. Standardization of sodium hydroxide solution. In future experiments, you will need standardized sodium hydroxide to determine the mass percent of acetic acid in vinegar, the formula mass of an unknown acid, the titration curves of HCl and acetic acid, the formula mass of a carbonate, the effectiveness of an antacid and the solubility of potassium hydrogen tartrate in water. As sodium hydroxide cannot be used as a primary standard, a titration must be performed to determine its concentration accurately. One of the most convenient primary standards for this purpose is **potassium hydrogen phthalate (KHP)**. KHP is a monoprotic organic acid with a formula mass of 204.22 g/mol. Note that the K and H stand for potassium and hydrogen respectively but the P stands for phthalate, not phosphorous.

1. Preparation of a 0.26 M NaOH solution. Do either option a or b at the discretion of the instructor. ***Note that sodium hydroxide pellets (option a) are dangerously corrosive and should be handled with great care. The 6 M sodium hydroxide solution used in option b avoids the use of the pellets but is also a corrosive solution. If any sodium hydroxide solution is spilled on skin, it should be immediately washed off with a large quantity of water. The presence of sodium hydroxide (or other bases) on skin can usually be detected by its slippery feeling.***

   a. Weigh out about 10 g of NaOH pellets and transfer them to a 1 L bottle (preferably plastic as NaOH etches glass) [Note: 5 g of NaOH in a 500 mL bottle will suffice if only some of the ***Experiments 22 - 25*** are to be performed]. Add about 600 mL of deionized water and mix until the NaOH dissolves. Add another 350 mL of deionized water and mix again. Stopper, label and save this solution for the experiments today and next few weeks.

   b. Using a 50 mL graduated cylinder, transfer about 41 mL of the laboratory stock solution of 6 M NaOH to a 1 L bottle (preferably plastic as NaOH etches glass) [Note: 20 mL of NaOH in a 500 mL bottle will suffice if only some of the ***Experiments 22 - 25*** are to be performed]. Add about 900 mL of water and thoroughly mix the contents. Stopper, label and save this solution for the experiments today and next few weeks.

2. Label three clean Erlenmeyer flasks 1, 2, and 3 and weigh into each of them 1 to 1.2 g of dry KHP to at least the nearest 0.001 g. Add about 50 mL of water to each flask and swirl to dissolve the KHP. Rinse and fill a 50 mL buret with the NaOH solution. Add 2 or 3 drops of phenolphthalein to flask #1 and titrate (see *Figures 21-1* and *21-2*) until the first tinge of pink appears. As you approach the end point, the pink will persist for longer periods of time before fading to colorless. At this point add the NaOH in half drop quantities by rapidly turning the closed stopcock 180° (see *Figure 21-1*). Repeat the titrations on flasks 2 and 3.

C. Four acid and base unknowns. For this challenge, you will have available a set of 4 bottles labeled only A, B, C and D. The bottles will contain 0.1 M HCl, 0.4 M HCl, 0.1 M NaOH and 0.4 M NaOH. Each of the acid unknowns also contains phenolphthalein (colorless in acid and pink in base). No other reagents or test papers may be used but the solutions can be mixed with each other. Before you attack this problem, think about the various possibilities when two solutions are mixed including the order of mixing and develop a scheme that will enable you to identify the contents of the 4 bottles.

Fig. 21-2

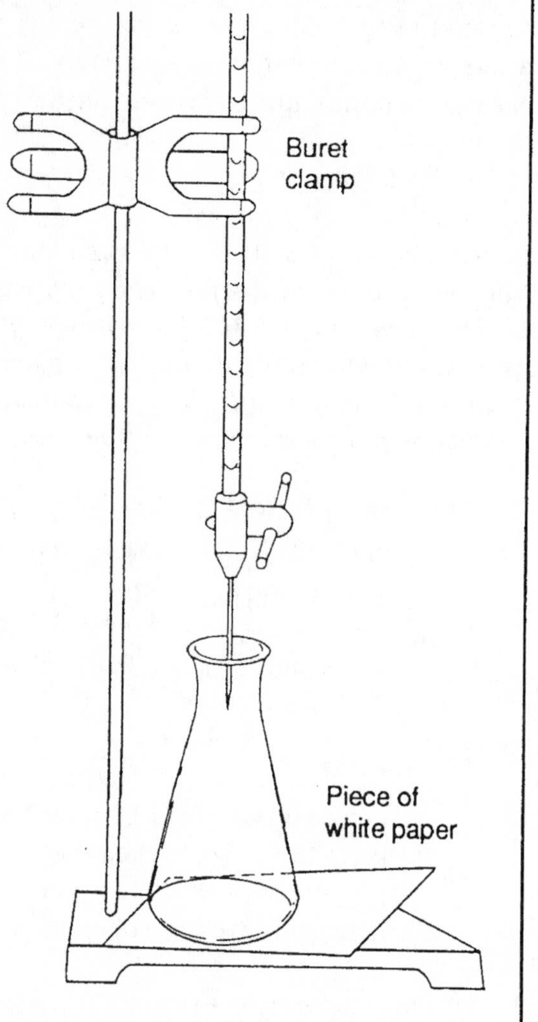

Fig. 21-1

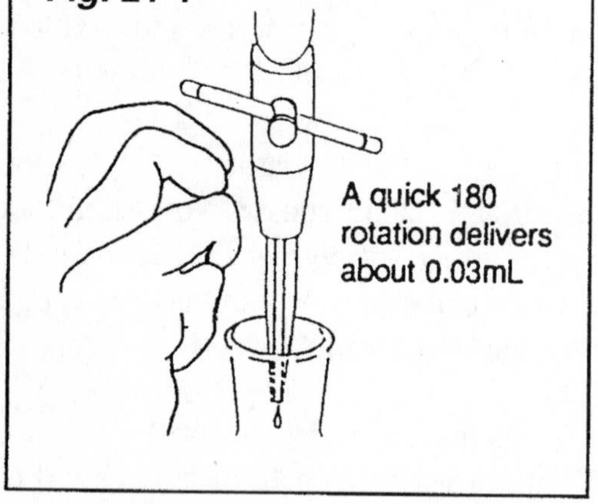

Name________________________________Date__________Lab Section__________

Prelaboratory Problems - *Experiment 21* - Acids and Bases: Reactions and Standardization

The solutions to the starred problems are in ***Appendix 4***.

1. a. Option 1-a of the instructions suggests the dissolving of about 10 g of NaOH in 0.95 L of water to prepare the solution to be standardized. Calculate the approximate concentration of this solution.

____

b. Option 1-b of the instructions suggests that about 41 mL of 6 M NaOH be diluted to about 0.95 L of water to prepare the solution to be standardized. Calculate the approximate concentration of this solution.

____

c.* 1 to 1.2 gram samples of KHP (204.22 g/mol, potassium hydrogen phthalate, not potassium hydrogen phosphorous) will be used to standardize the NaOH solution. Approximately what volume of the NaOH solution will be required to titrate the KHP?

____

d. Why would it have been very inaccurate to have just weighed out 10.000 g of NaOH, diluted it to 1.000 L and calculated the molarity without standardizing it (in other words, can NaOH be used as a primary standard?)?

2.* A 0.4904 g sample of KHP requires 23.82 mL of NaOH to reach the end point. What is the concentration of NaOH?

3. A 0.3535 g sample of KHP requires 19.27 mL of NaOH to reach the end point. What is the concentration of NaOH?

4. What properties should the indicator have for titration of KHP with NaOH?

5. Give a detailed scheme (preferably illustrated with a matrix - see ***Experiment 9***) with predicted observations for the analysis of the acids and bases in *Part C*. Be sure to consider order of mixing. Also enter your plan on page 263 for *Problem C*. What is the minimum number of mixtures needed to solve the system?

Name________________________________Date_________Lab Section_________

Results and Discussion - *Experiment 21* - Acids and Bases: Reactions and Standardization

A. Reactions and indicators.

For numbers 1 - 3 below, report your observations and write balanced equations for each reaction.

1. hydrochloric acid + sodium hydroxide

temperature of HCl _________

temperature after mixing _________

2. hydrochloric acid + sodium carbonate solution

3. sodium hydroxide solution + aluminum

evolved gas + oxygen

4. Observations for A + B

time elapsed _________

Observations for A + C

time elapsed _________

Does it appear that hydrogen ion catalyzes the process that is responsible for the observed change? Explain your answer.

5. Color chart of indicators in 0.01 M HCl, pH 7 buffer, 0.01 M NaOH

	0.01 M HCl	pH 7 buffer	0.01 M NaOH
phenolphthalein	________	________	________
methyl orange	________	________	________
bromothymol blue	________	________	________
red cabbage extract	________	________	________

B. Standardization of sodium hydroxide solution.

	flask 1	flask 2	flask 3
1. Mass of flask + KHP[1]	________	________	________
2. Mass of flask	________	________	________
3. Mass of KHP	________	________	________
4. Moles of KHP (204.22 g/mole)	________	________	________
5. Final buret reading	________	________	________
6. Initial buret reading	________	________	________
7. Volume of NaOH soln.	________	________	________
8. Molarity of NaOH soln.	________	________	________
9. Average molarity of NaOH solution (Record also in next 4 experiments)			________
10. Deviation from average	________	________	________
11. Average deviation of molarity (see ***Expt. 4***)			________

[1]KHP stands for potassium hydrogen phthalate, 204.22 g/mol, not potassium hydrogen phosphous.

C. Four acid and base unknowns.

1. Give your scheme (see *Prelaboratory Exercise 5*) for identifying the four solutions and the observations. Include prediction and observation matrices. Be sure to consider the order of mixing.

2. Give the compound and the concentration for each bottle.

A = ____________ B = ____________ C = ____________ D = ____________

3. Suggest any ways you can think of to improve any part(s) of this experiment.

4. Some of the *Learning Objectives* of this experiment are listed on the first page of this experiment. Did you achieve the *Learning Objectives*? Explain your answer.

Experiment 22

ACIDS AND BASES: ANALYSIS

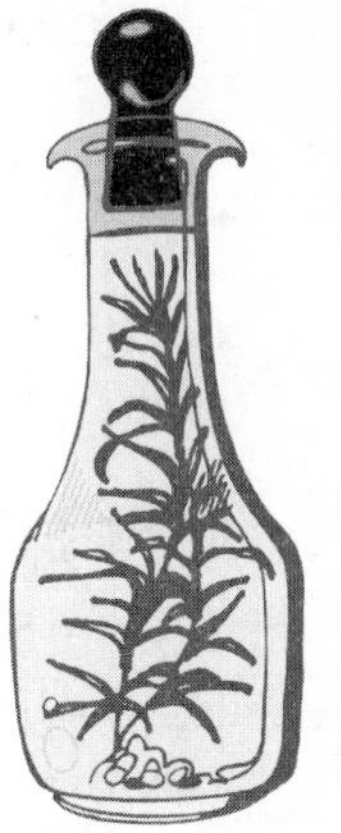

Learning Objectives

Upon completion of this experiment, students will have experienced:
1. The determination of the percent by mass of acetic acid in vinegar.
2. The determination of the molecular mass of an unknown acid.

Text Topics

Acids and bases, indicators, titrations, (for correlation to some texts, see page ix).

Notes to Students and Instructor

The solution of sodium hydroxide prepared last week will be used to determine the percent by mass of acetic acid in vinegar and to determine the molecular mass of an unknown acid.

Discussion

Have you ever tasted wine that has gone sour? The taste is due to the presence of acetic acid which results from the oxidation of the ethanol in the wine. In fact vinegar is produced by fermentation of sugar to ethanol followed by bacteria catalyzed oxidation of the ethanol to acetic acid. One question that a chemist might be confronted with is how much acetic acid is present in the vinegar. This is just one of the questions commonly encountered by chemists working in analytical laboratories. Another challenge might be the identification of an acid in a sample. The determination of its molecular mass can go a long way toward facilitating this identification. The amount of acetic acid in vinegar and the molecular mass of an unknown acid can be determined by titration with a standardized base. These are the topics for today's experiment.

Procedure

A. Titration of vinegar. In ***Experiment 21***, you prepared and standardized an approximately 0.2 M NaOH solution. Because vinegar is close to 1 M acetic acid, it is advisable either to quantitatively dilute the vinegar or titrate only 5.00 mL quantities to avoid using more than 1 buret full of sodium hydroxide. The use of 5.00 mL delivered with a volumetric pipet is suggested.

Obtain a vinegar unknown from your instructor. Weigh a 125 or 250 mL Erlenmeyer flask and pipet 5.00 mL of vinegar into the flask. Reweigh the flask and calculate the density of the vinegar solution. Add about 50 mL of water and 3 drops of phenolphthalein and titrate with the standardized sodium hydroxide until the first tinge of pink persists. Repeat the titration two more times and calculate the molarity and mass percent of acetic acid in vinegar.

B. Molecular mass of an unknown acid. Obtain an unknown acid from your instructor. Weigh approximately 0.5 g samples to at least the nearest 0.001 g into three Erlenmeyer flasks. Dissolve each of the samples in about 50 mL of water, add 3 drops of phenolphthalein indicator and titrate with standardized sodium hydroxide. Repeat for the second and third samples. Calculate the molecular mass for each trial and the average molecular mass.

Fig. 22-1

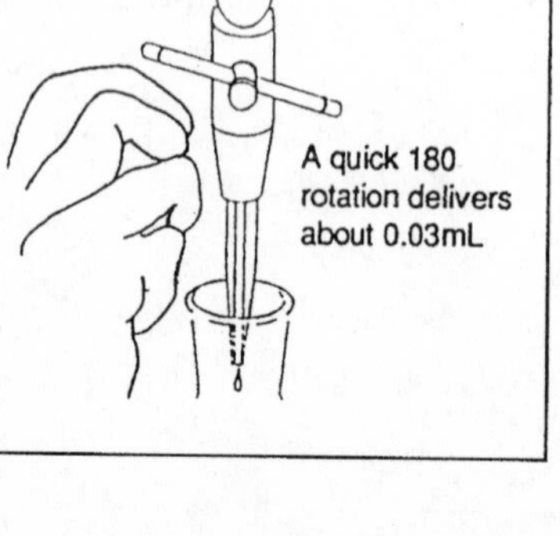

Fig. 22-2

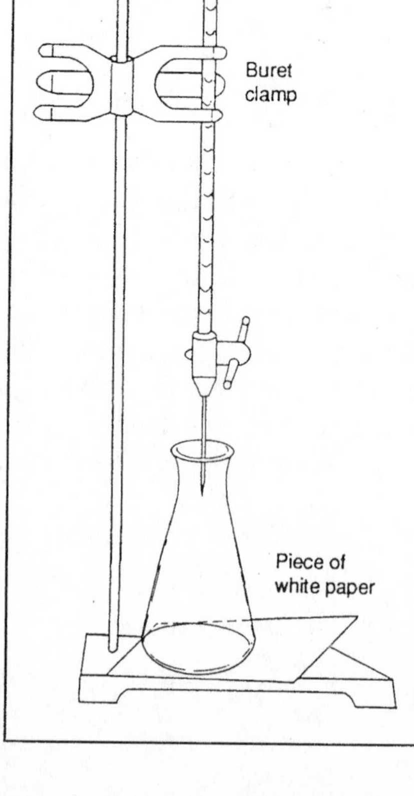

Name________________________________Date__________Lab Section_________

Prelaboratory Problems - ***Experiment 22*** **- Acids and Bases: Analysis**

The solutions to the starred problems are in ***Appendix 4***.

1.* The titration of 25.00 mL of a sulfuric acid solution of unknown concentration requires 31.22 mL of a 0.1234 M NaOH solution. What is the concentration of the sulfuric acid solution?

2.* 10.00 mL of vinegar (mass = 10.05 g) requires 16.28 mL of 0.5120 M NaOH to reach the end point. Calculate the molarity and mass percent of the acetic acid in the vinegar.

3.* A 0.1936 g sample of an unknown monoprotic acid requires 15.56 mL of 0.1020 M NaOH solution to reach the end point. What is the molecular mass of the acid?

4. The titration of 10.00 mL of a diprotic acid solution of unknown concentration requires 21.37 mL of a 0.1432 M NaOH solution. What is the concentration of the diprotic acid solution?

5. 10.00 mL of vinegar (mass = 10.05 g) requires 14.77 mL of 0.4926 M NaOH to reach the end point. Calculate the molarity and mass percent of the acetic acid in the vinegar.

6. A 0.2602 g sample of an unknown monoprotic acid requires 12.23 mL of 0.1298 M NaOH solution to reach the end point. What is the molecular mass of the acid?

Name________________________________Date__________Lab Section__________

Results and Discussion - *Experiment 22* - Acids and Bases: Analysis

A. Analysis of vinegar

1. Unknown vinegar number ________
2. Mass of flask ________
3. Mass of flask + 5.00 mL of vinegar ________
4. Mass of 5.00 mL of vinegar ________
5. Density of vinegar ________
6. Molarity of sodium hydroxide solution ________

Titrations	flask 1	flask 2	flask 3
7. Final buret reading	________	________	________
8. Initial buret reading	________	________	________
9. Volume of NaOH soln.	________	________	________
10. Moles of NaOH	________	________	________
11. Moles of acetic acid	________	________	________
12. Molarity of acetic acid	________	________	________
13. Average molarity of acetic acid in vinegar			________
14. Deviation of each molarity from average	________	________	________

15. Average deviation of molarity (see *Expt. 4*) ________
16. Mass percent of acetic acid in vinegar (show calculations below) ________

B. Molecular mass of an unknown acid.

1. Unknown number ________

2. Molarity of sodium hydroxide solution ________

Titrations	flask 1	flask 2	flask 3
3. Mass of flask + unk.	________	________	________
4. Mass of flask	________	________	________
5. Mass of unknown	________	________	________
6. Final buret reading	________	________	________
8. Initial buret reading	________	________	________
9. Volume of NaOH soln.	________	________	________
10. Moles of unknown acid (assume monoprotic)	________	________	________
11. Molecular mass of acid	________	________	________
12. Average molecular mass of acid			________
13. Deviation of each mol. mass from average	________	________	________

14. Average deviation of molecular mass ________

15. Suggest experimental modifications you could make if the acid to be titrated has low solubility in water.

16. Some of the *Learning Objectives* of this experiment are listed on the first page of this experiment. Did you achieve the *Learning Objectives*? Explain your answer.

Experiment 23

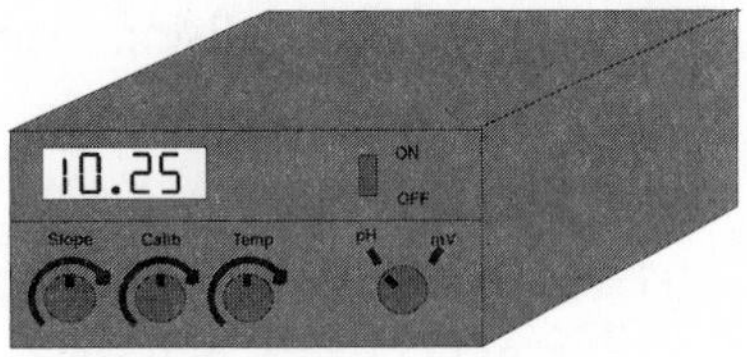

ACIDS AND BASES: pH, pK_a MEASUREMENTS

Learning Objectives

Upon completion of this experiment, students will have experienced:

1. The use of a pH meter.
2. The determination of the approximate acid dissociation constant of acetic acid using four different techniques.
3. The determination of titration curves.
4. The concepts of hydrolysis and buffers.

Text Topics

Acids and bases, pH, acid dissociation constants, titration curves, buffers (for correlation to some textbooks, see page ix).

Notes to Students and Instructor

The solution of sodium hydroxide prepared in ***Experiment 21*** will be used to determine the titration curves of hydrochloric acid and acetic acid. Because of the convenience of having a partner to either read or record results, it is recommended that a partner be allowed for this experiment. A pH probe interfaced to a computer can considerably facilitate this experiment.

Discussion

Most of us have heard of or used the pH scale as a measure of the acidity of water, soil or shampoo. In fact, one of the most valuable uses of pH is as a measure of the acidity of a solution. The concept of pH was developed because the hydrogen ion concentration in solutions can vary by many orders of magnitude and may have values as low as $1x10^{-14}$ M. To avoid the nuisance of writing these tiny numbers, a logarithmic or pH scale was established which more conveniently communicates acidity values. The letter p is defined as $-\log_{10}$ thus pH means $-\log_{10}[H^+]$ where $[H^+]$ is the concentration of hydrogen ions. The definition includes a negative sign to avoid negative signs in the pH values. Because of the negative sign, the lower the pH is, the higher the acidity. Unfortunately, pH electrodes measure $-\log_{10}a_{H+}$ where a_{H+} is the activity rather than the concentration of hydrogen ion. A discussion of activity is beyond the scope of this text but please recognize that the use of pH to mean $-\log_{10}[H^+]$ should be considered as an **approximation** and will have a substantial discrepancy at low pH. The approximation should increase in accuracy as the hydrogen ion concentration decreases.

For a solution of a strong acid such as HCl which almost totally ionizes in water, a 0.010 M solution contains 0.010 M H^+. For this solution, the pH is approximately 2.0. For concentrations such as 3.75×10^{-4} M H^+, the pH must be estimated using logarithm tables or a scientific calculator (Enter the number in scientific notation, push the log button and record the negative of the reading. The pH in this case would be approximately 3.43). It is also relatively simple to convert back from pH to hydrogen ion concentration by taking antilogs. Instrumentation has been developed to read out pH values. A pH of 9.21 would mean that $[H^+] = 6.2 \times 10^{-10}$ M. On a scientific calculator, enter 9.21, hit the +/- key and then the inverse or second function key followed by the log key.

The reaction and equilibrium expressions for the dissociation of water are:

$$H_2O(l) = H^+ + OH^- \qquad K_w = [H^+][OH^-]$$

An experimental determination of K_w has yielded the result $K_w = 1 \times 10^{-14}$ consequently, the equilibrium expression for the dissociation of water is:

$$K_w = [H^+][OH^-] = 1 \times 10^{-14} \qquad \textit{equation 23-1}$$

Rearrangement of *equation 23-1 leads to:*

$$[OH^-] = \frac{1 \times 10^{-14}}{[H^+]} \quad \textit{equation 23-2} \quad \text{and} \quad [H^+] = \frac{1 \times 10^{-14}}{[OH^-]} \quad \textit{equation 23-3}$$

It is important to recognize, as the above equations indicate that the base strength is inversely proportional to the acid strength. As one goes up, the other must go down. Another very useful equation results if logarithms are taken of *equation 23-1*. The base strength, or hydroxide ion concentration, can also be expressed logarithmically using the symbol pOH.

$$\log_{10}K_w = \log_{10}[H^+][OH^-] = \log_{10}1 \times 10^{-14}$$

$$\log_{10}[H^+] + \log_{10}[OH^-] = -14$$

$$pH + pOH = 14 \qquad \textit{equation 23-4}$$

Equations 23-1 to 23-4 can be used to calculate or approximate the remaining three values of $[H^+]$, pH, pOH, $[OH^-]$ if one of the values such as pH is known or has been measured. For a neutral solution, $[H^+] = [OH^-]$. Solving *equation 23-1* for a neutral solution, $[H^+] = [OH^-] = 1 \times 10^{-7}$ or pH = pOH = 7.

pH measurements provide valuable information for many different kinds of solution studies. Comparison of acid strength, determination of acid, base and hydrolysis equilibrium constants and equivalent point determinations are a few of the many applications. The relative strength of an acid may be determined by comparing the pH values of acids of equal concentration. Alternatively and better is a comparison of the acid dissociation constants. For a monoprotic acid, the general reaction for ionization can be simplified to:

$$HA(aq) = H^+ + A^- \qquad \textit{reaction 23-1}$$

It is possible to show that the concentrations of reactants and products are related to an equilibrium constant by the equation:

$$K_a = \frac{[H^+][A^-]}{[HA]} \qquad \textit{equation 23-5}$$

Notice from *equation 23-5* that the greater the amount of ionization (or the stronger the acid), the larger the K_a value is. For strong acids such as hydrochloric acid, which are almost completely ionized, K_a values are typically between 1 and 10^{10}. Acids of moderate strength such as phosphoric acid have K_a values between 10^{-3} and 1. Weak acids such as acetic acid have K_a values below 10^{-3}. Compounds with K_a values below 10^{-11} are usually not called acids.

In this experiment, the acid dissociation constant of acetic acid will be determined four different ways.

1. For a weak acid, measurement of the pH of a solution of known concentration leads to a straightforward approximation of the K_a value. For a weak acid the amount of ionization is small compared to the total concentration of the acid. Insertion of the initial concentration of the acid for [HA] simplifies the calculation. However, always check the completed calculations to be sure the assumption was justified. The pH is used to approximate $[H^+]$ and $[A^-]$. For the pure acid, these two values are the same because each time an acid molecule ionizes, one hydrogen ion is produced and one anion is produced.

2. The K_a for acetic acid may be determined by making a buffer solution in which the concentrations of HA (acetic acid) and A^- (acetate ion) are equal. Since their concentrations will not change significantly as a result of ionization, their concentrations cancel in *equation 23-5* leaving $K_a = [H^+]$. Thus the antilog of the negative of the pH gives the approximate K_a value directly.

3. Many ions react with water in what is called a hydrolysis reaction. The most commonly encountered compounds that undergo hydrolysis are the salts of weak acids such as sodium acetate. Sodium acetate completely ionizes when dissolved in water but then a small fraction of the acetate ions react with water as in the reaction below with A^- representing the acetate.

$$A^- + H_2O(l) = HA(aq) + OH^- \qquad \textit{reaction 23-2}$$

The equilibrium constants for hydrolysis are generally very small but it takes only tiny amounts of hydroxide for the pH to be significantly changed. Remember that in a neutral solution the hydrogen and hydroxide ion concentrations have the minuscule values of 1×10^{-7} M. Any reaction that produces even small amounts of hydrogen or hydroxide ions will affect the very sensitive pH measurement. Because of hydrolysis, ions such as acetate, carbonate and phosphate tend to make their aqueous solutions basic. The equilibrium constant expression for the hydrolysis reaction above is:

$$K_H = \frac{[HA][OH^-]}{[A^-]} = \frac{[HA][OH^-][H^+]}{[A^-][H^+]} = \frac{K_w}{K_a}$$

In the second step above, both the numerator and denominator have been multiplied by $[H^+]$. By careful inspection, we notice that the hydrolysis constant K_H is equal to K_w/K_a and $K_a = K_w/K_H$. If the pH is measured for a solution of known concentration of a salt (such as sodium acetate), the $[OH^-]$ can quickly be calculated by applying *equation 23-4* and taking the antilogarithm of the negative of the result. According to *reaction 23-2*, for every hydroxide produced, there will also be one HA produced thus the hydroxide concentration is the same as the HA concentration. As the amount of hydrolysis of A^- is very low, the original concentration of A^- can be inserted into the expression for K_H and the value for K_H can now be calculated. K_a is $10^{-14}/K_H$.

4. The fourth method you will use to determine K_a for acetic acid is more complex and involves the determination of the titration curve for acetic acid. This technique will be discussed in the ***Procedure*** section. Basically, the titration curve is produced by taking pH measurements at small intervals throughout the neutralization of acetic acid by sodium hydroxide. The equivalence point will be easily obtainable by inspection of a graph of pH vs the volume of NaOH solution added. At the half equivalence point (divide the volume of NaOH used to reach the equivalence point by 2), the concentration of HA will be very close to the concentration of A^- as half of the HA will have been neutralized or converted to product. At this point, as with the buffer solution described in #2, the concentrations of HA and A^- cancel and K_a is equal to $[H^+]$. Thus the antilogarithm of the negative of the pH reading at the half titration point should be a good value for K_a.

The second and fourth methods described above for the determination of the K_a of acetic acid involved a pH measurement on a buffer system. Buffers contain a weak acid and its conjugate base (such as acetic acid and sodium acetate) or a weak base and its conjugate acid (such as ammonia and ammonium chloride) and are designed to resist pH change. Consider your blood for example. It is buffered to maintain a pH of about 7.4. The pH of 7.4 means the hydrogen and hydroxide ion concentrations are very low. If your blood were not buffered, any influx of an acid or base would drastically change the pH and kill you. However, the presence of the buffer prevents significant pH changes unless very large amounts of acids or bases are added.

Procedure

A. pH of acids, bases and salts. Follow the instructions that accompany your pH meter to calibrate the pH meter with 2 buffer solutions. For each solution listed in the ***Results and Discussion*** section, predict the pH before performing the measurement. Use the pH meter to read the pH of each solution to the nearest 0.1 pH unit. In addition to checking each solution with the pH meter, also determine the pH using pH 1-11 paper.

B. Buffers. Two solutions, ***A*** and ***B*** will be provided.

A = $1x10^{-3}$ M HCl   **B** = 0.1 M $HC_2H_3O_2$ + 0.1 M $NaC_2H_3O_2$

Transfer about 25 mL of solution ***A*** to a small beaker, insert the calibrated pH electrode and read the pH. Leave the pH electrode in the solution, add 1 drop of 1 M NaOH, stir and read the pH. Add 9 more drops of 1 M NaOH, stir and read again. Add 15 more drops and repeat reading. Clean out the beaker and repeat the experiment with solution ***B***.

C. Titration curves.

1. Pipet 10.00 mL of an HCl solution of unknown concentration (obtain from instructor) into a 150 mL beaker. Insert into the beaker a clamped buret containing your standardized NaOH and a calibrated pH electrode.

2. Read the pH and then add standardized NaOH solution until either the pH changes by 0.5 units or you have added 1.00 mL of base (whichever is smaller). Record both the new pH and the amount of NaOH solution added, in Part C of the ***Results and Discussion*** section. Repeat this process until you observe a very large pH change (about 4 pH units in less than 0.4 mL). Be sure to add only very small amounts (half or single drops) between readings when the pH starts changing rapidly. Continue the process until you have added about 5 mL of NaOH solution beyond the point where you observed the abrupt pH change. Graph the pH on the vertical axis and volume of NaOH solution on the horizontal axis. The equivalence point should be near a pH value of 7. From the volume of NaOH solution at the pH = 7 point, calculate the concentration of the HCl solution.

3. Pipet 10.00 mL of an acetic acid solution of unknown concentration (obtain from your instructor) into a 150 mL beaker. Repeat the manipulations given in #1 and 2 above except that the equivalence pH should be close (why?) to the pH read for 0.1 M $NaC_2H_3O_2$ in Part A above (#9). From the volume of NaOH solution at the appropriate pH, calculate the concentration of the acetic acid solution. Also determine from the graph the pH when half of the volume of NaOH necessary to reach the equivlaence point has been added. Calculate the K_a value for acetic acid from the pH at this half equivalence point (see *Figure 23-1)*.

4. (Optional) Repeat the above by pipetting 10.00 mL of a maleic acid solution into a 150 mL beaker. As maleic acid is a diprotic acid, the pH curve should reveal not one but two inflections. Be sure to proceed slowly so that you don't miss the first one. From the titration curve determine the values of both acid dissociation constants and compare your values to those in the *Handbook of Chemistry and Physics.*

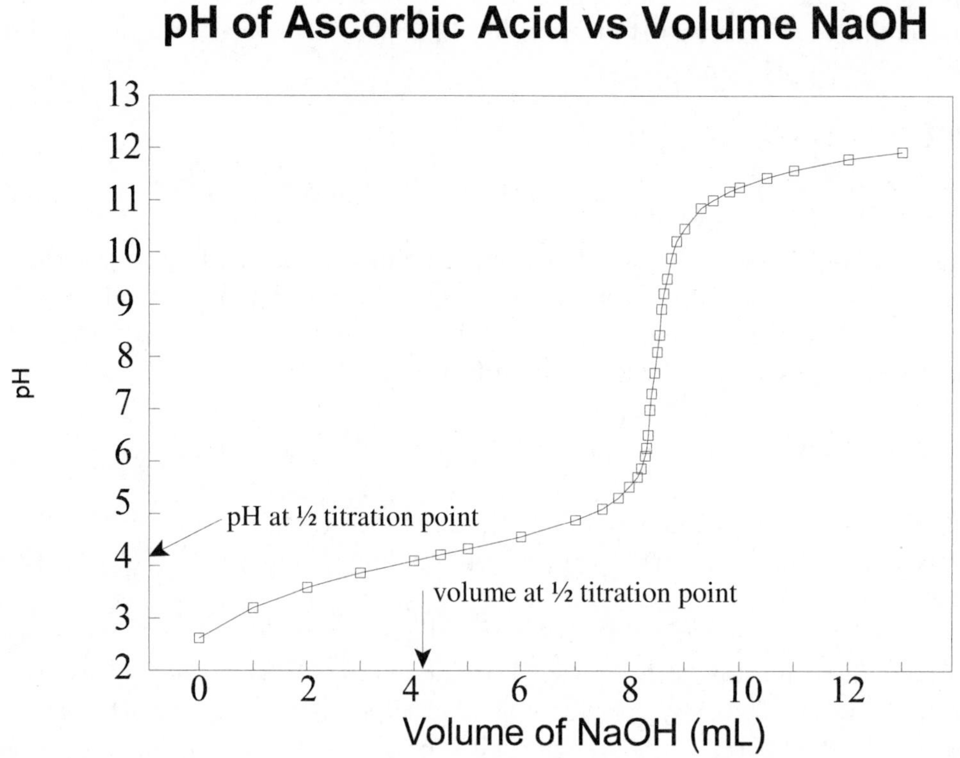
pH of Ascorbic Acid vs Volume NaOH
pH
13
12
11
10
9
8
7
6
5
4
3
2
pH at ½ titration point
volume at ½ titration point
0
2
4
6
8
10
12
Volume of NaOH (mL)

Fig. 23 - 1

Name________________________________Date__________Lab Section_________

Prelaboratory Problems - *Experiment 23* - Acids and Bases: pH Measurements

The solutions to the starred problems are in ***Appendix 4***.

1. Fill in the blanks in the following table:

	pH	$[H^+]_{(mol/L)}$	$[OH^-]_{(mol/L)}$	pOH
a.*	4.7	________	________	________
b.*	________	$7.7x10^{-3}$	________	________
c.	________	________	$2.3x10^{-8}$	________
d.	________	________	________	3.6

2. Make predictions and fill in the first column of *Part A* of this experiment (#1 - 13)

3. What is the approximate expected pH of a 0.030 M HNO_3 solution? ________

4. What is the final pH after 1 drop (0.05 mL) of 6 M HCl is added to 1.0 L of freshly prepared pure water that was originally at a pH of 7.0. Is there a significant pH change? ________

5. a. What is the molarity of pure water at 20°C? ________

   b. Does the concentration of water change significantly (from the value in 4-a) when enough acetic acid is added to make the solution 0.1 M acetic acid?

c. Although the dissociation of an acid is commonly written as indicated in *reaction 23-1*, better representations of the process and the resulting equilibrium expression are:

$$HA(aq) + H_2O(l) = H_3O^+ + A^- \qquad K = \frac{[H_3O^+][A^-]}{[HA][H_2O]}$$

The equilibrium expression is usually simplified to *equation 23-5* by multiplying both sides of the equation by the concentration of water and defining a new constant, the acid dissociation constant, K_a as $[K][H_2O]$. Is it a valid procedure to incorporate the water concentration in to the acid dissociation constant. Base your on the results of parts *a* and *b* of this problem.

6.* A 0.10 M H_2S solution has a pH of about 4.0. What are the approximate concentrations of H^+, SH^-, and H_2S in the solution and what is K_a for the extremely toxic substance, H_2S? (Assume that ionization of the second hydrogen can be neglected).

$[H^+]$ = ________

$[SH^-]$ = ________

$[H_2S]$ = ________

K_a = ________

7. An aqueous solution containing 0.10 M HF and 0.10 M KF has a pH of 3.45. What is the approximate value of K_a for HF?

K_a = ________

8. A 0.10 M solution of sodium cyanide gives a pH reading of 11.1. Write a net ionic equation that accounts for this pH and calculate the approximate concentrations of CN^-, HCN and OH^- in the solution and the value of K_a for HCN.

$[CN^-]$ ________

[HCN]________

$[OH^-]$ ________

K_a ________

Name________________________________Date__________Lab Section_________

Results and Discussion - *Experiment 23* - Acids and Bases: pH Measurements

A. pH of acids, bases and salts.

#	Solution	pH (predicted)	pH (paper)	pH (electronic)
1.	0.1 M HCl	________	________	________
2.	0.01 M HCl	________	________	________
3.	0.001 M HCl	________	________	________
4.	0.1 M $HC_2H_3O_2$	________	________	________
5.	0.1 M NH_4Cl	________	________	________
6.	deionized water	________	________	________
7.	tap water	________	________	________
8.	0.1 M NaCl	________	________	________
9.	0.1 M $NaC_2H_3O_2$	________	________	________
10.	0.1 M $NaHCO_3$	________	________	________
11.	0.1 M Na_2CO_3	________	________	________
12.	0.1 M NH_3	________	________	________
13.	0.1 M NaOH	________	________	________

14. Compare and explain the pH differences between solutions 1 and 4.

15. Do the pH values for solutions 1, 2 and 3 indicate that pH approaches $-\log_{10}[H^+]$ as the concentration of H^+ decreases? Explain your answer.

16. Use the concentrations and pH values for solution 4 to approximate K_a for acetic acid. ________

17. Use the concentrations and pH value from solution 9 and the K_w value to approximate the K_a for acetic acid. ________

18. Write net ionic equations that account for the pH values observed in numbers 5, 9, 11, and 12.

#5

#9

#11

#12

B. Buffers

A = 1×10^{-3} M HCl		B = 0.1 M $HC_2H_3O_2$ + 0.1 M $NaC_2H_3O_2$	
total drops of 1 M NaOH	pH	total drops of 1 M NaOH	pH
0	____	0	____
1	____	1	____
10	____	10	____
25	____	25	____

1. Compare your observations for the addition of 1 M NaOH to solutions A and B and account for any differences.

2. Use the concentrations and pH value for the first pH measurement of solution B to calculate K_a for acetic acid. ________

C. Titration curves.

HCl unknown #___		acetic acid unknown #___	
mL NaOH	pH	mL NaOH	pH
________	____	________	____
________	____	________	____
________	____	________	____
________	____	________	____
________	____	________	____
________	____	________	____
________	____	________	____
________	____	________	____
________	____	________	____
________	____	________	____
________	____	________	____
________	____	________	____
________	____	________	____
________	____	________	____
________	____	________	____
________	____	________	____
________	____	________	____
________	____	________	____
________	____	________	____
________	____	________	____
________	____	________	____
________	____	________	____
________	____	________	____

Maleic acid (optional) unknown #___

mL NaOH	pH	mL NaOH	pH
________	____	________	____
________	____	________	____
________	____	________	____
________	____	________	____
________	____	________	____
________	____	________	____
________	____	________	____
________	____	________	____
________	____	________	____
________	____	________	____
________	____	________	____
________	____	________	____
________	____	________	____
________	____	________	____
________	____	________	____
________	____	________	____
________	____	________	____
________	____	________	____
________	____	________	____
________	____	________	____
________	____	________	____
________	____	________	____
________	____	________	____

1. Plot pH (vertical axis) versus volume of NaOH (horizontal axis) added for each of the titrations performed.

2. Write balanced molecular and net ionic equations for the reaction that occurs when NaOH is added to hydrochloric acid.

3. Write balanced molecular and net ionic equations for the reaction that occurs when NaOH is added to acetic acid.

4. If the maleic acid option was performed, write balanced molecular and net ionic equations for the reactions (2 sets) that occur when NaOH is added to maleic acid.

5. From the titration curve and the concentration of your standardized NaOH, calculate the concentration of your unknown HCl solution.

   a. Unknown # ________

   b. Concentration of standard NaOH solution ________

   c. Volume of NaOH at equivalence point ________

   d. Moles of NaOH ________

   e. Moles of HCl ________

   f. Concentration of HCl ________

6. Repeat the above for acetic acid.

   a. Unknown # ________

   b. Volume of NaOH at equivalence point ________

   c. Moles of NaOH ________

   d. Moles of acetic acid ________

   e. Concentration of acetic acid ________

7. (optional) Repeat the above for maleic acid. Use the sharper of the two end points to calculate the concentration of the maleic acid. Be sure to use the correct mole ratio.

   a. Unknown # ________

   b. Volume of NaOH at equivalence point ________

   c. Moles of NaOH ________

   d. Moles of maleic acid ________

   e. Concentration of maleic acid ________

8. From the titration curve for acetic acid, determine the value of K_a for acetic acid.

   K_a = ________

9. (optional) Repeat the above to determine the two K_a values of maleic acid. Compare your values to the values in the *Handbook of Chemistry and Physics*.

   K_{a1} (experimental) = ________

   K_{a1} (literature) = ________

   K_{a2} (experimental) = ________

   K_{a2} (literature) = ________

10. Some of the *Learning Objectives* of this experiment are listed on the first page of this experiment. Did you achieve the *Learning Objectives*? Explain your answer.

Name___________________________________Date__________Lab Section_________

Postlaboratory Problems - *Experiment 23* - Acids and Bases: pH Measurements

1. Compare the four values of K_a that you have obtained for acetic acid. Which one would you have the most confidence in and why?

   K_a = ______ ______ ______ ______

2. Compare the titration curves of hydrochloric acid and acetic acid. Were there any significant differences? If so, why?

3. a. If you were choosing indicators for the hydrochloric acid and acetic acid titrations, at what pH would you want a sharp color change?

   HCl ________

   acetic acid ________

   b. Look up indicators in the *Handbook of Chemistry and Physics* and list the best indicators for each acid.

HCl	acetic acid
____________	____________
____________	____________
____________	____________

4. Discuss criteria you would use to choose between the use of indicators and a pH meter to determine the endpoint of an acid base titration.

5. Suggest reasons why you might want to buffer a system. In particular, comment on the fact that biological systems are usually buffered.

6. Suggest any ways you can think of to improve any part(s) of this experiment.

Experiment 24

ACIDS AND BASES: CARBONATE ANALYSIS

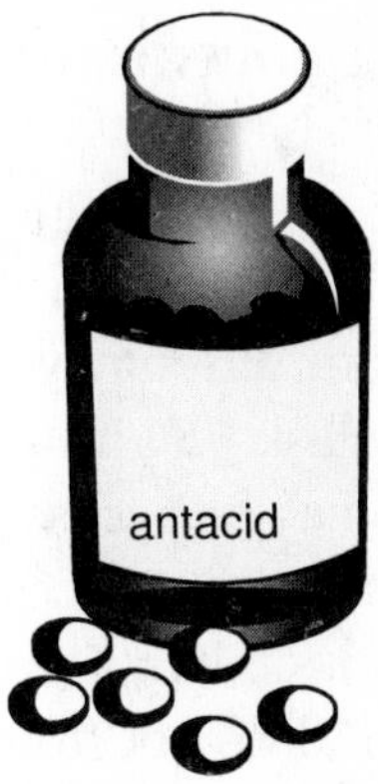

Learning Objectives

Upon completion of this experiment, students will have experienced:
1. The technique of back titration.
2. The determination of the formula mass of a carbonate.
3. The determination of the effectiveness of an antacid.

Text Topics

Acids and bases, pH, titration curves, (for correlation to some textbooks, see page ix).

Notes to Students and Instructor

The solution of sodium hydroxide prepared in ***Experiment 21*** will be used to determine the formula mass of a carbonate and the effectiveness of an antacid. The end point can be determined by using an indicator or a pH meter. A pH probe interfaced to a computer can considerably facilitate this experiment. In addition to standardized sodium hydroxide, standardized hydrochloric acid is required for this experiment.

Discussion

Hydrogen carbonates (bicarbonates) and carbonates are commonly encountered anions. As both react with acid to form carbon dioxide and water, it would seem that it should be possible to titrate either one with hydrochloric acid. Several factors make direct titration difficult or impractical. One of the options for today's experiment is the titration of an antacid tablet but these tablets often contain binders and other additives that can interfere with the titration. Also because the carbonates in today's experiment are insoluble in water, direct titration is hindered. It is possible, however, to react the bicarbonate or carbonate with an excess of standardized acid and then to titrate the excess acid with standardized NaOH solution. This technique is called back titration. This procedure will be performed on an unknown alkaline earth (group IIA or 2) carbonate and on an antacid. The end point will be determined using either a pH meter or an indicator.

Procedure

A. Formula mass of a carbonate. Obtain about 1 g of an unknown group IIA (or 2) carbonate from your instructor and weigh a 0.15 g sample to at least the nearest milligram into a 250 mL Erlenmeyer flask. Pipet 10.00 mL of standardized 0.5 M HCl (supplied by instructor) into the flask. Swirl the mixture until the reaction is complete and the sample is completely dissolved. It may be necessary to break up the larger chunks of the carbonate with a stirring rod to speed up the solution process. If you do so, rinse the liquid remaining on the tip of the stirring rod back into the flask with a stream of deionized water. The fizzing should stop before the addition is complete. If it does not stop fizzing, add 10.00 mL more of the 0.5 M HCl. Insert a pH electrode or add 3 drops of an appropriate indicator (methyl orange or bromophenol blue). If the pH meter or indicator does not indicate that the solution is acidic, add another 5.00 mL of the 0.5 M HCl. Titrate with your standardized 0.26 M NaOH solution to the end point. If you are using a pH meter, record the pH at least at every 0.5 mL interval and 0.1 mL intervals or smaller near the end point. Repeat with at least one more sample. Calculate the formula mass of the carbonate and determine the identity of the carbonate.

B. Effectiveness of an antacid tablet. Obtain an antacid tablet and grind the tablet with a mortar and pestle. Transfer about 0.15 g to a preweighed (to at least the nearest milligram) 250 mL Erlenmeyer flask and weigh again. Now follow the procedure above beginning with the addition of 10.00 mL of standardized 0.5 M HCl. If time is available, repeat with as many brands of antacid tablets as possible.

Name____________________________________Date__________Lab Section__________

Prelaboratory Problems - *Experiment 24* - Acids and Bases: Carbonate Analysis

The solution to the starred problem is in ***Appendix 4***.

1. The instructions recommend the use of either methyl orange or bromophenol blue as an appropriate indicator for the back titration. Is either or both of these indicators appropriate and if so, why? Can you suggest any other appropriate indicators?

2. The discussion section states that one of the reasons that back titration is preferred to direct titration is because of the limited solubility of carbonates in water. Why does this make direct titration difficult and why does back titration tend to circumvent this difficulty?

3.* The mixing of barium chloride solution with sodium carbonate solution resulted in a white precipitate. 25.00 mL of 0.3000 M HCl was added to 0.250 g of the solid with resultant fizzing and dissolving of the sample. 24.80 mL of 0.2000 M NaOH was required to back titrate the excess HCl. What was the calculated molecular mass of the product and how likely is it that the product was barium carbonate?

4. A 1.37 gram antacid tablet is added to 25.00 mL of 0.4500 M HCl. After the resulting reaction was complete, back titration required 10.40 mL of 0.1212 M NaOH. Assuming that the tablet contained only calcium carbonate and an inert binder, write balanced formula and net ionic equations for the first and second reactions and calculate the mass and mass percent of calcium carbonate in the tablet.

   a. Formula equation for calcium carbonate + hydrochloric acid

   b. Net ionic equation for calcium carbonate + hydrochloric acid

   c. Formula equation for sodium hydroxide + hydrochloric acid

   d. Net ionic equation for sodium hydroxide + hydrochloric acid

   e. mass of calcium carbonate ________

      mass percent of calcium carbonate ________

Name______________________________ Date__________ Lab Section__________

Results and Discussion - *Experiment 24* - Acids and Bases: Carbonate Analysis

A. Formula mass of a carbonate.

1. Assuming a formula of MCO_3 for your carbonate, write balanced molecular and net ionic equations for the reaction between your carbonate and HCl.

2. Calculations

   a. Unknown number __________

   b. Molarity of HCl solution __________

   c. Molarity of NaOH solution __________

	flask 1	flask 2	flask 3
d. Mass of flask + carbonate	________	________	________
e. Mass of flask	________	________	________
f. Mass of carbonate	________	________	________
g. Volume of HCl solution	________	________	________
h. Moles of HCl	________	________	________
i. Final buret reading*	________	________	________
j. Initial buret reading*	________	________	________
k. Volume of NaOH solution	________	________	________
l. Moles of NaOH	________	________	________
m. Moles of carbonate (consider mole ratio)	________	________	________
n. Formula mass of carbonate	________	________	________

   o. Average formula mass of carbonate __________

3. Probable identity of metal carbonate __________

4. Percentage error of formula mass __________

*For data obtained with a pH meter, use next page.

Unknown carbonate titrations

titration 1		titration 2	
mL NaOH	pH	mL NaOH	pH

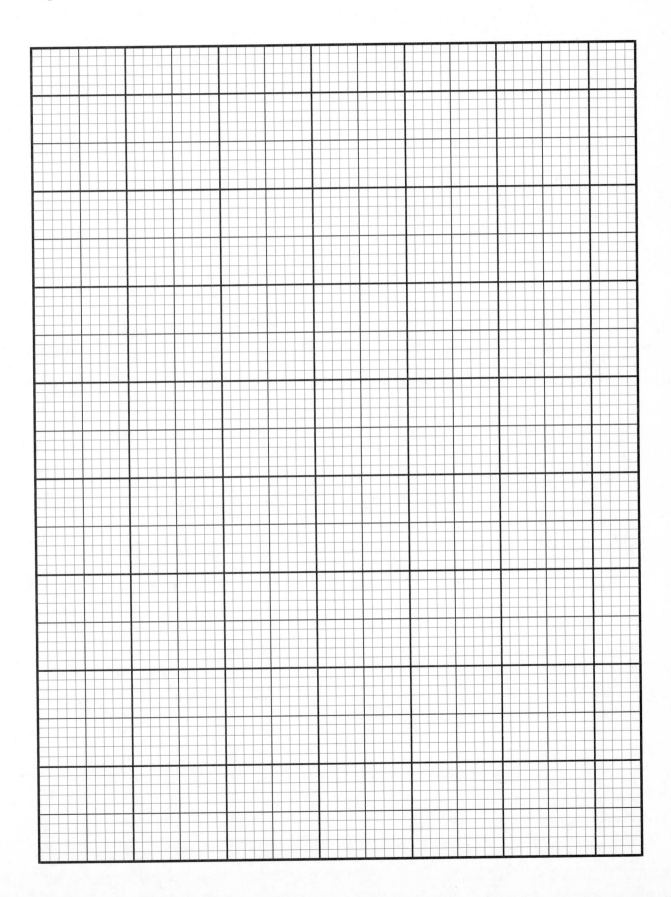

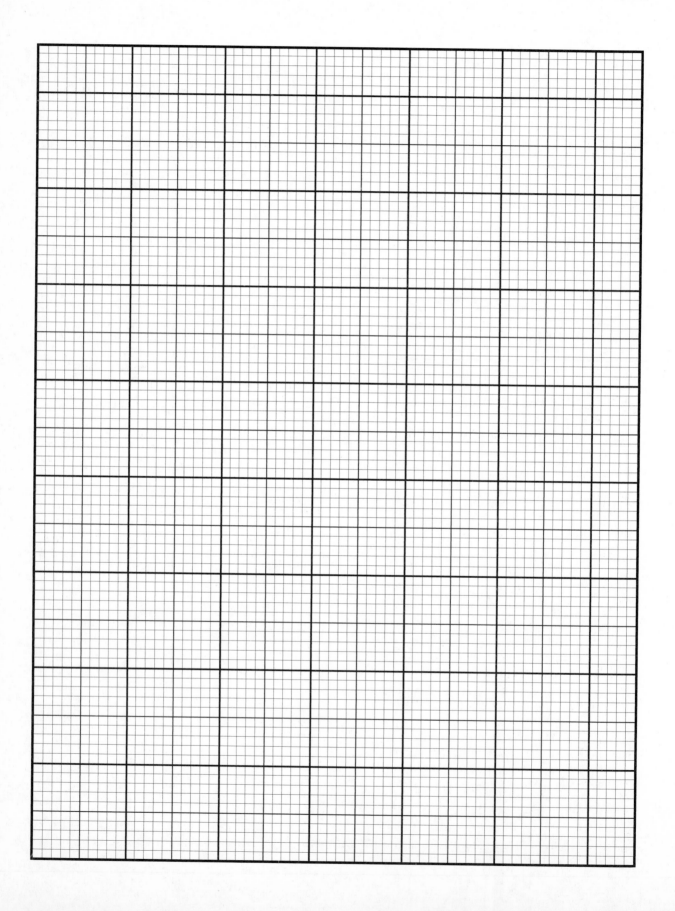

B. Effectiveness of an antacid tablet.

1. Calculations

a. Molarity of HCl solution ________

b. Molarity of NaOH solution ________

	brand 1	brand 2	brand 3
c. Brand name	______	______	______
d. Mass of flask + antacid	______	______	______
e. Mass of flask	______	______	______
f. Mass of antacid	______	______	______
g. Volume of HCl solution	______	______	______
h. Moles of HCl	______	______	______
i. Final buret reading*	______	______	______
j. Initial buret reading*	______	______	______
k. Volume of NaOH solution	______	______	______
l. Moles of NaOH	______	______	______
m. Moles of HCl consumed	______	______	______
n. Moles of HCl consumed per gram of tablet	______	______	______

* For data obtained with a pH meter, use next page.

2.

a. Calculate the number of moles of HCl that would be consumed by 1.0 g of $NaHCO_3$. ________

b. Compare the amount consumed by sodium bicarbonate and each of the brands tested. Which is the best antacid?

c. Considering the prices of the antacids, which antacid would you buy and why?

Antacid titrations

Brand 1		Brand 2	
mL NaOH	pH	mL NaOH	pH

Some of the *Learning Objectives* of this experiment are listed on the first page of this experiment. Did you achieve the *Learning Objectives*? Explain your answer.

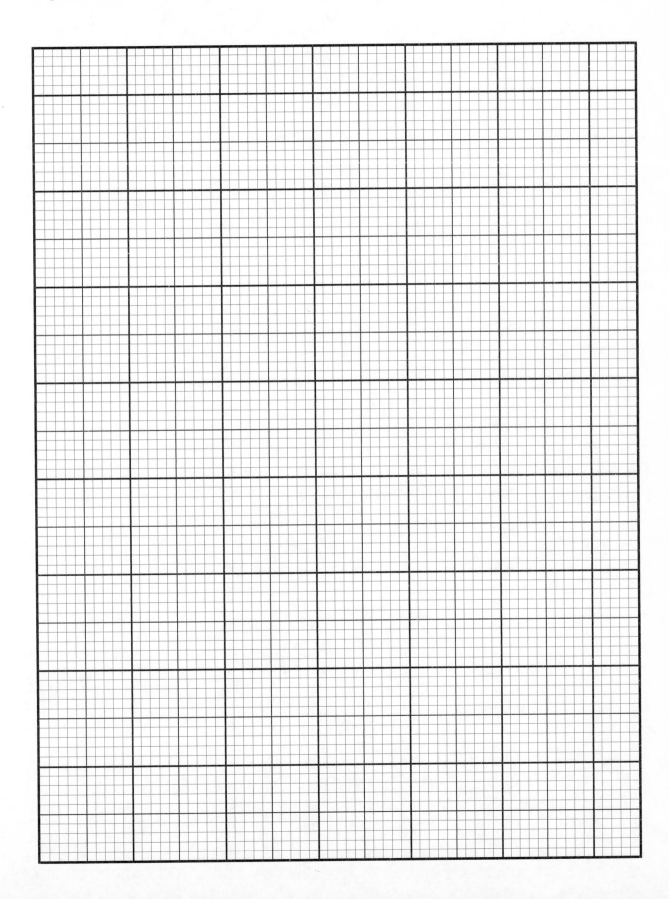

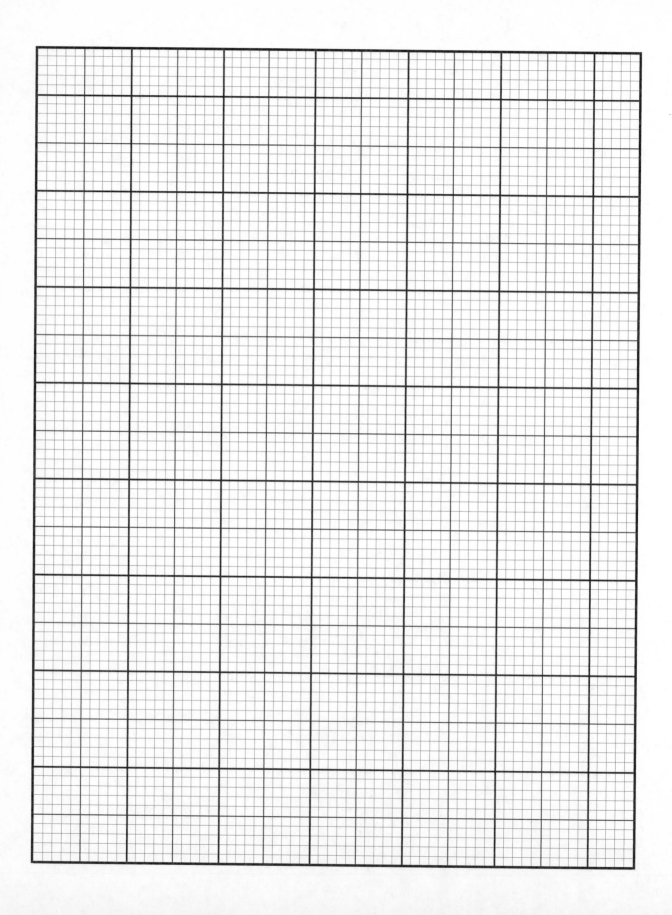

Experiment 25

EQUILIBRIUM - SOLUBILITY PRODUCT

Learning Objectives

Upon completion of this experiment, students will have experienced:
1. The determination of a solubility product at several temperatures.
2. The determination of the enthalpy, entropy and free energy of solution.

Text Topics

Solubility product, enthalpy, entropy, free energy (see page ix).

Notes to Students and Instructor

Because several measurements of K_{sp} are required to graphically determine the enthalpy and entropy of solution, a very quick method will be used. It should be recognized that this method is not very accurate and is intended as a learning experience rather than an exercise in research quality techniques. [**Note:** ***Experiment 26*** is an alternative experiment that covers similar concepts.] One measurement will be performed using a titration to demonstrate a better way of obtaining data. Standardized NaOH from ***Experiment 21*** or another source will be needed for the titration.

Discussion

The preparation of a good wine often involves a long period of storage at 25° to 28°F. The cooling procedure decreases the solubility (as with vanillin in ***Experiment 2***) of potassium hydrogen tartrate (KHT - cream of tartar) in the wine and the resulting crystals of KHT are filtered out. This helps to prevent formation of haziness in the final wine product that can sometimes result from precipitation of the KHT. The decrease in KHT concentration also often helps to adjust the acidity of the wine to a proper level. In today's experiment, the dependence of KHT concentration on temperature will be determined. The data will be used to calculate the solubility product for KHT and the enthalpy, entropy and free energy of solution for KHT.

Equilibrium Constants. Many chemical reactions are reversible. In other words, if two chemical species, in solution, are mixed and form new species, there is some tendency for the new species to react reforming the original species. When the rate of formation of the new species becomes equal to the rate of the reverse reaction, a state of equilibrium is said to have been reached.

Equilibrium equations are written with two arrows pointed in opposite directions between reactants and products indicating that both processes are taking place simultaneously.

Reactants $\rightleftarrows$ Products

For example, the equilibrium for saturated silver chloride is represented by the expression:

$AgCl(s) \rightleftarrows Ag^+ + Cl^-$

Solid silver chloride is breaking into aqueous silver and chloride ions at the same rate that silver ions and chloride ions are coming together to reform solid silver chloride.

For 0.1 M acetic acid at room temperature about 1% of the acetic acid in solution dissociates (percent ionization is concentration and temperature dependent) according to:

$HC_2H_3O_2(aq) \rightleftarrows H^+ + C_2H_3O_2^-$

Equilibrium is achieved when the rate of the forward reaction equals the rate of the reverse reaction. For systems for which a state of equilibrium exists, it can be shown that the product of the concentrations (concentration is used as an approximation in this text for the correct parameter, activity) of the products divided by the product of the concentrations of the reactants is equal to a constant. For example, for a system with two reactants, A and B, and two products, C and D:

$$aA + bB \rightleftarrows cC + dD \qquad K_{eq} = \frac{[C]^c[D]^d}{[A]^a[B]^b}$$

The mathematical relationship for K_{eq} demonstrates Le Chatelier's Principle that a stress placed on a system will cause the system to shift in a direction to reachieve equilibrium. If an aqueous acetic acid solution for which

$$HC_2H_3O_2(aq) \rightleftarrows H^+ + C_2H_3O_2^- \qquad K_a = \frac{[H^+][C_2H_3O_2^-]}{[HC_2H_3O_2]}$$

is disturbed by the addition of HCl, the system will shift to the left as the hydrogen ion concentration has increased. Looking at the equilibrium expression, addition of H^+ makes the numerator large relative to the denominator. Consequently, some H^+ must react with $C_2H_3O_2^-$ to decrease the numerator and increase the denominator until the hydrogen ion concentration multiplied by acetate concentration divided by the acetic acid concentration once again equals K_a. Addition of NaOH to the system would decrease the hydrogen ion concentration and the system would shift to the right to reestablish equilibrium.

Solubility Products. The equilibrium constant for compounds that have very low solubilities in water is called a solubility product. For example, only 2×10^{-4} g of silver chloride dissolves in 100 mL of water to give silver and chloride ions.

$AgCl(s) \rightleftarrows Ag^+ + Cl^-$

For this system, the equilibrium constant, $K_{eq} = \dfrac{[Ag^+][Cl^-]}{[AgCl]}$

However, since the concentration of solid AgCl is a constant, the equation can be written $K_{eq}[AgCl] = [Ag^+][Cl^-]$. The solubility product, K_{sp} is defined as the product:

$$K_{eq}[AgCl] = K_{sp} \qquad \text{or} \qquad K_{sp} = [Ag^+][Cl^-].$$

Calcium hydroxide also has a low solubility (0.1 g/100 mL) in water. The K_{sp} for the dissociation of calcium hydroxide into ions can be represented by:

$$Ca(OH)_2(s) \rightleftharpoons Ca^{2+} + 2\,OH^- \qquad K_{sp} = [Ca^{2+}][OH^-]^2$$

If Ca^{2+} (e.g., $CaCl_2$) or OH^- (e.g., NaOH) is added to a saturated calcium hydroxide solution, the product $[Ca^{2+}][OH^-]^2$ will exceed the value of K_{sp} and additional calcium hydroxide will precipitate out of the solution. This is called a **common ion effect.**

Enthalpy, entropy and free energy. It can be demonstrated that the equilibrium constant for a process is related to the free energy change for the process by the equation, $\Delta G^\circ = -RT\ln K_{eq}$ (ΔG° is the change in free energy at standard state). Consistent with the previous equation, the definition of ΔG and the second law of thermodynamics (the entropy of the universe must increase for a spontaneous process), we observe that a negative free energy or a $K_{eq} > 1$ is a spontaneous process (products favored) and a positive free energy or $K_{eq} < 1$ is a nonspontaneous process (reactants favored).

It is possible to develop the extremely useful relationship for the free energy change of a process, $\Delta G = \Delta H - T\Delta S$. Thus the spontaneity of a process is determined by the magnitude and signs of ΔH (change of enthalpy or heat of the system at constant pressure) and ΔS (change of entropy or disorder of the system). The system under consideration today will be the dissolving of KHT in water. The enthalpy of dissolution of an ionic compound is determined by the crystal lattice energy and the heat of hydration. If the latter is bigger than the former, the enthalpy of solution will be negative (this helps favor the products) and vice versa. The entropy of dissolution is a measure of the number of ways energy can be stored by the system. The number of ways energy can be stored often correlates with the disorder of the system. If the products are more randomly oriented (more disorder) than the reactants, ΔS is positive and this helps favor formation of the products.

The two equations presented for ΔG° can be combined into another useful relationship: $\Delta G^\circ = -RT\ln K_{eq} = \Delta H^\circ - T\Delta S^\circ$. Solving for $\ln K_{eq}$ yields:

$$\ln K_{eq} = -\Delta H^\circ/RT + \Delta S^\circ/R$$

In today's experiment, several values of K_{sp} for KHT as a function of temperature will be determined. Note from the equation that a graph of lnK (y axis) versus 1/T (x axis) is in the form $y = mx + b$ and should produce a straight line [Note: This applies over a limited range of temperature as entropy is not completely independent of temperature]. In this graph, $m = -\Delta H^\circ/R$ and $b = \Delta S^\circ/R$. Consequently the enthalpy and entropy should be derivable from the graph and the value of the gas constant, R (8.313×10^{-3} kJ/mol K). Once ΔH° and ΔS° are known, ΔG° can be calculated at a temperature of interest.

Procedure

A. K_{sp} vs Temperature. KHT dissolves in water according to the following equations:

$$KHT(s) = K^+ + HT^- \qquad K_{sp} = [K^+][HT^-]$$

Unsaturated solutions of known KHT concentration will be cooled and the temperature recorded when the saturation point is attained. As the mass of the KHT is known and the volume can be approximated fairly closely, the concentration of dissolved KHT ([KHT]) can be calculated. Since $[KHT] = [K^+] = [HT^-]$, K_{sp} can be calculated directly at the observed temperature. Saturation temperatures will be determined for 6 different concentrations.

1. Add to a 125 mL clean, dry Erlenmeyer flask (preferably equipped with a magnetic stirring bar), 1.0 gram of KHT weighed to the nearest 0.001 g.

2. Add 25.00 mL of water to the flask with a buret, and insert a 0 - 110°C thermometer.

3. Heat the flask on a heater-stirrer unit with stirring until all the solid dissolves (about 80°C).

4. Keep the heater-stirrer unit on medium heat but move the flask to a room temperature stirrer unit. Stir continuously while it cools and record the temperature at the first evidence of cloudiness.

5. Using the buret, add 5.00 mL more water to the flask and repeat #'s 3 and 4 above (about 75°C should be sufficient to dissolve the solid this time).

6. Repeat the above adding the 5.00 mL amounts of water 4 more times.

7. Add 1.0 gram (weighed to the nearest 0.001 g) of KCl to the final solution. Heat until the solution clarifies and determine the saturation temperature as above by cooling until cloudiness appears.

Fig. 25-1

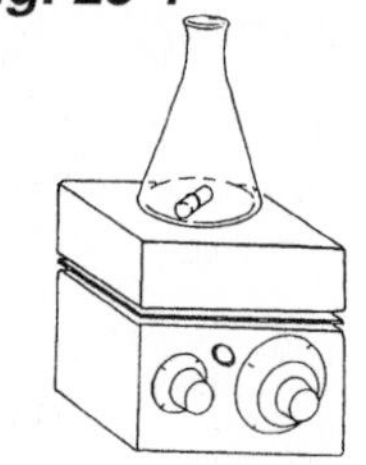

B. K_{sp} by titration. There are several small errors that enter into the total volume values in Part A above. In addition, it is not easy to accurately determine the saturation point temperature. In this part of the experiment, the KHT will be equilibrated in a water bath at fixed temperature for several hours. 10.00 mL of the saturated solution will be withdrawn with a pipet and titrated with standardized sodium hydroxide solution (from ***Experiment 21*** if available). If it is possible to have several water baths in the laboratory at different temperatures, then the entire experiment can be done using this titration method.

A flask containing at least 500 mL of a saturated KHT solution (about 25 g KHT/500 mL of water) should be placed in a 60.0°C water bath several hours before the laboratory period begins by the instructor. Rinse a 10 mL pipet with a little of the solution. Be sure not to disturb the solid KHT on the bottom of flask. Pipet 10.00 mL of the saturated KHT into a 250 mL Erlenmeyer flask. Add about 50 mL of water and 3 drops of phenolphthalein indicator to the flask and titrate with 0.1 or 0.2 M standardized NaOH. Calculate the concentration of KHT in the solution and the K_{sp} value.

C. The entropy of a reaction. In ***Experiment 6*** you ran a reaction between the two solids, ammonium chloride and strontium hydroxide octahydrate. Hopefully, it will now be possible for you to gain some more insight into this reaction by rerunning it. Add 3 grams of ammonium chloride and 7 grams of strontium hydroxide octahydrate to a clean, dry 125 mL Erlenmeyer flask. Swirl the crystals, listen while you are stirring and report all your observations. After a few minutes of swirling, insert a thermometer and read the temperature of the mixture.

Name________________________________Date__________Lab Section_________

Prelaboratory Problems - *Experiment 25* - Solubility Product

The solutions to the starred problems are in ***Appendix 4***.

1.* a. For the dissolving of ammonium nitrate in water, $\Delta H^\circ = 25.7$ kJ/mol and $\Delta S^\circ = 108.7$ J/mol K. Calculate ΔG° and K_{sp} for the process at 25°C.

ΔG° ________

K_{sp} ________

b. Is the process endothermic or exothermic? ________

c. Is the sign of the entropy change consistent with your expectations? Explain your answer.

2. 100 mL of a saturated aqueous $SrCrO_4$ solution contains 0.12 g of strontium chromate at 25°C. Calculate K_{sp} and ΔG° for the dissolving of strontium chromate in water.

K_{sp} ________

ΔG° ________

3. 16.5 mL of 0.1000 M HCl are required to titrate a 10.00 mL aliquot of a saturated strontium hydroxide solution (20°C). Calculate K_{sp} and ΔG° for the dissolving of strontium hydroxide in water.

K_{sp} ________

ΔG ________

Name________________________________Date__________Lab Section_________

Results and Discussion - *Experiment 25* - Solubility Product

A. K_{sp} vs Temperature.

1. Mass of flask + KHT __________

2. Mass of flask __________

3. Mass of KHT __________

4. Data Table

total. volume (mL)	saturation temperature (°C)	saturation temperature (K)	1/T (1/K)	$[KHT]^a = [K^+] = [HT^-]$ (mol/L)	K_{sp}	$\ln K_{sp}$
25						
30						
35						
40						
45						
50						

[a] [KHT] is the concentration of potassium hydrogen tartrate that has dissolved. The formula mass of KHT = 188.18 g/mol.

5. On the accompanying piece of graph paper, graph $\ln K_{sp}$ (y axis) vs 1/T (x axis) over the range of data gathered. From the graph determine the slope and evaluate $\Delta H°$.

   a. slope = $-\Delta H°/R$ __________

   b. $\Delta H°$ __________

6. Use the slope and one point on the graph to calculate the intercept.

   a. $\Delta S°/R = \ln K_{sp} + \Delta H°/RT$ __________

   b. $\Delta S°$ __________

7. Is the sign of ΔS^o consistent with your expectations for the dissolving of a salt in water? Explain your answer.

8.

a. Calculate ΔG^o for the dissolving of KHT in water at 60.0°C ________

b. What is the significance of the sign of ΔG^o_{333} (is dissolving of KHT at 60.0°C spontaneous or nonspontaneous)?

c. Is there a temperature for this system where $\Delta G^o_T = 0$?
If so calculate it. ________

9. a. The *International Critical Tables* give the heats of formation of solid and aqueous KHT as -1545 and -1497 kJ/mol respectively. From the values given, calculate ΔH of solution for KHT. ________

b. What is the percent difference between your experimental value (*5b*) and the literature value? ________

10. Common ion effect

a. Concentration of KHT final solution (50 mL) ________

b. Volume of solution 5.0×10^1 mL

c. Mass of KCl ________

d. Concentration of KCl ________

e. Total $[K^+] = ([KHT]_{(dissolved)} + [KCl])$ ________

f. Total $[HT^-] = [KHT]_{(dissolved)}$ ________

g. Saturation temperature (°C) ________

h. Saturation temperature (K) ________

i. $K_{sp} = [K^+][HT^-]$ ________

j. Why did the temperature have to be increased to dissolve the KHT when the KCl was added?

B. K_{sp} by titration.

1. Temperature of water bath ________
2. Volume of KHT aliquot ________
3. Molarity of NaOH solution ________
4. Final buret reading ________
5. Initial buret reading ________
6. Volume of NaOH ________
7. Moles of NaOH ________
8. Moles of KHT ________
9. $[K^+] = [HT^-]$ ________
10. K_{sp} ________

11. a. Using the graph in Part A, determine K_{sp} at the water bath temperature in Part B. ________

b. What is the percent difference between the K_{sp} values? ________

C. The entropy of a reaction.

1. Record all of your observations about the reaction between ammonium chloride and strontium hydroxide octahydrate.

temperature ________

2. Write a balanced equation for the reaction.

3. Was the reaction endothermic or exothermic? ________

4. Was the entropy positive or negative for the reaction? Explain your answer. ________

5. Was the free energy positive or negative? Explain your answer. ________

6. Some of the *Learning Objectives* of this experiment are listed on the first page of this experiment. Did you achieve the *Learning Objectives*? Explain your answer.

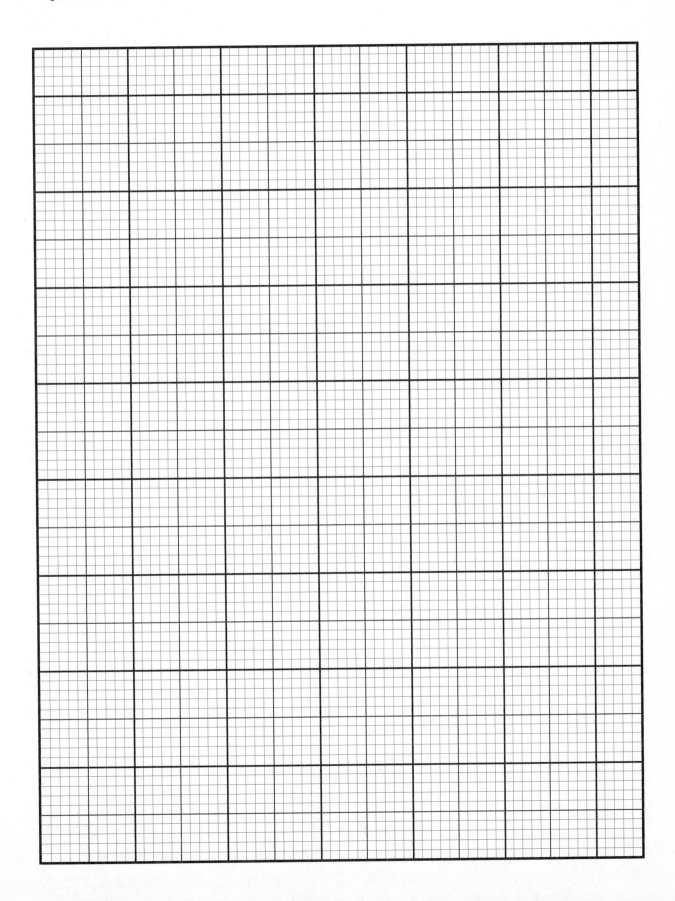

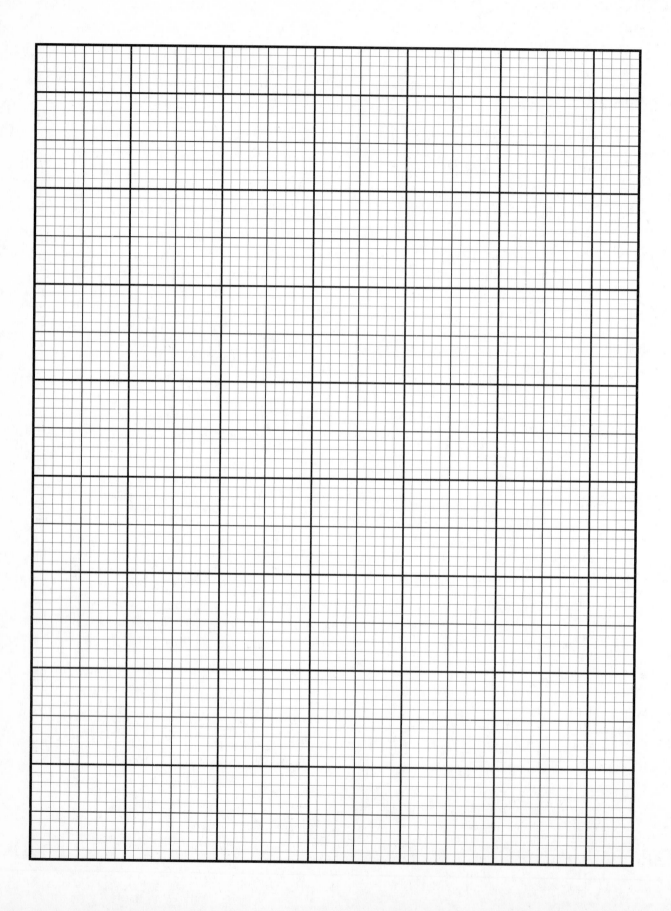

Name________________________________ Date__________ Lab Section__________

Postlaboratory Problems - *Experiment 25* - Solubility Product

1. What are some of the errors involved in the total volume measurements in Part A?

2. In our laboratories at Modesto Junior College, we generally obtain sizeable percent differences (about 100%) between the K_{sp} values obtained using the method of *Part A* and the titration method of *Part B*. We have not been able to determine the source of this discrepancy although we have considered complications from hydrolysis and/or dissociation of the hydrogen tartrate ion. If you also obtained significant differences, suggest possible explanations for them.

3. Concerning the applicability of the titration method, comment on if and how you could use it to determine K_{sp} for each of the following:

   a. barium hydroxide

   b. calcium sulfate

   c. copper(II) iodate

d. lead(II) chloride

e. lithium carbonate

f. silver acetate

g. potassium nitrate

h. silver chloride

4. Suggest any ways you can think of to improve any part(s) of this experiment.

Experiment 26

Equilibrium - Determination of K_{eq}

J. Willard Gibbs
1839 - 1903 (ΔG)

Learning Objectives

Upon completion of this experiment, students will have experienced:

1. The use of techniques of absorption spectroscopy.
2. The determination of an equilibrium constant at several temperatures.
3. The determination of the enthalpy, entropy and free energy of solution.

Text Topics

Equilibrium constant, enthalpy, entropy, free energy (for correlation with some textbooks, see page ix).

Notes to Students and Instructor

This experiment is offered as an alternative to ***Experiment 25***. If this experiment is performed instead of the previous one, it is still advisable to perform Part C of the previous experiment if time permits. It would be convenient to have students work with a partner as quick, multiple measurements are required.

Discussion

As water is a very polar solvent and iodine is non-polar, the solubility of iodine in water would be expected to be very low. This prediction is a correct one as an aqueous solution saturated with iodine contains about 0.03% by mass iodine. Since chemists have set an arbitrary guideline for solubility at 1% by mass, iodine is said to be insoluble in water. However, the addition of iodide to water considerably increases the solubility of iodine. Iodine and iodide react to form triiodide and the equilibrium is far enough to the right to substantially increase the solubility of iodine in water.

$$I_2(s) + I^- = I_3^-$$

For a discussion of equilibrium, equilibrium constants and the enthalpy, entropy and free energy of reactions, please refer to the discussion in ***Experiment 25***. In this experiment, the equilibrium constant for formation of triiodide from iodine and iodide will be determined at several temperatures. The enthalpy, entropy and free energy can be determined from an analysis of the resulting data.

For the iodine, iodide, triiodide system, it is necessary to determine the concentrations of the three species in order to calculate the equilibrium constant. In this experiment, the initial concentrations of iodine and iodide will be known and only the triiodide concentration will actually be measured after equilibrium is achieved. Once the triiodide concentration has been determined, the final iodine and iodide concentrations can be calculated from their initial concentrations and the stoichiometry of the reaction.

Procedure

A. Determination of the absorption spectrum of triiodide. Please review the discussion of absorption spectroscopy in ***Experiment 15***. Iodine and triiodide solutions are both colored and absorb light in the visible region of the spectrum. Iodine has an absorption maximum around 460 nm in the blue green region but fortunately for this experiment, the absorption tails off at higher energy to a value (extinction coefficient of about 16) that is low enough in the near ultraviolet (<400 nm) that it causes an insignificant error at the wavelength used for study in this experiment. Triiodide has two peaks in the ultraviolet (288 nm and one to be determined near 350 nm) and both are very strong absorptions. The absorption then tails off at lower energy into the visible giving rise to its yellowish color. Use of the peak in the near ultraviolet (~350 nm) avoids the use of expensive quartz cells needed at 288 nm (Pyrex glass does not transmit below about 300 nm) and it is just within reach of inexpensive spectrometers such as the Spectronic 20 (capable of reaching to 340 nm). The first step in this experiment will be to determine the optimum wavelength for the study by determining the absorption spectrum of triiodide. Alternatively, the instructor can run the spectrum on a recording spectrophotometer or provide the optimum wavelength. However, if this is done, the absorption of a solution of triiodide of known concentration should still be determined at the absorption maximum so that the results do not depend on a literature value of the extinction coefficient. Because round tubes will probably be used and other variables are difficult to exactly duplicate, extinction coefficients should generally be determined with the equipment and conditions at hand or an error will be introduced.

A solution of triiodide with an easily measured absorption throughout the range of interest (340 nm - 450 nm) needs to be prepared. Because the solution will be prepared by mixing solutions of iodine and iodide, the iodide concentration will be made very high to make sure the equilibrium strongly favors triiodide. In addition to providing sufficient triiodide, this technique makes the iodine concentration very low keeping the absorption of iodine in the region of interest to an insignificant level. Into a dry 125 mL Erlenmeyer, pipet 5.00 mL of a stock solution of 1.00×10^{-4} M aqueous iodine and 10.00 mL of a 0.5 M KI solution. Review the instructions for your spectrometer and determine the absorption spectrum of the solution every 10 nm from 340 nm to 420 nm using water as a blank. Plot the data and determine the wavelength and extinction coefficient at the wavelength of maximum absorption. Set the spectrometer on that wavelength.

B, C. Determination of the equilibrium constant for the iodine + iodide = triiodide reaction as a function of concentration and temperature. Depending on the time available, the constant should be determined at different concentrations of iodide and iodine. Instructions will be supplied for only two measurements (Steps 1, 3) and amounts for additional determinations will be left up to the students (Step 8). Remember that a high amount of iodine could lead to iodine contributing to the absorption measurement resulting in an error. The experiment could also be extended by measuring the iodine absorption at 460 nm where the extinction coefficient of iodine is about 750. Unfortunately, triiodide also absorbs at 460 nm with an extinction coefficient of about 1000 and this complicates the analysis.

Step 1. Stock solutions of 2.00×10^{-3} M KI and 1.00×10^{-4} M iodine should be prepared or provided for this experiment. Pipet 1.00 mL of water, 1.00 mL of the 2.00×10^{-3} M KI solution and 2.00 mL of the 1.00×10^{-4} M iodine solution into a dry cuvette, mix and measure its absorption at the wavelength selected in Part A above. After measuring the absorption, determine the temperature of the solution in the cuvette with a 110°C thermometer. Record your results in the table under **B-4**.

Step 2. Prepare an ice-water slush in a 250 mL beaker. Also heat a 400 mL beaker with about 200 mL of water on a hot plate to about 75°C.

Step 3. Pipet 2.00 mL of the 2.00×10^{-3} M KI solution and 2.00 mL of the 1.00×10^{-4} M iodine solution into a dry cuvette, mix and measure its absorption and the temperature as in Step 1 above. Record your results in the tables under **B-4** and **C-2**.

Step 4. Now immerse the cuvette from Step 3 in the ice bath and stir with the thermometer until the temperature dips below 10°C. Quickly reinsert the cuvette into the spectrometer, measure the absorption value and read its temperature. Record your results in the table under **C-2**.

Step 5. Now put the cuvette into the hot water and stir with the thermometer until the temperature reaches about 30°C. Again determine the absorption value and the temperature. Record your results in the table under **C-2**.

Step 6. Repeat the heating, absorption measurement and temperature measurements two more times increasing the temperature about 10°C each time until you reach about 50°C. Record your results in the table under **C-2**.

Step 7. Use the analysis in ***Experiment 25*** with the data from Steps 3 - 6 to determine the enthalpy, entropy and free energy for the reaction.

Step 8 (optional). Determine the equilibrium constant at room temperature for additional concentrations of iodine and KI. Record your results in the table under **B-4**.

Prelaboratory Problems - *Experiment 26* - Equilibrium - Determination of K_{eq}

Do the Prelaboratory Problems for ***Experiment 25***.

Name________________________________Date__________Lab Section__________

Results and Discussion - *Experiment 26* - Equilibrium - Determination of K_{eq}

A. Determination of the absorption spectrum of triiodide.

1. Concentration of iodine in stock solution ____________

2. Concentration of triiodide in diluted solution ____________

3. Absorption vs wavelength data and graph (plot A on the vertical axis and λ on the horizontal axis.

λ (nm)	340	350	360	370	380	390	400	410	420
A									

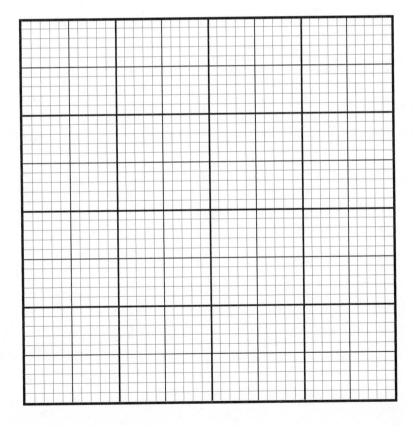

4. Wavelength of maximum absorption ____________

5. Extinction coefficient at absorption maximum ____________

6. The **Procedure** section stated that the absorption spectrum of triiodide tails off in to the visible giving rise to the yellowish color of its solutions. Explain this statement.

7. The equilibrium constant for triiodide formation from iodine and iodide is about 700. The extinction coefficients of iodine and triiodide at the absorption maximum of triiodide are about 16 and 25,000 respectively. Show mathematically that the absorption of iodine does not make a significant contribution to the absorption of the solution you prepared for Part A of the **Procedure** (5 mL of 1×10^{-4} M iodine + 10 mL of 0.5 M KI)

B. Determination of the equilibrium constant for the iodine + iodide = triiodide reaction as a function of concentration.

1. Concentration of KI in original solution ____________

2. Concentration of I_2 in original solution ____________

3. Equilibrium expression for reaction ____________

4. Results and calculations (Calculate the concentrations of iodine ($[I_2]_o$) and iodide ($[I^-]_o$) that would result from the dilutions if a reaction did not occur. Using the absorption measurement for triiodide (A) and its extinction coefficient, calculate the triiodide concentration ($[I_3^-]$) that results from the achievement of equilibrium, then calculate the final concentrations of iodine ($[I_2]$) and iodide ($[I^-]$) by subtraction. Finally, calculate the value of the equilibrium constant (K_{eq}).

#	mL I_2	mL KI	total vol. (mL)	$[I_2]_o$ (mol/L)	$[I^-]_o$ (mol/L)	A	Temp. (°C)	$[I_3^-]$ (mol/L)	$[I_2]$ (mol/L)	$[I^-]$ (mol/L)	K_{eq}
1	2.00	1.00	4.00								
2	2.00	2.00	4.00								

5. Average value of K_{eq} ____________

6. Are the values obtained for the equilibrium constants within experimental error of each other? Explain your answer.

7. Palmer, D. A., Ramette, R. W. and Mesmer, R. E., *J. Soln. Chem.*, **13**, 1984, pp 673 -691, report a best-fit molal equilibrium constant for this system at 25°C of 698. Comment on a comparison of your average value to the value of 698.

C. Determination of the equilibrium constant as function of temperature.

1. From Part B, run #2, $[I_2]_o$ = ____________ $[I^-]_o$ = ____________

2. Fill in the table and graph ln K_{eq} vs 1/T (or use a spreadsheet)

#	Temp. (°C)	Temp. (K)	A	$[I_3^-]$ (mol/L)	$[I_2]$ (mol/L)	$[I^-]$ (mol/L)	K_{eq}	1/T (1/K)	ln K_{eq}

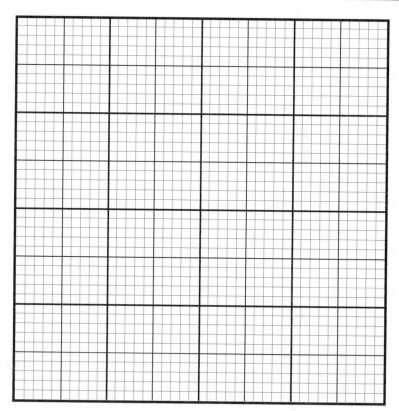

3. On the graph paper or using a spreadsheet, graph $\ln K_{eq}$ (y axis) vs 1/T (x axis) over the range of data gathered. From the graph (a spreadsheet can calculate these values with a couple of clicks) determine the slope and evaluate ΔH°.

   a. slope = $-\Delta H^\circ/R$ ____________

   b. ΔH° ____________

4. Use the slope and one point on the graph to calculate the intercept.

   a. $\Delta S^\circ/R = \ln K_{eq} + \Delta H^\circ/RT$ ____________

   b. ΔS° ____________

5. (Optional) Does the sign of the entropy value make sense (because this reaction is in solution and solvation probably plays an important and difficult to predict effect on the entropy, this question is probably a complex one).

6. Palmer, et. al. (see Part B, #7) report values of $\Delta H^\circ = -17.0$ kJ/mol and $\Delta S^\circ = -0.6$ $JK^{-1}mol^{-1}$ for the iodine + iodide = triodide system. Compare your results and attempt to account for differences.

7. The NBS Tables of Chemical Thermodynamic Properties in the *Journal of Physical & Chemical Reference Data*, **Vol. 11**, 1982, p 252, report the following data:

	ΔH_f^o kJ/mol	ΔG_f^o kJ/mol	S_f^o J/K-mol
I_2(aq)	-55.19	-51.57	111.3
I^-(aq)	22.6	16.46	137.2
I_3^-(aq)	-51.5	-51.4	239.3

a. Using the data in the table, use Hess's law to calculate the enthalpy and entropy for the triiodide formation.

ΔH^o = ____________

ΔS^o = ____________

b. Compare your results to the values calculated above and try to account for any differences.

c. Use the free energy data in the table above to calculate the free energy for the reaction and use this value to calculate an equilibrium constant for the reaction at 25°C. ($\Delta G_f^o = -RT\ln K_{eq}$)

ΔG_f^o = ____________

K_{eq} = ____________

d. Compare your value of the equilibrium constant to value in the question above and try to account for any differences.

8. Some of the *Learning Objectives* of this experiment are listed on the first page of this experiment. Did you achieve the *Learning Objectives*? Explain your answer.

Experiment 27

COMPLEXES

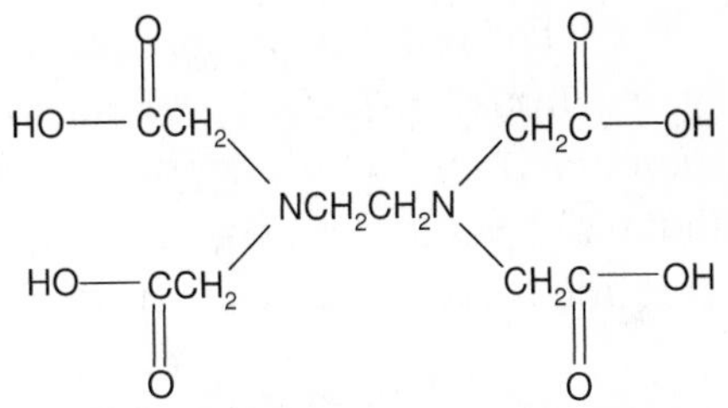

ethylenediaminetetraacetic acid (EDTA)

Learning Objectives

Upon completion of this experiment, students will have experienced:

1. The qualitative observation of complexes.
2. Calculations involving equilibrium constants for complex formation, acid dissociation and solubility.
3. The analysis of cations using precipitation reactions.

Text Topics

Complexes, solubility products, formation constants for complexes, qualitative analysis schemes (for correlation to some textbooks, see page ix).

Notes to Students and Instructor

This experiment should be studied and some calculations and predictions made before the laboratory period begins. It is helpful for these predictions if Part A is performed the first week and Parts B and C the second week.

Discussion

The melting point of sodium chloride is 801°C. The crystalline array with each sodium surrounded by 6 chlorides and each chloride surrounded by 6 sodium ions must have very strong attractive forces for the crystal to have such a high melting point. And yet when mixed with water, the salt dissolves and as we have learned, breaks up into sodium and chloride ions. Why does it do this when the crystal has such strong ionic attractions? Apparently the energetics are more favorable for the ions in water than in the crystalline state. This is due to the hydration of the ions by water molecules. The positively charged sodium ions find themselves surrounded by the partially negatively charged oxygens of several water molecules. The chlorides end up surrounded by the partially positively charged hydrogens of the water molecules. These hydrated ions are sometimes called aquo complexes. Whenever an ion combines with molecules or other ions to form new ions, we call the result a complex. In the distillation experiment (***Experiment 19***), the amount of Ca^{2+} and Mg^{2+} was determined by quantitatively complexing the ions with EDTA.

Today's experiment involves the qualitative observation of complex formation and utilization of reactions in qualitative analysis schemes. In the classical scheme to analyze for silver and mercury(I) ions, chloride ion is added to precipitate AgCl and Hg_2Cl_2 (mercurous ion is unusual in that it exists in solution as dimers, Hg_2^{2+} rather than as Hg^+). Addition of ammonia causes the silver chloride to dissolve as silver ion complexes strongly with ammonia:

$$AgCl(s) + 2\,NH_3(aq) = Ag(NH_3)_2^+ + Cl^-$$

Hg_2Cl_2 disproportionates in the presence of ammonia and is easily detected by the black color of mercury produced.

$$Hg_2Cl_2(s) + 2\,NH_3(aq) = HgNH_2Cl(s) + Hg(l) + NH_4Cl(aq)$$

The decantate containing the silver ammonia complex is acidified to reform the white precipitate of silver chloride.

$$Ag(NH_3)_2^+ + 2\,H^+ + Cl^- = AgCl(s) + 2\,NH_4^+$$

Many formation constants for complexes (K_f) have been determined. These may be used to calculate the amounts of complex that will be formed as a function of the concentrations of cations and complexing agents. By utilizing these constants, it is possible to determine if a precipitate will form in the presence of a complexing agent or an acid. Using the silver-ammonia system as an example, it is possible to calculate the concentration of ammonia needed to produce soluble 0.1 M $Ag(NH_3)_2Cl$ from insoluble AgCl.

$$AgCl(s) = Ag^+ + Cl^- \qquad K_{sp} = [Ag^+][Cl^-] = 1.8\text{x}10^{-10}$$

$$Ag^+ + 2\,NH_3(aq) = Ag(NH_3)_2^+ \qquad K_f = \frac{[Ag(NH_3)_2^+]}{[Ag^+][NH_3]^2} = 1.7\text{x}10^7$$

Combining the two expressions results in the cancellation of silver ion from the equation.

$$K_fK_{sp} = \frac{[Ag(NH_3)_2^+][Cl^-]}{[NH_3]^2}$$

Solving for $[NH_3]^2$ and assuming that virtually all the silver ion is complexed results in:

$$[NH_3]^2 = \frac{[Ag(NH_3)_2^+][Cl^-]}{K_fK_{sp}} = \frac{(0.10)(0.10)}{(1.7\text{x}10^7)(1.8\text{x}10^{-10})} = 3.3 \qquad [NH_3] = 1.8\text{ M}$$

For the above example, the ammonia concentration comes out 1.8 M. Because 0.2 M NH_3 is used for complexing with the AgCl, the solution should be made a total of 2.0 M in ammonia.

When the anion of a precipitate is the conjugate base of a weak acid (e.g., sulfide from hydrogen sulfide), the precipitate can sometimes be dissolved by the addition of a strong acid. H^+ decreases the concentration of the anion by shifting the equilibrium, $HA = H^+ + A^-$, to the left. As the anion concentration decreases, the cation concentration increases. Recall that [anion][cation] = constant. The additional cations are provided by the precipitate as it dissolves. The following example presents the calculations for the concentration of acid necessary to prevent precipitation of 0.1 M CaF_2.

$$CaF_2(s) = Ca^{+2} + 2\,F^- \qquad K_{sp} = [Ca^{+2}][F^-]^2 = 3.9\times10^{-11}$$

$$HF(aq) = H^+ + F^- \qquad K_a = \frac{[H^+][F^-]}{[HF]} = 3.5\times10^{-4}$$

It is now possible to solve for the concentration, $[H^+]$, necessary to prevent precipitation by combining the two equilibrium expressions in the appropriate way to eliminate fluoride concentration.

$$[H^+] = \frac{K_a[HF]}{[F^-]} = \frac{K_a[HF][Ca^{2+}]^{1/2}}{(K_{sp})^{1/2}} = \frac{(3.5\times10^{-4})(0.20)(.32)}{6.2\times10^{-6}} = 3.6\text{ M}$$

Thus 0.10 M CaF_2 should be soluble in dilute (6 M) HCl.

Procedure

A. Qualitative observations. A series of experiments will be performed to study the possible formation of precipitates and complexes. Experimental observations will be compared to the results of calculations using literature values of equilibrium constants.

1. Transfer about 2 mL (about 40 drops) of 0.1 M $MgCl_2$ solution to 5 test tubes.

2. Add 3 M NaOH dropwise to the first test tube being sure to stir and to make visual observations for each drop. Record any significant changes. Continue until about 20 drops of NaOH have been added.

3. Repeat #2 in the second tube but substitute 3 M NH_3 for the NaOH.

4. Repeat #2 in the third tube but substitute 1 M KSCN for the NaOH.

5. Add about 2 mL of 0.1 M $K_4Fe(CN)_6$ to the fourth tube.

6. Add 3 mL of 0.1 M Na_3PO_4 to the fifth tube. Record all observations. Continuing with the fifth tube, add 10 drops of 6 M HCl. Be sure to stir and record any significant changes as you proceed.

7. Repeat 1-6 with 0.1 M $CuSO_4$, 0.1 M $Fe(NO_3)_3$ and 0.1 M $Zn(NO_3)_2$.

The tables below contain the pertinent literature data for you to determine if your observations are consistent with the published values for the equilibrium constants. In the ***Results and Discussion*** section, you will be asked to perform a few of the possible calculations to check for consistency with your observations.

Solubility Product		Formation Constants		Acid Diss. Constants	
Substance	K_{sp}	Complex	K_f	Substance or ion	K_a
$Cu_3(PO_4)_2$	$1.3x10^{-37}$	$Cu(NH_3)_4^{2+}$	$4.8x10^{12}$	H_3PO_4	$7.5x10^{-3}$
$Mg_3(PO_4)_2$	$6.3x10^{-26}$	$Zn(NH_3)_4^{2+}$	$2.9x10^{9}$	$H_2PO_4^-$	$6.2x10^{-8}$
$Zn_3(PO_4)_2$	$9.1x10^{-33}$	$Cu(OH)_4^{2-}$	$1.3x10^{16}$	HPO_4^{-2}	$1x10^{-12}$
$Cu(OH)_2$	$1.3x10^{-20}$	$Zn(OH)_4^{2-}$	$2x10^{20}$		
$Fe(OH)_3$	$3x10^{-39}$	$FeSCN^{2+}$	$1.2x10^{2}$		
$Mg(OH)_2$	$7.1x10^{-12}$				
$Zn(OH)_2$	$1.2x10^{-17}$				

B. Unlabeled Bottles. In ***Experiment 9***, you were given the challenge of assigning identities to seven unlabeled bottles. You were provided with identities of the seven possible solutions and the information that when mixed together, they could undergo only double replacement reactions. This experiment is the same except that the mixtures could result in complex formation as well as double replacement reactions. Based on the information in Part A, you should be able to develop a prediction matrix. By mixing the possible pairs, you will be able to fill in an observation matrix. Comparison of the two should enable you to assign identities to the solutions. Be careful with your comparisons as **observations will depend on the order of mixing**. Addition of zinc nitrate to sodium hydroxide may yield different observations than when sodium hydroxide is added to zinc nitrate. Because the observations may depend on the order of mixing, the matrix provided on page 360 has been designed to allow you to enter predictions and observations for both mixing directions. The seven solutions that will be provided are:

0.1 M $CuSO_4$	3 M NH_3
0.1 M $Fe(NO_3)_3$	3 M NaOH
0.1 M $MgCl_2$	1 M KSCN
0.1 M $Zn(NO_3)_2$	

C. A Qualitative Analysis Scheme. For this part of the experiment, you will be given an unknown that could contain one or a mixture of any of the following cations: Cu^{2+}, Fe^{3+}, Na^+, Zn^{2+}. Read the section on qualitative analysis in ***Experiment 9*** and devise a scheme for the analysis of the mixture. The results of Part A of this experiment should help you devise this scheme. Fill in the flow diagram in the ***Results and Discussion*** section. Two of the possible ions are colored and at least one of the two can probably be identified or ruled out simply by observing the original mixture. However, you should develop a scheme that enables you to verify your preliminary conclusions from color and then perform the entire scheme both on a known mixture of all four ions and your unknown.

Name____________________________________Date____________Lab Section____________

Prelaboratory Problems - *Experiment 27* - Complexes

The solution to the starred problem is in ***Appendix 4***.

1. a.* Should the equivalent of 0.01 M AgBr [$K_{sp} = 5x10^{-13}$] dissolve in 6 M NH_3 [K_f for $Ag(NH_3)_2^+ = 1.7x10^7$]?

   b. What is the minimum concentration of ammonia that should dissolve AgBr to produce a 0.01 M $Ag(NH_3)_2^+$ solution?

2. What hydrogen ion concentration should be required to dissolve 0.10 M MgF_2 if the K_{sp} for magnesium fluoride is $6.8x10^{-9}$ and K_a for HF is $6.5x10^{-4}$?

Name________________________________Date__________Lab Section_________

Results and Discussion - *Experiment 27* - Complexes

A. **Qualitative observations.** Record any significant observations made during the addition of the reagent to the solution containing the cation. Write net ionic equations that account for your observations.

Reagent	Mg^{2+}	Cu^{2+}
NaOH		
NH_3		
KSCN		
$K_4Fe(CN)_6$		
Na_3PO_4		
Na_3PO_4 + 6 M HCl		

Reagent	Fe^{3+}	Zn^{2+}
NaOH		
NH_3		
KSCN		
$K_4Fe(CN)_6$		
Na_3PO_4		
Na_3PO_4 + 6 M HCl		

1. a. Use values of K_{sp} and K_f for copper(II) hydroxide and its complex to predict if a precipitate should form in a 0.1 M Cu^{2+} solution that is 3 M in OH^-.

   b. Are the results of 1a consistent with your observations? Explain your answer.

2. Repeat 1a and 1b for 0.1 M Zn^{2+} in 3 M OH^-.

   a.

   b.

3. a. Should $Cu(OH)_2$ precipitate in 3 M NH_3 if the original $[Cu^{2+}]$ is 0.1 M (the K_b for ammonia is 1.6×10^{-5})?

   b. Are the results of 3a consistent with your observations? Explain your answer.

4. What is the approximate percent of complexed iron in a 0.1 M Fe^{3+} solution containing 1 M KSCN?

5. Qualitatively account for your observations when 3 M HCl is added to the phosphates of magnesium, copper(II), iron(III) and zinc. Discuss this in terms of Le Chatelier's principle and the prevailing equilibria.

6. Refer to your observations on the $Cu^{2+} + SCN^-$ system. See if you can find any information in the literature on $Cu(SCN)_2$. Are your observations consistent with the literature information?

B. Unlabeled bottles. Seven unlabeled bottles containing the solutions below in some scrambled sequence will be provided.

0.1 M $CuSO_4$	3 M NH_3
0.1 M $Fe(NO_3)_3$	3 M NaOH
0.1 M $MgCl_2$	1 M KSCN
0.1 M $Zn(NO_3)_2$	

For the 21 possible mixtures of the seven solutions write net ionic equations (NIE). When no reaction is expected, write *NAR* for no apparent reaction. In some cases, a precipitate will form first and then dissolve as a complex is formed. In these cases, write net ionic equations for precipitation and dissolving reactions. Based on these equations, fill in the prediction matrix. Based on your experimental observations, fill in the observation matrix. Compare the two and assign identities to *A - G*. Some of the lines below may be extras.

1. copper(II) sulfate + iron(III) nitrate

   NIE____

2. copper(II) sulfate + magnesium chloride

   NIE____

3. copper(II) sulfate + zinc nitrate

   NIE____

4. copper(II) sulfate + ammonia

   NIE____

   NIE____

5. copper(II) sulfate + sodium hydroxide

   NIE____

   NIE____

6. copper(II) sulfate + potassium thiocyanate

   NIE____

7. iron(III) nitrate + magnesium chloride

   NIE____

8. iron(III) nitrate + zinc nitrate

NIE____

9. iron(III) nitrate + ammonia

NIE____

NIE____

10. iron(III) nitrate + sodium hydroxide

NIE____

NIE____

11. iron(III) nitrate + potassium thiocyanate

NIE____

NIE____

12. magnesium chloride + zinc nitrate

NIE____

13. magnesium chloride + ammonia

NIE____

NIE____

14. magnesium chloride + sodium hydroxide

NIE____

NIE____

15. magnesium chloride + potassium thiocyanate

NIE____

16. zinc nitrate + ammonia

NIE____

NIE____

17. zinc nitrate + sodium hydroxide

NIE____

NIE____

18. zinc nitrate + potassium thiocyanate

NIE____

19. ammonia + sodium hydroxide

NIE____

20. ammonia + potassium thiocyanate

NIE____

21. sodium hydroxide + potassium thiocyanate

NIE____

Prediction matrix (order of mixing - substance on top added to substance on right)

$CuSO_4$	$Fe(NO_3)_3$	$MgCl_2$	$Zn(NO_3)_2$	NH_3	NaOH	KSCN	
☺	1	2	3	4	5	6	$CuSO_4$
1	☺	7	8	9	10	11	$Fe(NO_3)_3$
2	7	☺	12	13	14	15	$MgCl_2$
3	8	12	☺	16	17	18	$Zn(NO_3)_2$
4	9	13	16	☺	19	20	NH_3
5	10	14	17	19	☺	21	NaOH
6	11	15	18	20	21	☺	KSCN

Experimental observation matrix

A	*B*	*C*	*D*	*E*	*F*	*G*	
☺							*A*
	☺						*B*
		☺					*C*
			☺				*D*
				☺			*E*
					☺		*F*
						☺	*G*

Label color______________________

A = _________ *C* = _________ *E* = _________ *G* = _________

B = _________ *D* = _________ *F* = _________

C. A Qualitative Analysis Scheme. 1. Prepare a flow diagram for the analysis of the cations, Cu^{2+}, Fe^{3+}, Na^+, and Zn^{2+}.

Cu^{2+}, Fe^{3+}, Na^+, Zn^{2+}

2. List the steps of your procedure for determining your known (a mixture of all four cations) followed by your observations and molecular and net ionic equations that account for your observations.

Procedure

<u>Observations</u> <u>Equations</u>

3. Give your observations and conclusions for your unknown.

4. Cations present in your unknown (# = ____), ____________________________

5. Some of the *Learning Objectives* of this experiment are listed on the first page of this experiment. Did you achieve the *Learning Objectives*? Explain your answer.

Experiment 28

RATES AND MECHANISMS OF REACTIONS - VISUAL AND/OR SPECTROSCOPIC MONITORING

Learning Objectives

Upon completion of this experiment, students will have experienced:

1. A study of concentration, temperature and a catalyst on reaction rates.
2. An investigation of a reaction mechanism.
3. The derivation of rate law expressions.

Text Topics

Kinetics, reaction rate expressions, reaction mechanisms, energy of activation, spectroscopy (for correlation with some textbooks, see page ix).

Notes to Students and Instructor

This experiment demonstrates the general principles of kinetics including the effects of concentrations and temperature on rates of reaction. Two different methods for following the reaction are presented. The first uses a built in clock that leads to a stunning visual indication that the reaction has progressed to a certain percentage of product formation. The second method follows the same reaction but uses spectroscopy in the near ultraviolet to follow the progress of the reaction. The second method is clearly closer to research quality but both approaches compromise on some techniques to allow for the limited time available for the experiment.

Discussion

Try to visualize a chemical reaction between hydrogen and chlorine to produce hydrogen chloride. When we write the equation on paper, $H_2(g) + Cl_2(g) = 2\,HCl(g)$, it appears there has been a sudden change from reactants to products. But upon further consideration, we realize that this reaction required several changes. The hydrogen-hydrogen bond and the chlorine-chlorine bond had to break and the hydrogen-chlorine bond had to form. Did these changes occur simultaneously or sequentially? If the latter is the case, in what sequence did they occur? Ideally we would like to be able to videotape the reaction to study the mechanism. Technology that approaches this capability might become available in the future. Currently, for most reactions, the best we can do is to observe and study the behavior of the reactants and products. We are able to obtain important information about the mechanism of the reaction by studying the effects of concentration and temperature on the rate of the reaction.

Reaction rates are also valuable for practical purposes. When adjusting conditions for running an industrial reaction, the chemical engineer wants the reaction to proceed at a reasonable rate. If it goes too fast, it might get out of control and overheat or even explode. On the other hand, if it is too slow, it may not be profitable to run the reaction. A knowledge of the dependence of the reaction rate on concentration and temperature helps the chemist to understand the mechanism and the engineer to optimize reaction conditions.

The rate of virtually every reaction increases as the temperature increases (as a rough rule of thumb, a 10°C increase doubles the reaction rate). Consider the potentially explosive reaction between hydrogen and oxygen to yield water. Although the reaction is highly exothermic, a mixture of hydrogen and oxygen will sit in a bottle indefinitely without noticeable formation of water. Before the oxygens can begin to form bonds with hydrogens, the hydrogen-hydrogen and oxygen-oxygen bonds must begin to break. The cleavage of bonds requires energy. This process requires more energy than most of the molecules have at ambient temperature. As the temperature of the system is increased, the fraction of molecules with sufficient energy to undergo bond cleavage upon a collision increases dramatically and the rate of reaction increases.

Compare the situation to a bowl containing several marbles. If the bowl is lifted off the ground, the marbles will have potential energy because of their position relative to the ground. But because of the lip of the bowl, the marbles are hindered from returning to the lower energy state. If energy is transferred to the system by shaking the bowl, eventually, if the shaking is hard enough, some of the marbles will attain enough energy to make it over the lip and fall to the ground. In this case, the energy of the marble is transferred to the ground. In the case of the molecules, the energy that is released when product bond formation occurs can often be transferred to unreacted molecules to help them break bonds. The energy barrier between the reactants and products that must be surmounted for the reactants to pass over to products is called the energy of activation. The activation energy can be determined from a study of the reaction rate constant as a function of temperature.

$$\ln k = -E_a/RT + \ln A$$ (k = rate constant at absolute temperature, E_a = energy of activation, R = gas constant, A = constant)

The mathematical dependence of the rate of reaction on concentration is more intuitively understandable. Consider again the $H_2 + Cl_2$ reaction. For the reaction to proceed, a collision between H_2 and Cl_2 apparently occurs. One would expect that a doubling of either the H_2 concentration or the Cl_2 concentration would double the number of collisions and the rate. A doubling of both H_2 and Cl_2 will quadruple the number of collisions and the rate. As the rate is expressed as a change in the amount of a reactant, $\Delta[H_2]$ per unit time, t, or rate = $\Delta[H_2]/\Delta t$, the rate should show a proportionality to the concentrations of H_2 and Cl_2 or

$$\Delta[H_2]/\Delta t = -k[H_2][Cl_2]$$

where k is the proportionality constant that is called the rate constant. The above rate expression is consistent with the conclusions above that a doubling of the concentration of either hydrogen or chlorine will double the rate. The rate constant, k, actually is temperature dependent (see previous paragraph) and increases markedly with temperature due to an increase in the number and more importantly, the effectiveness of the collisions. The rate expression then for **any step** of a reaction will be proportional to the products of the concentrations of the reacting species each raised to a power given by the coefficient of the species in the reaction step. Some examples are:

$$A \rightarrow P \qquad \Delta[A]/\Delta t = -k[A]$$

$$2A \rightarrow P \qquad \Delta[A]/\Delta t = -2k[A]^2$$

$$A + B \rightarrow P \qquad \Delta[A]/\Delta t = -k[A][B]$$

If the reaction is a multistep reaction, the determination of the rate expression involves a somewhat more complex analysis that is beyond the scope of this discussion. In fact, the system you will study today is a multistep process but the possible rate expressions will be provided to you.

Some additives to a reaction mixture markedly increase the reaction rate by providing alternate or lower energy pathways to products. Compounds that increase reaction rates without undergoing any ***net*** chemical change themselves are called catalysts.

Procedure

The stoichiometry of the reaction you will explore today is :

$$2I^- + S_2O_8^{2-} = I_2(aq) + 2SO_4^{2-}$$

This reaction apparently requires the simultaneous collision of three ions. The probability of such an occurrence is very small. Consider the probability of Dylan finding Hope and Carson at the same time. The chance of such a coincidence is small. However, Dylan could encounter one of them and then the two of them could go looking for the third person. Reactions that involve more than two molecules or ions also usually proceed in steps. For the reaction between two iodides and a persulfate, a possible sequence would be the following:

$$I^- + S_2O_8^{2-} = SO_4^{2-} + SO_4I^-$$

mechanism 1

$$SO_4I^- + I^- = I_2(aq) + SO_4^{2-}$$

One of these reactions would probably be slower than the other. The slower reaction in a multistep reaction is called the rate determining step. This means that the overall rate for the process is determined primarily by the rate of the slow step. By assuming which step is the rate determining step and making an approximation, it is possible to derive a rate expression for the reaction. For *mechanism 1*, if the first step is rate determining, then the rate expression is just the rate expression for the first step.

$$\text{rate} = k[I^-][S_2O_8^{2-}]$$ rate expression for mechanism 1-a

If the second step is slower than the first, the rate expression is complicated and beyond the scope of this treatment.

A second possible sequence is:

$$2I^- = I_2^{2-}$$

mechanism 2

$$I_2^{2-} + S_2O_8^{2-} = I_2(aq) + 2SO_4^{2-}$$

If the first step is the rate determining step:

$$\text{rate} = 2k[I^-]^2$$ rate expression for *mechanism 2-a*

If the second step is the rate determining step, the rate expression is essentially the same as if the reaction occurred in one step.

$$\text{rate} = k[I^-]^2[S_2O_8^{2-}]$$ rate expression for *mechanism 2-b* and one step rxn.

Your experimental results should enable you to distinguish among the three rate expressions and eliminate some of the possible mechanisms. Rate studies cannot prove that a proposed mechanism is operative but only disprove possible mechanisms.

If you performed ***Experiment 26***, you observed that iodine reacts in the presence of iodide to form triiodide ion. This is a fast secondary reaction and was not included in the discussion above about the mechanism of the reaction. However, the triiodide formation is an extremely important part of the second method (spectroscopic) for following the reaction.

METHOD 1 - Visual clock reaction. The solutions for each of the kinetic runs are given in the table on the next page. The sodium thiosulfate ($Na_2S_2O_3$) is used as part of an indicator system (a purple color will form) to enable you to determine the time required for the reaction to occur for each kinetic run. As the amount of product formed in each case (SO_4^{2-}) will be the same, **the relative rates of reaction will be inversely proportional to the reaction time**. For example, a reaction that takes 25 seconds has twice the rate of one that takes 50 seconds. Because of the proportionality, for the analysis today, it will be possible to substitute the inverse of the time (1/t) for the actual rate of the reaction.

To study the effect of reactant (I^- and $S_2O_8^{2-}$) concentrations on reaction rates, the reactant concentrations will be decreased. The simplest method for achieving this goal is to use a smaller amount of one of the reactants while keeping the total volume constant by replacement with water containing a substance that will not affect the reaction but that will maintain the same concentration of ions. The solutions of KCl and K_2SO_4 are used for this purpose.

A. Concentration effects on reaction rates. For convenience, five 50 mL burets should be cleaned and filled with solutions a - e. If 5 groups of 2 students each set up one buret, considerable time should be saved. A dropper bottle should be available containing solution f. Perform run 1 three times. For runs 1 - 8, add with a buret the required amounts of solutions a - d to a 125 or 250 mL Erlenmeyer flask (for run 8 also add 1 drop of the copper(II) sulfate solution). Measure the required amount of solution e into a second flask. **Quickly add the contents of the second flask to the first. Simultaneously begin timing and swirl until thorough mixing is achieved.** Record the time elapsed until the purple color appears. For each run after the purple appears, record the temperature.

	Run Number (mL of solution)									
Solution	1a	1b	1c	2	3	4	5	6	7	
0.20 M KI	20	20	20	15	10	5	20	20	20	20
0.0050 M $Na_2S_2O_3$ in a 0.4% starch soln.	10	10	10	10	10	10	10	10	10	10
0.2 M KCl	-	-	-	5	10	15	-	-	-	-
0.1 M K_2SO_4	-	-	-	-	-	-	5	10	15	-
0.10 M $K_2S_2O_8$	20	20	20	20	20	20	15	10	5	20
0.1 M $CuSO_4$	-	-	-	-	-	-	-	-	-	1 drop

B. Temperature effects on reaction rates. To study the effect of temperature on the reaction rate, run 1 will be repeated at several temperatures. The flask containing solutions a and b will be heated or cooled and then solution e (at room temperature) will be added. It will be assumed that quick mixing will result in the same temperature for the whole solution. The temperature will be measured immediately after the purple color appears.

For each run, add 20 mL of solution a and 10 mL of solution b to the first flask. Add 20 mL of solution e to the second flask. Change the temperature of the first flask to about the value indicated below. Quickly add solution e, commence timing and stir. Record the time elapsed for the purple to appear and measure the temperature of the solution. Repeat the procedure for the remaining runs. To raise the temperature of a flask, suspend the flask in a beaker of water and heat the beaker over a wire gauze with a Bunsen burner. To lower the temperature of a flask, swirl the solution in an ice bath until the desired temperature is achieved.

run number	approximate temperature for flask 1 (°C)
1a, 1b, 1c	room temperature (already performed above)
2′	35
3′	50
4′	5

METHOD 2 - Spectroscopic Monitoring. The solutions for each of the kinetic runs are the same as for the Method 1 (except the sodium thiosulfate ($Na_2S_2O_3$) solution is not needed). You will need a spectrometer capable of doing measurements down to at least 350 nm and a couple of cuvettes. You will also either need the uv absorption spectrum of triiodide or you will need to perform ***Experiment 26*** - *Part A* to determine the optimal wavelength for this experiment.

A. Concentration effects on reaction rates. It is necessary in this experiment to add and mix solutions quickly ***(especially the addition of KI)***. To accomplish the speed needed, 1 mL calibrated Beral pipets (calibrated at 0.25, 0.50, 0.75 and 1 mL) are ideally suited. The 3.00 mL of water can be delivered with multiple transfers with a 1 mL Beral pipe or better with a buret or a 3 mL pipet. Transfer all of the amounts except the KI into a cuvette and stir. Load another Beral pipet with the proper amount of KI solution, quickly squirt it into the cuvette, start a timer, **mix thoroughly and quickly,** and insert the cuvette into a precalibrated spectrometer. Measure the absorption starting at 10.0 seconds and then every 10.0 seconds up to100 seconds. Withdraw the tube and measure the temperature of the contents.

solution	run 1	run 2	run 3	run 4	run 5	run 6
water (mL)	3.00	3.00	3.00	3.00	3.00	3.00
0.20 M KI (mL)	0.25	0.50	1.00	0.25	0.25	0.25
0.2 M KCl (mL)	0.75	0.50	0	0.75	0.75	0.75
0.1 M K_2SO_4 (mL)	0.75	0.75	0.75	0.50	0	0.75
0.10 M $K_2S_2O_8$ (mL)	0.25	0.25	0.25	0.50	1.00	0.25
0.1 M $CuSO_4$ (mL)	0	0	0	0	0	1 drop

The analysis of the rate data could in theory be performed using an integrated rate law. However, it is rather complex for this treatment and suffers from a significant figure problem. As the percent conversions are very small for these reactions, the method of initial rate analysis is valid and much easier to perform. All that is needed to determine the dependence of the reaction rate on the concentrations of iodide and persulfate is the relative rate of the initial reaction. To obtain this data, the absorption data is plotted vs time and the slope of the line is proportional to the rate. Use of a spreadsheet considerably facilitates this analysis.

B. Temperature effects on reaction rates. To study the effect of temperature on the reaction rate, run 2 will be repeated at three temperatures. The data for room temperature from Part A above is applicable for run 2' below. Load the cuvette with all of the solutions except the KI solution. Cool the cuvette to a little below 10°C and record the temperature. Squirt the KI solution into the cuvette, **mix thoroughly and quickly**, start the timer and insert the cuvette into the spectrometer. Take absorption measurements starting at the 10 second mark every 10 seconds up through 100 seconds. The temperature of the tube should be measured again immediately after removal of the tube from the spectrometer. The average of the temperatures recorded before and after the run will be taken as the temperature of the run. The process above should be repeated twice more with the tube preheated in a water bath to about 30°C and 40°C prior to the addition of the KI solution. Remember to read the temperature of the cuvette before and after each kinetic run. The absorptions are plotted versus time to determine the relative rates and then $\log_{10}$(relative rate) is plotted versus the reciprocal of the average Kelvin temperature to obtain the activation energy.

Name____________________________________Date___________Lab Section__________

Prelaboratory Problems - *Experiment 28* - Rates and Mechanisms of Reactions

The solutions to the starred problems are in ***Appendix 4***.

1. a.* For a reaction with the stoichiometry, 2 A = P, write an expression that relates the rate of disappearance of A to the of appearance of P.

   b. For a reaction with the stoichiometry, A + 2 B = 3 P, write an expression that relates the rate of disappearance of A to the rate of disappearance of B and another expression that relates the rate of disappearance of A to the rate of appearance of P.

2. a.* For a reaction with the stoichiometry, A + 2 B = P, a study of the initial rate of reaction revealed that the amount of time for formation of 1% of the total amount of possible product halved when the concentration of A was doubled, and doubled when the concentration of B was halved. Write a rate expression for this reaction and suggest a stepwise mechanism for the reaction.

   b. For a reaction with the stoichiometry, A + B + C = P, a study of the initial rate of reaction revealed that the amount of time for formation of 1% of the total amount of possible product quartered when the concentration of A was quadrupled, did not change when the concentration of B was halved, and doubled when the concentration of C was halved. Write a rate expression for this reaction and suggest a stepwise mechanism for the reaction.

3. The *Procedure section* for Method 1 states "As the amount of product formed in each case (SO_4^{2-}) will be the same, **the relative rates of reaction will be inversely proportional to the reaction time.**" Explain this statement using math if appropriate.

4. What is the function of the $Na_2S_2O_3$ solution in the first part of this experiment?

5. What are the functions of solutions KCl and K_2SO_4 in this experiment?

6. What is the purpose of using the $CuSO_4$ solution in this experiment?

Name________________________________Date__________Lab Section__________

Results and Discussion - *Experiment 28* - Rates and Mechanisms of Reactions

METHOD 1 - Visual clock reaction A. Concentration effects on reaction rates.

1. Concentrations and results.

run #	$[I^-]$ (mol/L)	$[S_2O_8^{2-}]$ (mol/L)	time (sec.)	1/time (sec.$^{-1}$)	temp. (°C)	$\log[I^-]$	$\log[S_2O_8^{2-}]$	log(time)
1a								
1b								
1c								
1 ave.								
2								
3								
4								
5								
6								
7								
8								

2. Considering the results of 1a, 1b, and 1c, how reproducible are the results of this experiment?

3.

   a. Plot the iodide concentration for runs 1 ave., 2, 3, and 4 vs the inverse of the time. Is the rate of the reaction proportional to the iodide concentration? Explain your answer. __________

b. Plot the persulfate concentration for runs 1 ave., 5, 6, and 7 versus the inverse of the time. Is the rate of the reaction proportional to the persulfate concentration? Explain your answer.

4. (optional)

a. Assuming the rate expression for this system is in the form, rate = $k[I^-]^x[S_2O_8^{2-}]^y$, it should be possible to determine the values of x and y from the data obtained. Show why a plot of log(time) vs log$[I^-]$ for runs 1 - 4 will give a slope of -x and log(time) vs log$[S_2O_8^{2-}]$ for runs 1, 5, 6, 7, will give a slope of -y.

b. Perform the graphing exercises indicated in *4-a* above and determine the values of x and y.

x = ________

y = ________

5. Based on your answer to *3-a,b* and/or *4-b* and the information in the ***Procedure*** section, write down a mechanism consistent with your results and indicate which step is the rate determining step.

6. Did the copper(II) sulfate catalyze the reaction? Explain your answer. ________

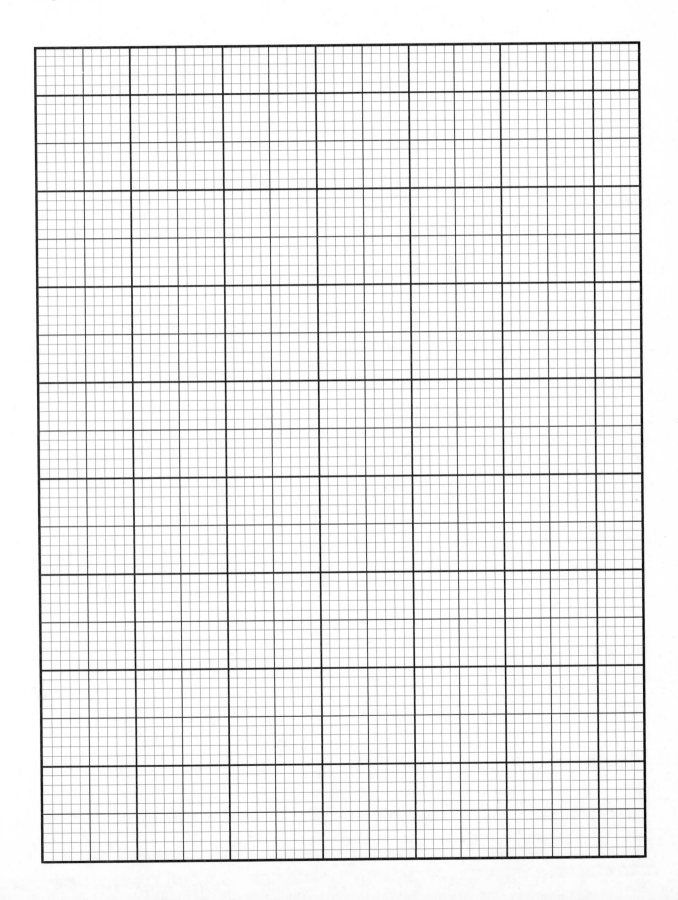

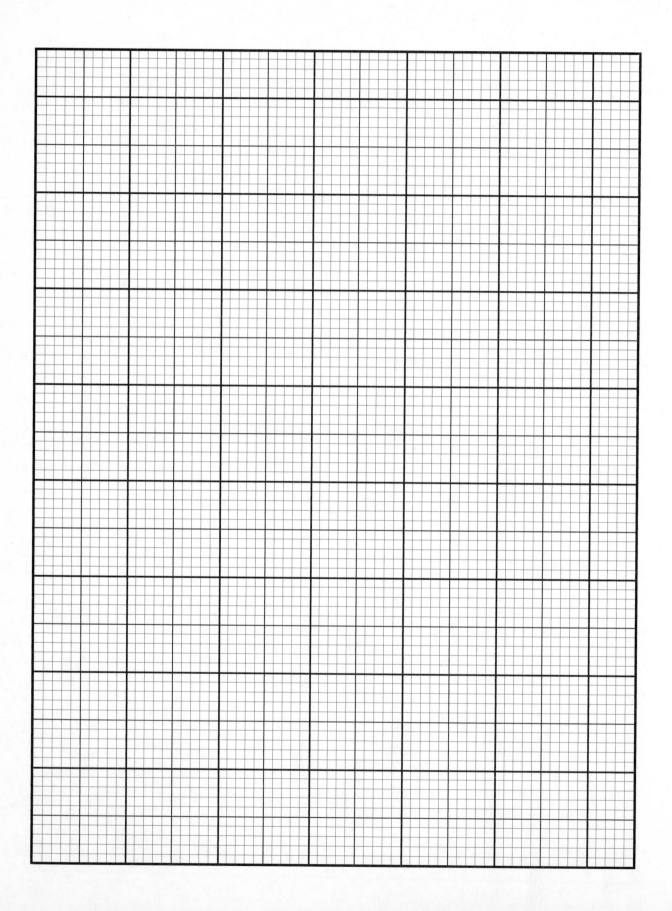

B. Temperature effects on reaction rates.

1. Results

run #	1 (average)	2′	3′	4′
time (sec.)	______	______	______	______
1/time (1/sec.)	______	______	______	______
ln[time]	______	______	______	______
Temp. (°C)	______	______	______	______
Temp. (K)	______	______	______	______
1/Temp. (1/K)	______	______	______	______

2. a. Show that a plot of ln(time) versus 1/Temp. should give a straight line with a slope of E_a/R.

   b. Graph ln(time) vs 1/Temp. and determine the slope. From the slope calculate the activation energy of the reaction.

   slope ________

   E_a ________

3. a. From the graph, determine the reaction time at 20°C and 30°C.

   20°C ________

   30°C ________

   b. For this particular reaction, how good is the rule of thumb that "a 10°C increase should double the reaction rate"? Explain your answer.

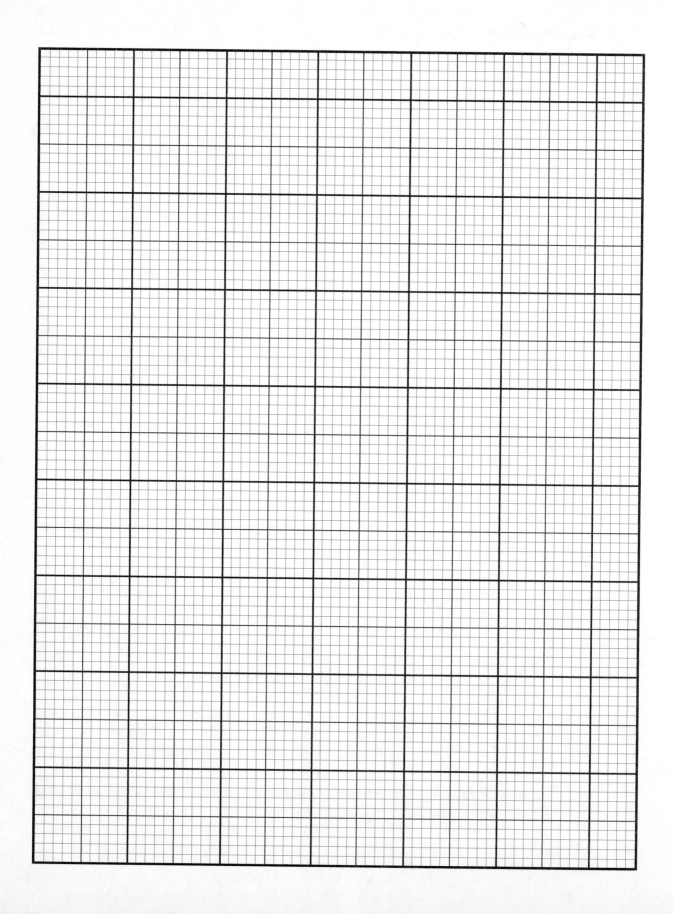

METHOD 2 - Spectroscopic Monitoring A. Concentration effects on reaction rates.

1. Absorption vs time results.

time (sec.)	run 1 A	run 2 A	run 3 A	run 4 A	run 5 A	run 6 A
10						
20						
30						
40						
50						
60						
70						
80						
90						
100						
temp. (°C)						

2. Concentrations and relative rates. Calculate the concentrations that were used for each run. For each run, plot the A value on the vertical axis and the time on the horizontal axis and determine the slope. A spreadsheet considerably facilitates this procedure.

run #	$[I^-]$ (mol/L)	$[S_2O_8^{2-}]$ (mol/L)	slope = relative rate (units/sec.)	temp. (°C)
1				
2				
3				
4				
5				
6				

3. Assuming the rate law for the system is rate = $k[I^-]^x[S_2O_8^{2-}]^y$, what is the value of x for the system? Explain your answer.

4. Assuming the rate law for the system is rate = $k[I^-]^x[S_2O_8^{2-}]^y$, what is the value of y for the system? Explain your answer.

5. Based on your answers to *3* and *4* above and the information in the ***Procedure*** section, write down a mechanism consistent with your results and indicate which step is the rate determining step.

6. a. Does copper(II) catalyze the reaction? Explain your answer.

   b. The copper(II) solution is colored. Did this interfere with the analysis method? Explain your answer.

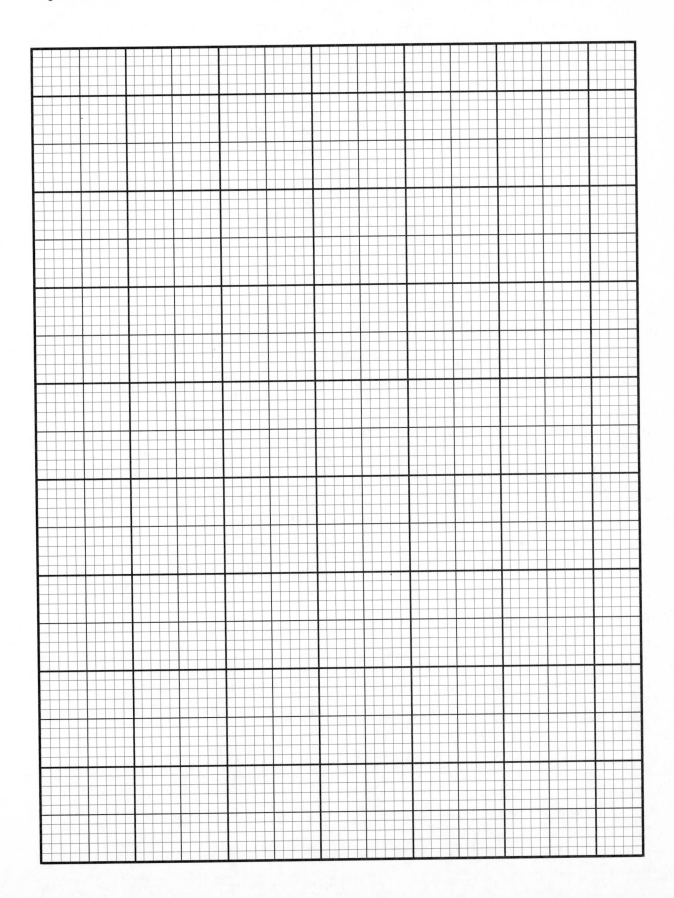

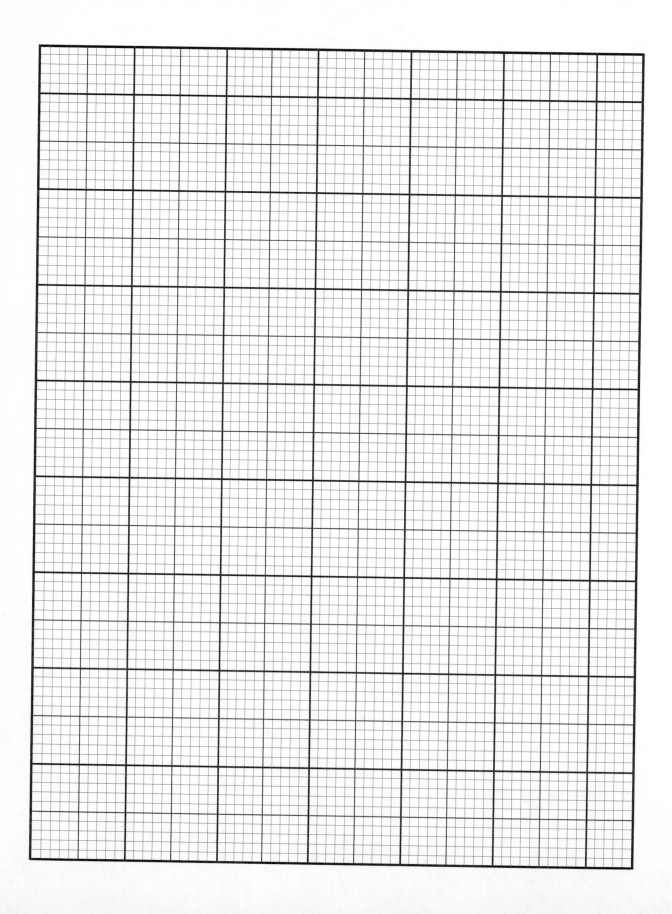

B. Temperature effects on reaction rates.

1. Concentrations.

$[I^-]$ (mol/L) = ____________

$[S_2O_8^{2-}]$ (mol/L) = ____________

2. Absorption vs time at different temperatures.

time (sec.)	run 1' A	run 2' = run 2 A	run 3' A	run 4' A
10				
20				
30				
40				
50				
60				
70				
80				
90				
100				
Temp. (°C) before				
Temp. (°C) after				
Temp. (°C) ave.				
Temp. (K) ave.				
1/Temp. (1/K)				
relative rate (units/sec.)				
ln(relative rate)				

3. a. Show that a plot of ln(relative rate) versus 1/Temp. should give a straight line with a slope of E_a/R.

b. Graph ln(relative rate) vs 1/Temp. and determine the slope. From the slope calculate the activation energy of the reaction.

slope ________

E_a ________

4. a. From the graph, determine the relative reaction rates at 20°C and 30°C.

20°C ________

30°C ________

b. For this particular reaction, how good is the rule of thumb that "a 10°C increase should double the reaction rate"? Explain your answer.

5. Suggest any ways you can think of to improve any part(s) of this experiment.

6. Some of the *Learning Objectives* of this experiment are listed on the first page of this experiment. Did you achieve the *Learning Objectives*? Explain your answer.

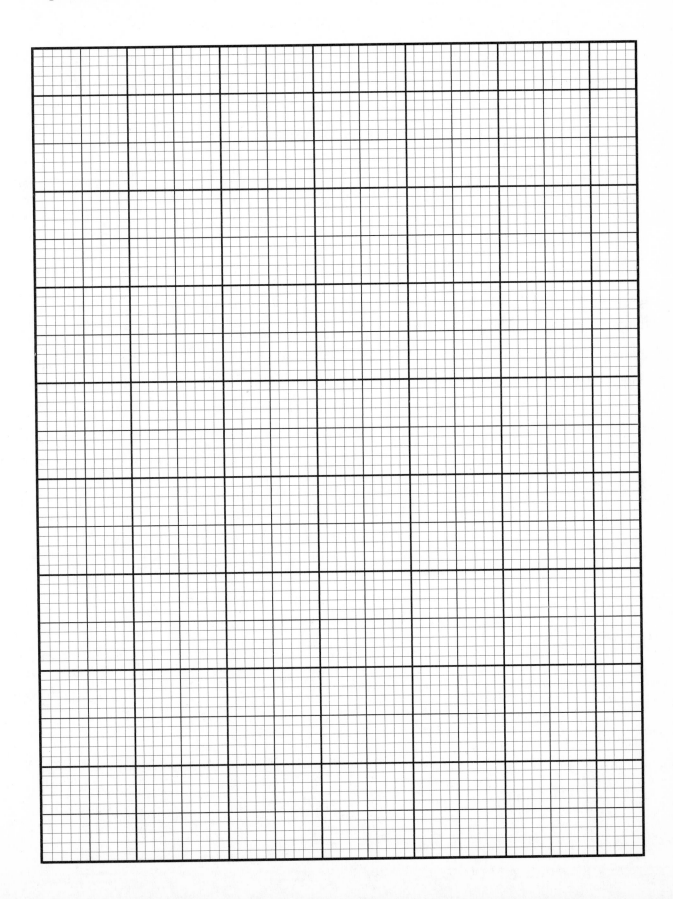

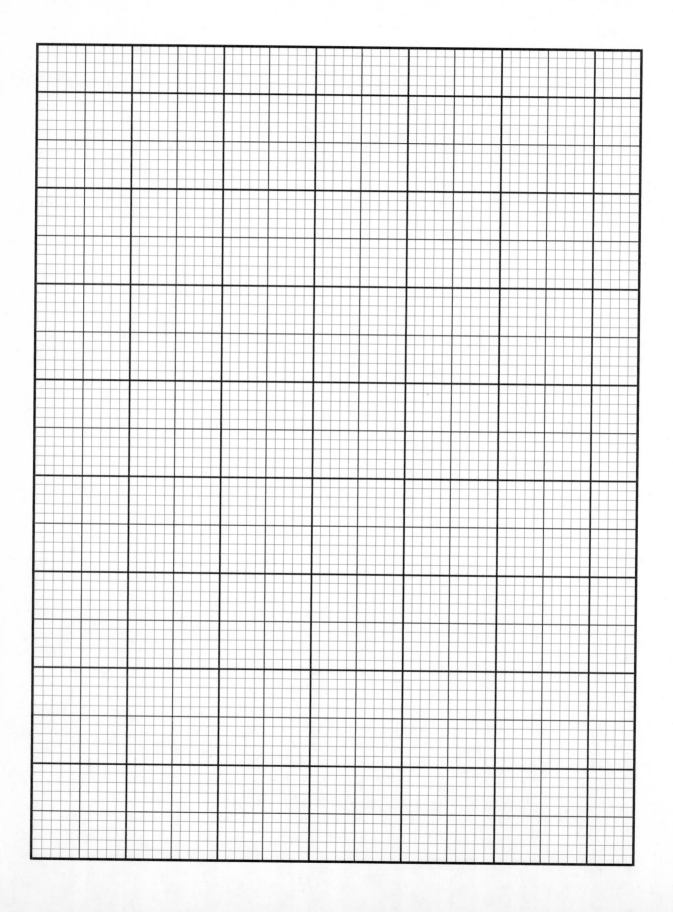

Experiment 29

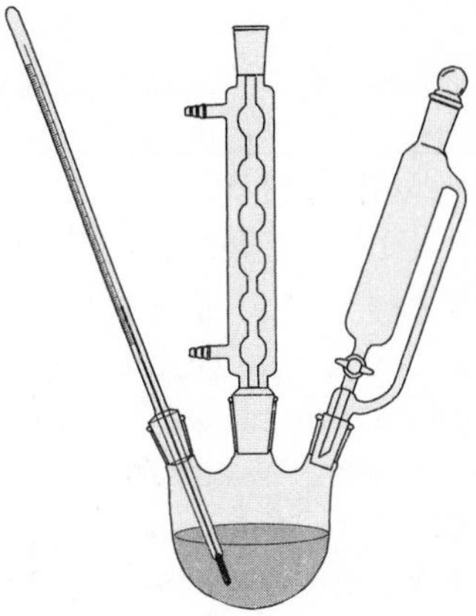

SYNTHESIS OF COPPER(II) GLYCINATE

Learning Objectives

Upon completion of this experiment, students will have experienced:

1. Techniques of chemical synthesis.
2. Concepts of geometric isomerism.
3. Determination of percent yield.

Text Topics

Stoichiometry, percent yield, isomerism, coordination compounds (for correlation with some textbooks, see page ix).

Notes to Students and Instructor

This experiment is short (about 1 hour) and can be conveniently performed the same day as ***Experiment 30***. The product is analyzed in ***Experiment 31***.

Discussion

Imagine the mindset of the great Leonardo Da Vinci about to paint the "The Last Supper" or of Frank Lloyd Wright about to design one of his unique and beautiful buildings. The challenges faced by these artisans are similar to those confronted by an organic chemist when he/she sets out to synthesize a complex molecule. For complex molecules, one approach involves the synthesis of fragments followed by the piecing together of the fragments. A series of reactions must be planned from start to finish that delicately puts various pieces of the molecule together without disturbing other parts of the molecule. For example, in 1972, R. B. Woodward and a team of postdoctoral fellows and graduate students were finally able after a sequence of many steps to complete the total synthesis of Vitamin B_{12}. Consider how many different parts there are, just as in a building. Work on one section can easily influence other sections unless the work is carefully planned and performed. The synthesis of Vitamin B_{12} could clearly be called a work of art.

One of the major goals of chemistry is synthesis. Chemists use their creative talents to make better polymers, medicines and pesticides. Pencillin G (the first penicillin discovered and tested) was not very effective when taken orally. It had a tendency to decompose rapidly in the acidic media of the stomach. As syringes are inconvenient and feared by most, organic chemists set out to synthesize a penicillin like structure that would maintain its antibiotic properties but be stable in the stomach. Ampicillin was one of the successful synthetic results (what happens to ampicillin in acidic media?).

Fig. 29 - 1

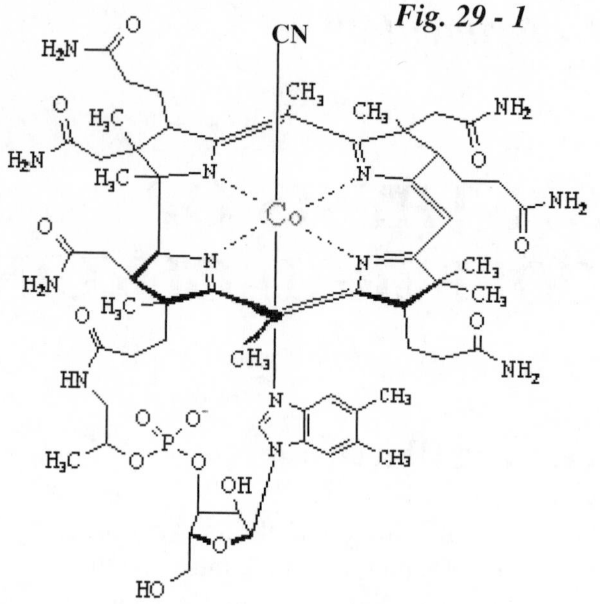

Pencillin G

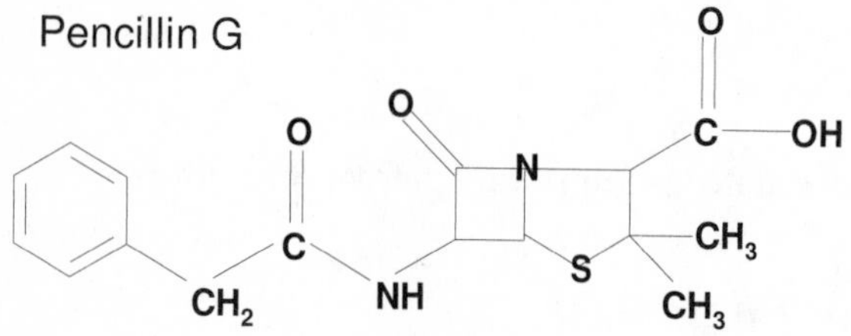

Ampicillin

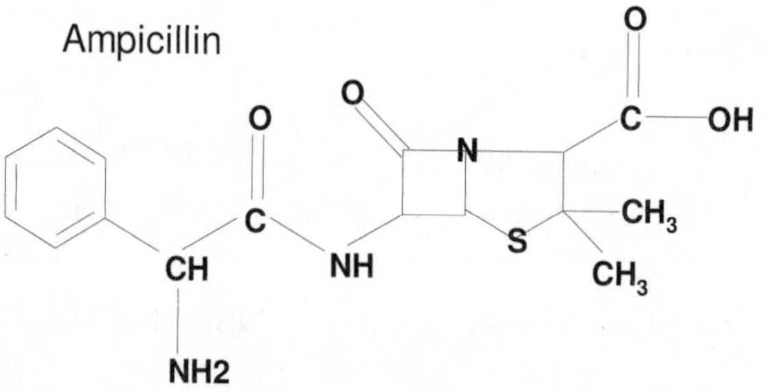

Fig. 29 -2

Morphine is another interesting challenge faced by pharmaceutical chemists. It has excellent pain killing (analgesic) properties but unfortunately is addictive. As morphine has several possible sites where it can interact with physiologically active sites in the body, it was (and still is) hoped that a molecule could be synthesized that would have the analgesic site but not the site responsible for addiction. To date only partial success has been achieved with such drugs as demerol and darvon (unfortunately both are still addictive).

Fig. 29 - 3

Morphine

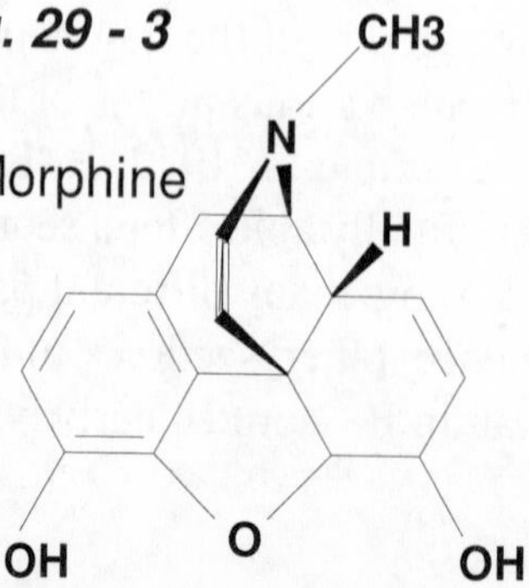

Demerol

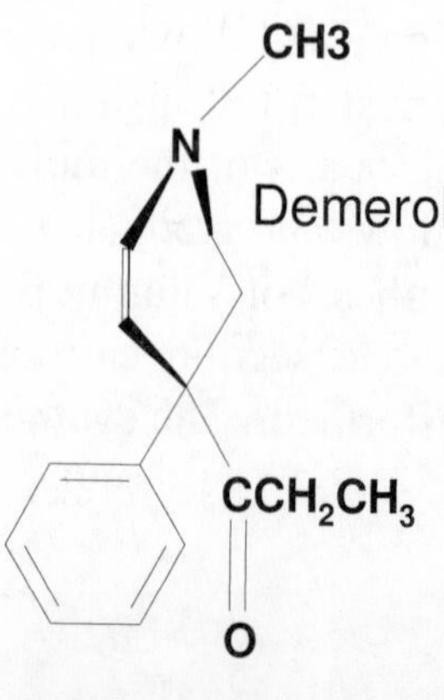

To the chemical industry, a synthesis should be economically beneficial. Costs of chemicals, reaction efficiency and waste disposal costs are a few of the economic considerations. Reaction efficiency is related to the percent yield which you can determine today by dividing the amount experimentally obtained by the theoretical amount and multiplying the result by 100%.

A chemical synthesis is usually followed by appropriate purification and identification techniques. In ***Experiment 31*** you will perform analytical techniques to determine the success of your synthesis.

Today's synthesis, unlike the vitamin B_{12} synthesis, involves only one step. The desired product, copper(II) glycinate monohydrate, like vitamin B_{12}, is a coordination compound between a metal and an organic substrate. The synthesis of copper(II) glycinate monohydrate can be viewed as a double replacement reaction between copper(II) acetate and the simplest of the amino acids, glycine.

$$Cu(C_2H_3O_2)_2(aq) + 2\,HC_2H_4NO_2(aq) = Cu(C_2H_4NO_2)_2(s) + 2\,HC_2H_3O_2(aq)$$

Glycine is one of the 20 amino acids that can link together to form proteins. It is actually related very closely to the acetic acid it replaces in the reaction. In the coordination compound that results, the copper is literally surrounded by the two glycines and one water.

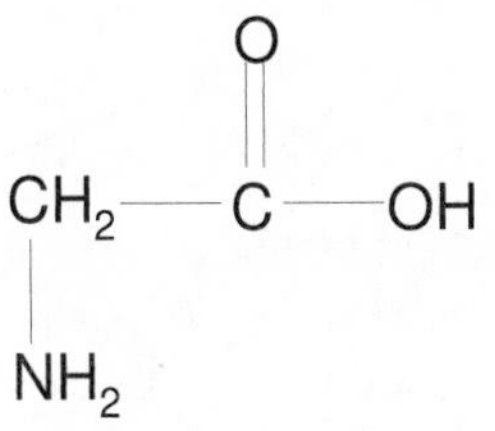

glycine

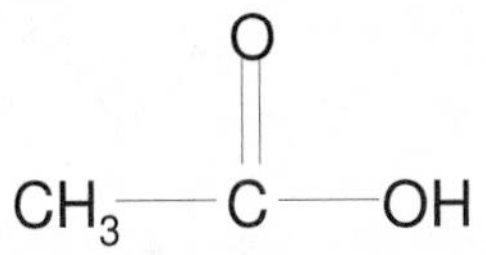

acetic acid

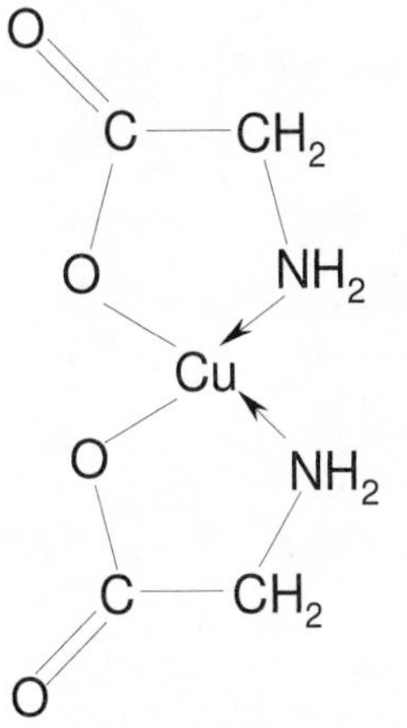

cis-copper(II) glycinate

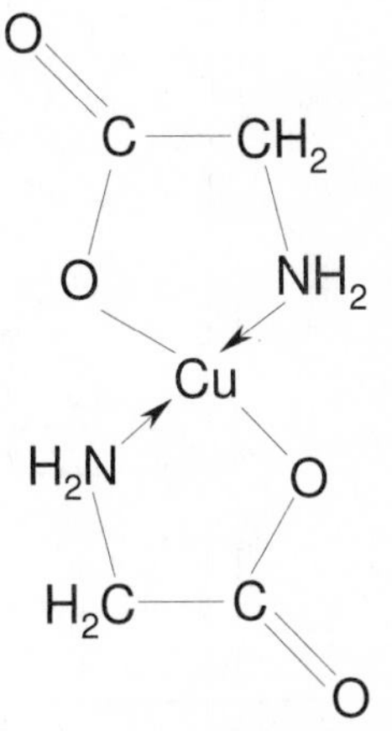

trans-copper(II) glycinate

Notice that there is a square planar configuration of the four atoms around the copper and there are two geometrically different ways of positioning the glycines. Compounds which have the same formula but different structures are called isomers and the special type of isomerism present here is called geometric isomerism. Your synthesis could yield one or the other or a mixture of both. Experimental evidence indicates that in this synthesis the product is exclusively the *cis* isomer. The *cis* isomer can be converted to the *trans* isomer by heating.

Procedure

Weigh about 1.6 grams of copper(II) acetate monohydrate to the limits of your balance into a 250 mL beaker and add 20 mL of deionized water. Weigh out about 1.3 g of glycine into a 150 mL beaker and add 15 mL of deionized water. Place both beakers on a hot plate and heat to the boiling point. Stir occasionally while heating until both solids dissolve. After the solids dissolve and the solutions are near the boiling point, **carefully** remove the beakers (use beaker tongs) from the hot plate and add the glycine solution to the copper(II) acetate solution and stir. Allow the solution to cool for several minutes and place the beaker in an ice bath. After crystals begin to form, add 20 mL of 1-propanol with continuous stirring. Continue to cool for several minutes and vacuum filter (use a Buchner funnel and a filter flask - see *Figure 29-4*) to collect the product. Wash the beaker (make sure all solid is transferred to the funnel) with acetone (**keep away from flames**) twice and add the wash liquid to the funnel. Pull air through the sample for a few minutes and then transfer the solid to a weighed piece of filter paper. Allow the sample to dry for a few days and then weigh it to determine the experimental and percent yields. Save the sample for analysis in ***Experiment 31***.

Fig. 29-4

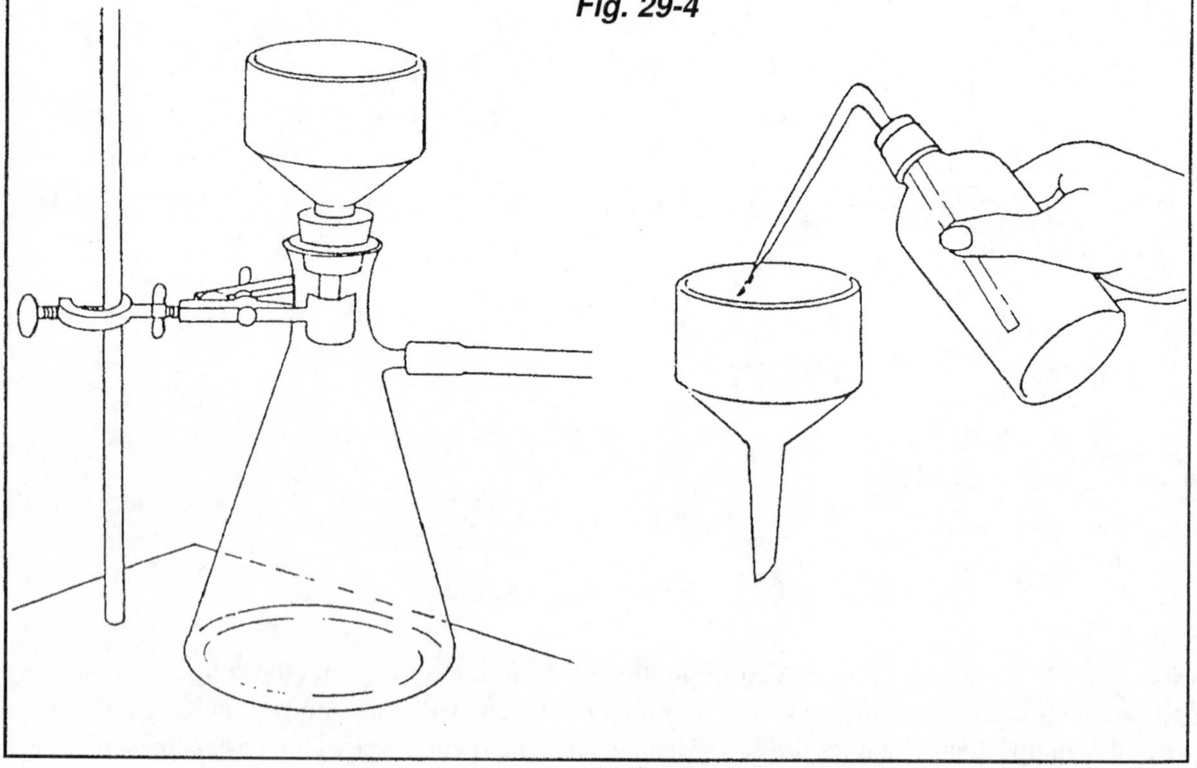

Name________________________________Date_________Lab Section_________

Prelaboratory Problems - *Experiment 29* - Synthesis of Copper(II) Glycinate

The solutions to the starred problems are in ***Appendix 4***.

1.* 12.0 grams of salicylic acid were reacted with an excess of acetic anhydride and 11.0 grams of aspirin (acetylsalicylic acid) were obtained. Calculate the theoretical and percent yields of aspirin.

$$C_7H_6O_3 \quad + \quad C_4H_6O_3 \quad = \quad C_9H_8O_4 \quad + \quad C_2H_4O_2$$

salicylic acid	acetic anhydride	aspirin	acetic acid

2. 25 mL of a 0.50 M $BaCl_2$ solution is mixed with 25 mL of a 0.50 M Na_2CO_3 solution and 2.0 g of $BaCO_3$ is collected. What is the percent yield of $BaCO_3$?

3. Double replacement reactions can be used in several ways to synthesize desired products.

If the desired product is insoluble in water and two soluble reactants can be obtained that give the desired substance as one of the products, filtration can be used to isolate the product. For example, silver chloride could be prepared using the following reaction:

$$AgNO_3(aq) \quad + \quad NaCl(aq) \quad = \quad AgCl(s) \quad + \quad NaNO_3(aq)$$

If the desired product is soluble in water and two soluble reactants can be obtained that yield an insoluble product plus the desired product, evaporation of the filtrate will give the desired product. Stoichiometric amounts of the reactants must be used. For example, sodium nitrate could be prepared using the preceding reaction.

If an acid and a carbonate can be used, carbon dioxide, water and the desired product will be formed. If the product is soluble in water, stoichiometric amounts of the reactants must be used. For example, sodium sulfate could be prepared using the following reaction:

$$Na_2CO_3(aq) + H_2SO_4(aq) = H_2O(l) + CO_2(g) + Na_2SO_4(aq)$$

If an acid and a base can be used, water and the desired product will be formed. If the product is soluble in water, stoichiometric amounts of the reactants must be used. For example, sodium sulfate could be prepared using the following reaction:

$$2\,NaOH(aq) + H_2SO_4(aq) = 2\,H_2O(l) + Na_2SO_4(aq)$$

a. Why must stoichiometric amounts be used for the cases described above where the product is soluble in water?

b. Why is the synthesis of sodium nitrate suggested above of little value?

Suggest pairs of reactants you could use in double replacement reactions to synthesize the compounds below. Include sufficient detail to make it clear how the product would be isolated (e.g., mix equal molar solutions of A and B, filter, evaporate the filtrate). ***Appendix 3*** might be useful for this exercise.

c.* calcium oxalate

d. lithium iodide

e. strontium acetate

Name____________________________________Date__________Lab Section__________

Results and Discussion - *Experiment 29* - Synthesis of Copper(II) Glycinate

$$Cu(C_2H_3O_2)_2 \cdot H_2O + 2\,HC_2H_4NO_2 = Cu(C_2H_4NO_2)_2 \cdot H_2O + 2\,HC_2H_3O_2$$

1. Mass of copper(II) acetate monohydrate __________
2. Formula mass of copper(II) acetate monohydrate __________
3. Moles of copper(II) acetate monohydrate __________
4. Mass of glycine __________
5. Formula mass of glycine __________
6. Moles of glycine __________
7. Formula mass of copper(II) glycinate monohydrate __________
8. Theoretical yield of copper(II) glycinate monohydrate based on copper(II) acetate monohydrate (show series of unit conversions) __________
9. Theoretical yield of copper(II) glycinate monohydrate based on glycine (show series of unit conversions) __________
10. Limiting reagent __________
11. Mass of filter paper __________
12. Mass of filter paper + product __________
13. Mass of copper(II) glycinate monohydrate __________
14. Percent yield of copper(II) glycinate monohydrate __________

15. Suggest any ways you can think of to improve any part(s) of this experiment.

16. Some of the *Learning Objectives* of this experiment are listed on the first page of this experiment. Did you achieve the *Learning Objectives*? Explain your answer.

Experiment 30

STANDARDIZATION OF THIOSULFATE

Learning Objectives

Upon completion of this experiment, students will have experienced:
1. Standardization of a sodium thiosulfate solution.
2. Stoichiometric calculations.

Text Topics

Stoichiometry, molarity, oxidation-reduction reactions, titrations (for correlation with some textbooks, see page ix).

Notes to Students and Instructor

This experiment should take less than two hours and may conveniently be performed in the same laboratory period as ***Experiment 29***.

Discussion

In ***Experiment 11***, you utilized titration techniques to determine the stoichiometry of a reaction and the concentration of potassium ferrocyanide in an unknown solution. Review the discussion, in that experiment, of titration techniques and preparation of standard solutions. Today's experiment utilizes an oxidation-reduction reaction to standardize a solution of sodium thiosulfate. This solution will be used in ***Experiment 31*** to determine the concentration of the active chemical in bleach and the mass percent of copper in the copper(II) glycinate monohydrate prepared in ***Experiment 29***.

Many compounds used for titration, such as sodium hydroxide are either too impure or too hygroscopic to be weighed accurately. In order to find the concentration of a sodium hydroxide solution accurately, it is necessary to titrate it against a primary standard. Primary standards are compounds that are certified to be at least 99.9% pure and are not hygroscopic. Sodium thiosulfate is not a primary standard so in order to determine the concentration of a sodium thiosulfate solution accurately, it must be titrated against one of several primary standards. In the procedure below, potassium ferricyanide is used because of its relatively high equivalent mass. Ferricyanide is reduced to ferrocyanide while iodide is oxidized to iodine (the iodine produced complexes with excess iodide present to form the more soluble triiodide ion).

$$2\,Fe(CN)_6^{3-} + 3\,I^- = 2\,Fe(CN)_6^{4-} + I_3^-$$

The triodide produced in the reaction is titrated with the sodium thiosulfate solution.

$$I_3^- + 2\,S_2O_3^{2-} = S_4O_6^{2-} + 3\,I^-$$

Zinc sulfate solution is added to remove ferrocyanide as it is produced (the presence of appreciable quantities of ferrocyanide ion reduces the accuracy of the determination). A starch solution is added to make the endpoint much more distinct. The overall net ionic equation is:

$$2\,Fe(CN)_6^{3-} + 2\,S_2O_3^{2-} = S_4O_6^{2-} + 2\,Fe(CN)_6^{4-}$$

Note that the iodide is converted to iodine (in the form of triiodide) and back to iodide. Thus the role of iodide is important but it does not show up in the overall reaction or the stoichiometric calculations. The net change shows that there is a one-to-one mole ratio between thiosulfate and ferricyanide.

Procedure

Calculate to the nearest 0.1 gram the amount of sodium thiosulfate pentahydrate needed to prepare 300 mL of a 0.11 M solution. Dissolve the calculated amount in about 300 mL of deionized water in a 500 mL Erlenmeyer flask, swirl and stopper the flask.

Weigh into a 250 mL Erlenmeyer flask to at least the nearest 0.001 g, about 0.6 g of primary standard potassium ferricyanide. Dissolve it in 25 mL of deionized water. In a beaker, dissolve about 1.5 g of potassium iodide in 10 mL of deionized water and add 1 mL of 6 M hydrochloric acid. Add this solution to the potassium ferricyanide solution, stopper the flask and put it in the dark (a closed locker will suffice) for a couple of minutes. Now add 12 mL of 1.0 M zinc sulfate and swirl the mixture. Titrate with the sodium thiosulfate solution you prepared until the iodine color begins to fade. Then add 2 mL of 0.4% starch indicator and titrate until the color changes abruptly from gray to white.

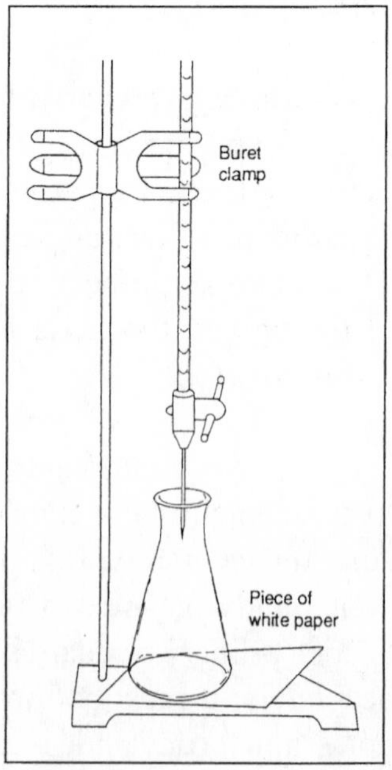

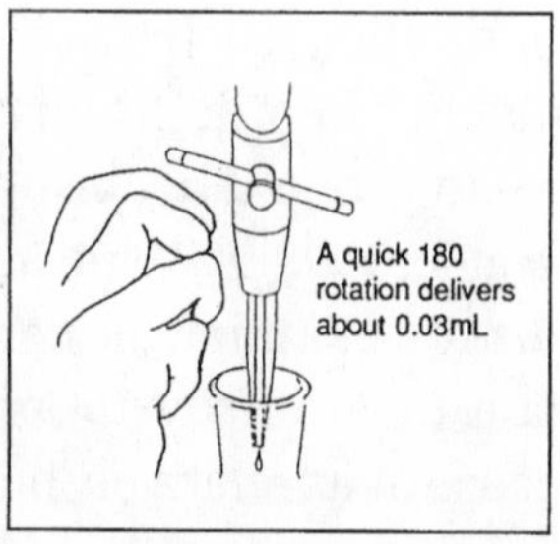

Repeat the titration with two more samples of potassium ferricyanide. Calculate the concentration of the sodium thiosulfate solution and label the flask containing the sodium thiosulfate solution with the identity and concentration of its contents. Save the solution for the titrations in ***Experiment 31***.

Name__________________________________Date__________Lab Section_________

Prelaboratory Problems - *Experiment 30* - Standardization of Thiosulfate

The solutions to the starred problems are in ***Appendix 4***.

1. Determine the oxidation numbers of the elements indicated.

   a.* manganese in $MnCl_2$, MnO_2, MnO_4^-

   _____ $MnCl_2$ _____ MnO_2 _____ MnO_4^-

   b. nitrogen in nitrate, nitrogen dioxide, nitrite, nitrogen monoxide, dinitrogen monoxide, nitrogen and ammonia

   _____ nitrate   _____ nitrogen dioxide   _____ nitrite   _____ nitrogen monoxide   _____ dinitrogen monoxide   _____ nitrogen   _____ ammonia

   c. arsenic in As_2O_3, As_2O_5, K_3AsO_4

   _____ As_2O_3 _____ As_2O_5 _____ K_3AsO_4

   d. oxygen in H_2O, H_2O_2

   _____ H_2O _____ H_2O_2

2. Write balanced half reactions and complete reactions for the following oxidation-reduction reactions.

   a.* aluminum(s) + hydrogen ion(aq) → aluminum ion(aq) + hydrogen(g)

   oxid. half rxn.:

   red. half rxn.:

   overall rxn.:

   b. chloride + permanganate → manganese(II) ion + chlorine(g) (aqueous acid)

   oxid. half rxn.:

   red. half rxn.:

   overall rxn.:

c. tin(II) ion + dichromate → tin(IV) ion + chromium(III) ion (aqueous acid)

oxid. half rxn.:

red. half rxn.:

overall rxn.:

3. For problem *2-c*, give answers to the following:

What is oxidized? ________

What is reduced? ________

What is the oxidizing agent? ________

What is the reducing agent? ________

4. Calculate the approximate amount of sodium thiosulfate pentahydrate needed to prepare 500 mL of a 0.20 M solution.

5.* A 0.800 gram sample of potassium ferricyanide is titrated with a sodium thiosulfate solution according to the procedure given above. 19.00 mL of the sodium thiosulfate solution are required to reach the starch end point. What was the concentration of the sodium thiosulfate solution?

6. For the titration in this experiment, explain how the indicator functions.

Name________________________________Date__________Lab Section_________

Results and Discussion - *Experiment 30* - Standardization of Thiosulfate

1. Write half reactions for the following:

   a. Oxidation of iodide to triiodide

   b. Reduction of ferricyanide to ferrocyanide

   c. Oxidation of thiosulfate to tetrathionate

2. Calculations

   a. Formula mass of $Na_2S_2O_3 \cdot 5H_2O$ ________

   b. Mass of $Na_2S_2O_3 \cdot 5H_2O$ (to prepare 300 mL of 0.11 M solution) ________

   c. Formula mass of $K_3Fe(CN)_6$ ________

	1st	2nd	3rd
d. Mass of $K_3Fe(CN)_6$	________	________	________
e. Moles of $K_3Fe(CN)_6$	________	________	________
f. Final buret reading	________	________	________
g. Initial buret reading	________	________	________
h. Volume $Na_2S_2O_3$	________	________	________
i. Molarity of $Na_2S_2O_3$	________	________	________

   j. Average molarity of $Na_2S_2O_3$ ________

   k. Show calculation for one of the values in "i"

[Note: Record the molarity from "j" in the appropriate blanks on pages 385 and 387 in ***Experiment 31***]

3. Suggest any ways you can think of to improve any part(s) of this experiment.

4. Some of the *Learning Objectives* of this experiment are listed on the first page of this experiment. Did you achieve the *Learning Objectives*? Explain your answer.

Experiment 31

ANALYSIS OF BLEACH AND COPPER(II) GLYCINATE

Learning Objectives

Upon completion of this experiment, students will have experienced:

1. Iodometric techniques in quantitative analysis.
2. Stoichiometric calculations involving solutions.

Text Topics

Stoichiometry of solution reactions, oxidation-reduction reactions, titrations (for correlation with some textbooks, see page ix)

Discussion

Do you believe the information supplied on the labels of commercial products? Today you will measure the percentage by mass of sodium hypochlorite (NaClO) in commercial household bleaches and compare your experimental values to the values on the label of the bleaches. The values will be measured by titrating the bleaches with the standardized sodium thiosulfate that you prepared in ***Experiment 30***. The same titration technique will also be used to analyze the product of your attempted synthesis of copper(II) glycinate from ***Experiment 29.*** An additional option is the determination of the percent by mass of copper in an unknown salt.

A. Bleach analysis. The oxidation number of the chlorine in the active ingredient in household bleach, NaClO, is +1. Because a change of oxidation state from +1 to -1 is energetically favorable for chlorine, the ClO^- is a strong oxidizing agent. It is the ability of ClO^- to remove electrons from colored compounds that results in its effectiveness as a bleaching agent. The determination of the concentration of NaClO in bleach also depends on its drive to be reduced. The hypochlorite ion will be reacted with iodide to yield chloride and triiodide.

$$H_2O(l) + OCl^- + 3I^- = I_3^- + Cl^- + 2OH^-$$

The triiodide is then reduced back to iodide by titration with the standardized thiosulfate solution.

$$I_3^- + 2S_2O_3^{2-} = 3I^- + S_4O_6^{2-}$$

The iodide ion thus does not undergo any net change in the process but assures that the correct stoichiometry occurs and provides a visually detectable endpoint. The overall reaction for calculation purposes is:

$$H_2O(l) + 2\,S_2O_3^{2-} + OCl^- = S_4O_6^{2-} + Cl^- + 2\,OH^-$$

Although the gradual disappearance of the triiodide color can be used to approximate the endpoint for the titration, a much sharper endpoint is achieved by the addition of a starch solution after the iodine color begins to fade. Iodine complexes with starch to form a dark purple color. When this color disappears, all of the triiodide has been reduced back to iodide. Using the molarity and volume of the sodium thiosulfate solution used to reach the endpoint, and the volume of bleach solution titrated, it is possible to calculate the percent of NaClO by mass in the bleach.

B. Copper(II) glycinate monohydrate analysis. A very similar procedure will be used to determine the percent by mass of copper in the product of the attempted synthesis of copper(II) glycinate monohydrate from ***Experiment 29***. In addition, or alternatively, the mass percent of copper in an unknown can be determined using the technique below. The copper(II) compound is dissolved in dilute acid and reacted with excess potassium iodide. This results in the production of triiodide according to the following equation:

$$2\,Cu^{2+} + 5\,I^- = 2\,CuI(s) + I_3^-$$

The triiodide produced in this reaction is titrated with standardized thiosulfate as in the bleach titration,

$$I_3^- + 2\,S_2O_3^{2-} = 3\,I^- + S_4O_6^{2-}$$

and the overall reactions is:

$$2\,Cu^{2+} + 2\,I^- + 2\,S_2O_3^{2-} = 2\,CuI(s) + S_4O_6^{2-}$$

Procedure

A. Bleach analysis. This analysis should be performed on two different brands of bleach; preferably one name brand and one generic. Weigh a 100 mL volumetric flask to at least the nearest 0.01 g. Pipet 10.00 mL of the first bleach into the 100 mL volumetric flask and reweigh the flask. The masses will be used to determine the density of bleach. Dilute to the 100 mL mark with deionized water and mix well. Rinse the pipet out with a little of the diluted bleach solution twice and pipet a 10.00 mL aliquot of it into a 250 mL Erlenmeyer flask. Add 1.0 g of potassium iodide and swirl the resulting mixture. Now add 5.0 mL of 6 M HCl to the mixture. ***[Caution: Make sure you add potassium iodide to the bleach solution before you add the hydrochloric acid. Addition of the acid first results in liberation of poisonous chlorine gas.]***

Titrate with the standardized sodium thiosulfate (~0.11 M) solution until the amber iodine color begins to fade. At this point add 2 mL of 0.4% starch solution and resume titrating until the dark color of the starch-iodine complex just disappears. Be sure to use ½ drop quantities as you near the endpoint by swiftly turning the stopcock 180°. An abrupt color change from dark blue to colorless marks the endpoint. Repeat the titration on a second 10.00 mL aliquot of the diluted bleach solution. Repeat the analysis with a second brand of bleach.

B. Copper(II) glycinate monohydrate analysis. Accurately weigh about 0.3 g to at least the nearest 0.001 g of the product from ***Experiment 29*** [hopefully copper(II) glycinate monohydrate] or a copper(II) unknown into a 250 mL Erlenmeyer flask and dissolve it in 35 mL of 0.05 M sulfuric acid. Add 1.0 g of potassium iodide and swirl the mixture. Titrate the mixture with standardized sodium thiosulfate until the triiodide color begins to fade. At this point, add about 2 mL of 0.4% starch solution. Continue to titrate until the gray starch-iodine color disappears. An abrupt transition from a gray-yellow color to a light pink color will mark the endpoint. Repeat the titration on a second 0.3 g sample of the complex or unknown. Calculate the percent by mass of copper in your sample. If you analyzed the synthesis product, copper(II) glycinate monohydrate, calculate the mass percent of copper from the formula and compare the theoretical and experimental values. Was your synthesis successful?

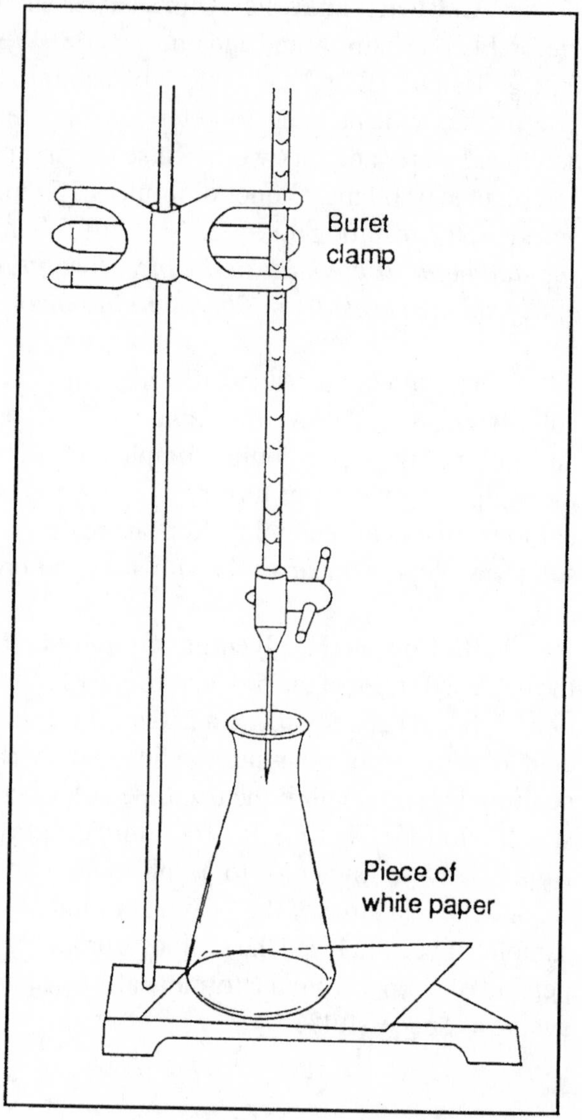
Buret
clamp
Piece of
white paper

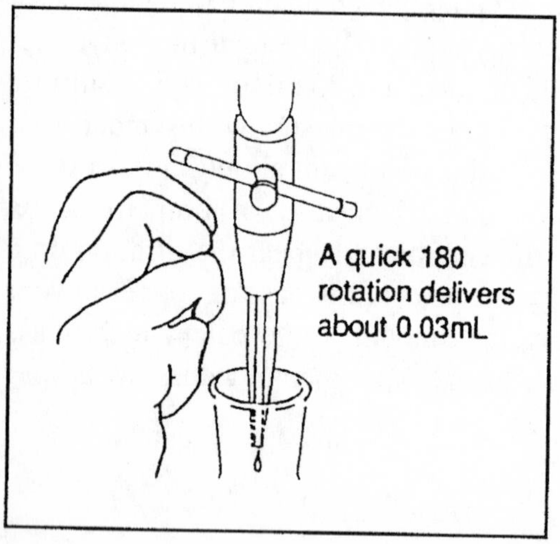
A quick 180
rotation delivers
about 0.03mL

Name________________________________Date__________Lab Section__________

Prelaboratory Problems - *Experiment 31* - Analysis of Bleach and Copper(II) Glycinate

The solutions to the starred problems are in ***Appendix 4***.

1.* Standardized potassium permanganate solutions are often used as the oxidizing agent in titrations. They are often standardized by titrating weighed out amounts of sodium oxalate. The products in acidic solution are manganese(II) and carbon dioxide.

a. Write oxidation and reduction half reactions and the net ionic equation for the above system.

b. 0.2600 g of sodium oxalate require 22.50 mL of a potassium permanganate solution to reach the endpoint. What is the concentration of the permanganate solution?

____

c. One of the advantages of permanganate titrations is that the titrating agent has its own built in indicator. Explain this statement.

2. The concentration of hydrogen peroxide in a solution can be determined by titration with permanganate solution in an acidic solution. The products are manganese(II) and oxygen.

a. Write oxidation and reduction half reactions and the net ionic reaction for the above system.

b. 1.900 grams of a hydrogen peroxide solution requires 14.65 mL of a 0.02500 M potassium permanganate solution to reach the endpoint. Calculate the mass percentage of hydrogen peroxide in the sample.

3.* A 1.00 mL sample of pool "chlorine" (NaClO, density = 1.18 g/mL) required 21.12 mL of 0.150 M $Na_2S_2O_3$ to titrate it to the endpoint using the iodometric technique. What was the molarity and percent of sodium hypochlorite in the pool chlorine?

4. A 0.250 g sample believed to be copper(II) acetate monohydrate required 11.84 mL of 0.105 M $Na_2S_2O_3$ to reach the endpoint in an iodometric titration. Calculate the experimental percent by mass of copper in the sample and the theoretical mass percent of copper in copper(II) acetate monohydrate.

Name________________________________Date__________Lab Section_________

Results and Discussion - *Experiment 31* - Analysis of Bleach and Copper(II) Glycinate

A. Bleach Analysis

1. Molarity of $Na_2S_2O_3$ solution (from ***Expt. 30***) ________

	Bleach 1		*Bleach 2*	
2. Brand	________		________	
3. Mass of vol. flask + 10 mL bleach	________		________	
4. Mass of vol. flask	________		________	
5. Mass of 10 mL bleach	________		________	
6. Density of bleach	______	______	______	______
7. Final buret reading	______	______	______	______
8. Initial buret reading	______	______	______	______
9. Volume $Na_2S_2O_3$ solution	______	______	______	______
10. Moles of $Na_2S_2O_3$	______	______	______	______
11. Moles NaClO in 10 mL aliquot	______	______	______	______
12. Molarity of NaClO in 10 mL aliquot (= molarity of NaClO in 100 mL flask)	______	______	______	______
13. Molarity of NaClO in bleach	______	______	______	______
14. Average molarity of NaClO in bleach	________		________	
15. Grams/L of NaClO in bleach	________		________	
16. Mass percent of NaClO in bleach	________		________	
17. Mass percent of NaClO from manufacturer's label	________		________	
18. Percent deviation between exptl. and label values	________		________	

19. Show the series of unit conversions you could use to calculate the molarity of the NaClO in the 100 mL flask from the volume and molarity of the $Na_2S_2O_3$ solution, the molar ratio and the volume of the NaClO for titration 1 of the first bleach.

20. Show the series of unit conversions you could use to calculate the mass percent of NaClO in bleach 1 from the average molarity, the formula mass of NaClO and the density of bleach.

B. Analysis of copper(II) glycinate or an unknown

1. Molarity of $Na_2S_2O_3$ solution (from ***Expt. 30***) ________

	Sample 1	Sample 2
2. Mass of copper sample	________	________
3. Final buret reading	________	________
4. Initial buret reading	________	________
5. Volume $Na_2S_2O_3$ solution	________	________
6. Moles of $Na_2S_2O_3$	________	________
7. Moles of Cu^{2+}	________	________
8. Grams of Cu^{2+}	________	________
9. Mass percent of copper in sample	________	________

10. Average mass percent of copper in sample ________

11. Theoretical mass percent of copper in sample ________
(formula mass of copper(II) glycinate monohydrate = 229.66 g/mol)

12. Percent error ________

13. Show the series of unit conversions you could use to calculate the mass percent of copper in copper(II) glycinate monohydrate for sample 1 from the volume and molarity of the $Na_2S_2O_3$ solution, the mass of the copper(II) glycinate monohydrate, the molar ratio and the atomic mass of copper.

14. In a previous experiment, copper(II) was analyzed by a quantitative precipitation method (see ***Experiment 7***). As copper(II) is a colored ion, it also should be possible to use visible spectroscopy to determine the concentration of copper(II). Compare the method used in this experiment (titration) to the other two methods (quantitative precipitation and spectroscopy) and discuss the applicability (e.g., under what conditions would each be preferable) of each method..

15. Suggest any ways you can think of to improve any part(s) of this experiment.

16. Some of the *Learning Objectives* of this experiment are listed on the first page of this experiment. Did you achieve the *Learning Objectives*? Explain your answer.

Experiment 32

REDOX REACTIONS

Learning Objectives

Upon completion of this experiment, students will have experienced:

1. The application of reduction potentials for the prediction of redox reactions.
2. The observation of several commonly encountered oxidation-reduction reactions.

Text Topics

Balancing of oxidation-reduction reactions, reduction potentials, half reactions (for correlations to some textbooks, see page ix).

Notes to Students and Instructor

It is important that students prepare for this experiment by determining electrochemical potentials for each reaction that will be tested. For those reactions that are predicted to be spontaneous, predictions of expected observations should be made. The lab manipulations of this experiment should only consume about 1 hour.

Discussion

Rusting, bleaching, electroplating, photosynthesizing and burning: what do these chemical changes have in common? They are all examples of processes that involve oxidation-reduction reactions. Electron transfer is the characteristic feature that each of the examples has in common. Identifying oxidation-reduction or redox reactions is a relatively straightforward process. If any of the reactants changes oxidation number during the reaction, the reaction is a redox reaction. As oxidation is a loss of electrons; the element that undergoes an increase in oxidation number (e.g., 2 to 3, 0 to 1, -2 to 0) is oxidized. Reduction is characterized by a decrease in oxidation number (e.g., 2 to 1, 2 to 0, 0 to -1). In some cases, redox is detectable by quick inspection. **If either a reactant or product is in elemental form, the reaction is of necessity a redox reaction.** The only way an element can react or be formed is with a transfer of electrons.

A special class of redox reactions, called single replacement reactions, was studied in ***Experiment 10.*** In that experiment, a series of reactions were performed to develop an activity list. In actuality, this list represented the relative ease of oxidation of a selected group of metals. Today's experiment will utilize a more quantitative method for ranking activity.

Redox reactions can be considered to result from the coupling of an oxidation half reaction with a reduction half reaction. For example, the replacement of silver ion by copper can be broken down into the following oxidation and reduction half reactions:

oxidation:	$Cu(s)$	=	$Cu^{2+} + 2\,e^-$
reduction:	$Ag^+ + e^-$	=	$Ag(s)$
net ionic:	$Cu(s) + 2\,Ag^+$	=	$Cu^{2+} + 2\,Ag(s)$

To arrive at the net ionic equation above, the reduction half reaction was multiplied by 2 to remove electrons from the overall equation when the two half reactions were added together. It would have been fairly simple to write the net ionic equation for this system without using half reactions. However, more complex redox reactions can be very difficult to balance without using the half reaction method. In addition, half reactions are useful in their own right as these are the processes that take place at the electrodes in electrochemical cells and batteries. A second example of the use of half reactions is the oxidation of Fe^{2+} to Fe^{3+} by dichromate ion.

oxidation:	Fe^{2+}	=	$Fe^{3+} + e^-$
reduction:	$Cr_2O_7^{2-} + 14\,H^+ + 6\,e^-$	=	$2\,Cr^{3+} + 7\,H_2O(l)$
net ionic:	$Cr_2O_7^{2-} + 14\,H^+ + 6\,Fe^{2+}$	=	$2\,Cr^{3+} + 7\,H_2O(l) + 6\,Fe^{3+}$

It is possible to determine electrochemical potentials for half reactions. These values are directly related to the driving force for the reaction, and potentials for the two half reactions can be combined algebraically to determine if the overall reaction is spontaneous. The potential E^o is related to the standard free energy of a process by $\Delta G^o = -nFE^o$ (where F is the charge on 1 mol of electrons, $1\,F = 96{,}485$ coulombs/mol e^- and n is the number of electrons transferred in the reaction). A reaction is spontaneous when the free energy change is negative so the potential for a spontaneous reaction must be positive. It should be noted that reversal of the direction of a reaction simply changes the sign of the potential (if it wants to go in one direction, it won't want to go in the reverse direction). As extensive tabulations of half reaction potentials exist, it is possible to calculate potentials and predict spontaneity for a great number of reactions.

To determine if the two examples above would go spontaneously, locate a table of standard reduction potentials. These tables may be found in many chemistry textbooks or handbooks such as the CRC *Handbook of Chemistry and Physics*.

reaction			E^o (volts)
$Ag^+ + e^-$	=	$Ag(s)$	0.80
$Cu^{2+} + 2\ e^-$	=	$Cu(s)$	0.34
$Cu(s) + 2\ Ag^+$	=	$Cu^{2+} + Ag(s)$	0.46
$Cr_2O_7^{2-} + 14\ H^+ + 6\ e^-$	=	$2\ Cr^{3+} + 7\ H_2O(l)$	1.33
$Fe^{3+} + e-$	=	Fe^{2+}	0.77
$Cr_2O_7^{2-} + 14\ H^+ + 6\ Fe^{2+}$	=	$2\ Cr^{3+} + 7\ H_2O(l) + 6\ Fe^{3+}$	0.56

The positive potentials for both net ionic equations (0.46 volts and 0.56 volts) indicate both are spontaneous reactions.

Procedure

In today's experiment, you will use a short table of half reaction potentials to calculate standard potentials for several reactions. For those that have positive potentials, you should also predict possible observations for the reaction when it is tried in the laboratory. For the silver ion - copper system, the potential indicates a spontaneous reaction and we expect to see silver metal plate out on the copper surface while the copper ion entering the solution should turn the solution from colorless to blue.

For each of the reactions listed in the ***Results and Discussion*** section, write down the net ionic equation and use half reaction potentials to determine the standard potential for the reaction. For those that should be spontaneous, predict the expected observations. Now, using 2 mL of each solution, try all the reactions including those predicted to be nonspontaneous to check for consistency with predictions. Be very careful with the half reactions for hydrogen peroxide as H_2O_2 is capable of undergoing either oxidation or reduction (or even both if a catalyst is present). Select the appropriate half reaction for H_2O_2 by considering the type of reaction that the other reactant is capable of undergoing. For example, if the reaction calls for the mixing of copper sulfate with hydrogen peroxide, consider only the oxidation of hydrogen peroxide as copper(II) can be reduced but not further oxidized.

reaction			E^o (volts)
$H_2O_2(aq) + 2H^+ + 2e^-$	=	$2H_2O(l)$	1.776
$Ce^{4+} + e^-$	=	Ce^{3+}	1.61
$MnO_4^- + 8H^+ + 5e^-$	=	$Mn^{2+} + 4H_2O(l)$	1.51
$2IO_3^- + 12H^+ + 10e^-$	=	$I_2(aq) + 6H_2O(l)$	1.195
$Br_2(l) + 2e^-$	=	$2Br^-$	1.066
$NO_3^- + 3H^+ + 2e^-$	=	$HNO_2(aq) + H_2O(l)$	0.94
$ClO^- + H_2O(l) + 2e^-$	=	$Cl^- + 2OH^-$	0.89
$Cu^{2+} + I^- + e^-$	=	$CuI(s)$	0.86
$Fe^{3+} + e^-$	=	Fe^{2+}	0.771
$O_2(g) + 2H^+ + 2e^-$	=	$H_2O_2(aq)$	0.683
$MnO_4^- + 2H_2O(l) + 3e^-$	=	$MnO_2(s) + 4OH^-$	0.588
$I_2(aq) + 2e^-$	=	$2I^-$	0.536
$[Fe(CN)_6]^{3-} + e^-$	=	$[Fe(CN)_6]^{4-}$	0.36
$Cu^{2+} + 2e^-$	=	$Cu(s)$	0.337
$SO_4^{2-} + 4H^+ + 2e^-$	=	$H_2SO_3(aq) + H_2O(l)$	0.17
$Cu^{2+} + e^-$	=	Cu^+	0.153
$Sn^{4+} + 2e^-$	=	Sn^{2+}	0.15
$S_4O_6^{2-} + 2e^-$	=	$2S_2O_3^{2-}$	0.090
$2H^+ + 2e^-$	=	$H_2(g)$	0.000
$O_2(g) + 2H_2O(l) + 2e^-$	=	$H_2O_2(aq) + 2OH^-$	-0.146
$2CO_2(g) + 2H^+ + 2e^-$	=	$H_2C_2O_4(aq)$	-0.49
$2SO_3^{2-} + 3H_2O(l) + 4e^-$	=	$S_2O_3^{2-} + 6OH^-$	-0.58
$Zn^{2+} + 2e^-$	=	$Zn(s)$	-0.763
$2H_2O(l) + 2e^-$	=	$H_2(g) + 2OH^-$	-0.828
$SO_4^{2-} + H_2O(l) + 2e^-$	=	$SO_3^{2-} + 2OH^-$	-0.92
$Mg^{2+} + 2e^-$	=	$Mg(s)$	-2.363
$Na^+ + e^-$	=	$Na(s)$	-2.714
$K^+ + e^-$	=	$K(s)$	-2.925

Name______________________________Date__________Lab Section__________

Prelaboratory Problems - *Experiment 32* - Redox Reactions

The solution to the starred problem is in ***Appendix 4***.

1. Write the net ionic equations and determine the standard potentials for the reaction of copper or zinc with tin(II). Which of the two reactions should be spontaneous?

   a.* copper + tin(IV)

   NIE____ E^o = ____________

   b. zinc + tin(IV)

   NIE____ E^o = ____________

   c. spontaneous reaction(s)____

2. Write the net ionic equation and determine the standard potential for the reaction between permanganate and tin(II). From the potential, calculate the free energy of the reaction.

   a. permanganate + tin(II)

   NIE____ E^o = ____________

   b. ΔG^o = ____________

3. For each of the reactions listed on the following pages (even if it shouldn't go), write down the net ionic equation (NIE), and determine its standard potential. For those with positive potentials, write down your expected observations. Use the following pages for your answers.

Name________________________________Date__________Lab Section_________

Results and Discussion - *Experiment 32* - Redox Reactions

For each of the reactions listed below (even if it doesn't go), write down the net ionic equation (NIE), and determine its standard potential. For those with positive potentials, write down your expected observations. Attempt to run all the reactions and record your experimental observations.

1. magnesium + 0.1 M zinc sulfate

oxidation half reaction______________________________ E^o = __________

reduction half reaction______________________________ E^o = __________

NIE______________________________ E^o = __________

Expected observations______________________________

Experimental observations______________________________

2. copper + 0.1 M zinc sulfate

oxidation half reaction______________________________ E^o = __________

reduction half reaction______________________________ E^o = __________

NIE______________________________ E^o = __________

Expected observations______________________________

Experimental observations______________________________

3. zinc + 0.1 M copper(II) sulfate

oxidation half reaction______________________________ E^o = __________

reduction half reaction______________________________ E^o = __________

NIE______________________________ E^o = __________

Expected observations______________________________

Experimental observations______________________________

4. zinc + 3 M hydrochloric acid

oxidation half reaction______________________________ E^o = __________

reduction half reaction______________________________ E^o = __________

NIE______________________________________ E^o = __________

Expected observations____

Experimental observations____

5. copper + 3 M hydrochloric acid

oxidation half reaction______________________________ E^o = __________

reduction half reaction______________________________ E^o = __________

NIE______________________________________ E^o = __________

Expected observations____

Experimental observations____

6. 0.1 M potassium iodide + 0.1 M copper(II) sulfate

oxidation half reaction______________________________ E^o = __________

reduction half reaction______________________________ E^o = __________

NIE______________________________________ E^o = __________

Expected observations____

Experimental observations____

7.

a. 0.1 M potassium iodide + 0.02 M potassium iodate
[Note: Do not discard after performing *7-a*. Proceed to *7-b*.]

oxidation half reaction____________________ E^o = ________

reduction half reaction____________________ E^o = ________

NIE____________________ E^o = ________

Expected observations____________________

Experimental observations____________________

b. Add 1 drop of 6 M hydrochloric acid to the solution from *7-a*. Report and explain your observations.

8. 0.1 M potassium iodide + 0.1 M potassium ferricyanide

oxidation half reaction____________________ E^o = ________

reduction half reaction____________________ E^o = ________

NIE____________________ E^o = ________

Expected observations____________________

Experimental observations____________________

9. 0.1 M iron(III) chloride + 0.1 M potassium iodide

oxidation half reaction____________________ E^o = ________

reduction half reaction____________________ E^o = ________

NIE____________________ E^o = ________

Expected observations____________________

Experimental observations____________________

10. 0.1 M iron(III) chloride + 0.1 M potassium bromide

oxidation half reaction____________________ E^o = ________

reduction half reaction____________________ E^o = ________

NIE____________________________ E^o = ________

Expected observations______________________________

Experimental observations____________________________

11. 3% hydrogen peroxide + 0.1 M cerium(IV) ammonium nitrate (acidic solution)

oxidation half reaction____________________ E^o = ________

reduction half reaction____________________ E^o = ________

NIE____________________________ E^o = ________

Expected observations______________________________

Experimental observations____________________________

12. aqueous iodine + 1 M sodium thiosulfate (acidic solution)

oxidation half reaction____________________ E^o = ________

reduction half reaction____________________ E^o = ________

NIE____________________________ E^o = ________

Expected observations______________________________

Experimental observations____________________________

Optional challenge (check with instructor)

13. The following reaction is probably going to be performed under basic conditions. To select appropriate half reactions, you should first run the reaction and devise some tests to identify the product anion that contains sulfur.

    3% hydrogen peroxide + 1 M sodium thiosulfate

    [Note: Insert a thermometer into the H_2O_2 solution and read it before adding the $Na_2S_2O_3$ solution. After addition, stir with the thermometer and occasionally read the thermometer.]

Describe any qualitative analysis experiments and the results.

List key half reactions (you may have to couple more than 2 half reactions)

NIE____ E^o = ____________

Expected observations____

Experimental observations____

Temperature reading before addition ____________

Temperature reading after addition ____________

Calculate $\Delta G°$ for this reaction ____________

Was any temperature change observed for the system consistent with the sign of $\Delta G°$? Explain your answer.

14. Like the previous reaction, you will need to do some qualitative analysis on the products before trying to figure out the net ionic equation for this reaction. Also consider taking the pH of the solution. This is probably a very complex system and may require that you look up the potentials for half reactions that are not listed in the table.

iron(III) chloride + sodium thiosulfate

Describe any qualitative analysis experiments and the results.

List key half reactions (you may have to couple more than 2 half reactions)

NIE____ E^o = ____________

Expected observations____

Experimental observations______________________________________

15. Suggest any way you can think of to improve any part(s) of this experiment.

16. Some of the *Learning Objectives* of this experiment are listed on the first page of this experiment. Did you achieve the *Learning Objectives*? Explain your answer.

Experiment 33

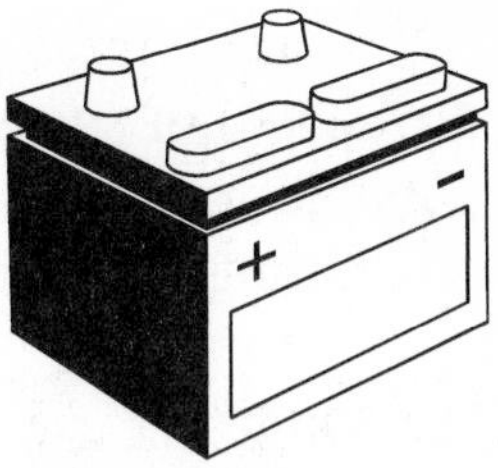

ELECTROCHEMISTRY

Learning Objectives

Upon completion of this experiment, students will have experienced:
1. The preparation of a galvanic cell.
2. The determination of Avogadro's number.
3. Electroplating.

Text Topics

Reduction potentials for half reactions, batteries, electroplating, the Faraday (for correlation with some textbooks, see page ix).

Notes to Students and Instructor

The determination of Avogadro's number requires the use of DC power supplies and ammeters. Unless you choose to do some of the suggested options, this should be a very short experiment.

Discussion

What do your watch, calculator, remote control, automobile ignition system, flashlight and portable cassette player have in common? They all derive their energy from the chemical reactions that occur in batteries. In ***Experiment 32***, we combined half reaction reduction potentials to determine the spontaneity of the resulting overall reaction. The complete reaction potential is the value of the voltage that could in theory be obtained if the two half reactions are appropriately coupled in a "battery".

Reactions with negative potentials will only go if driven by an input of energy. When electricity is used to drive a reaction uphill, the process is called electrolysis. Electrolysis is commonly utilized to prepare elements such as sodium and to electroplate metals with other metals. If a current is passed through a solution containing copper(II) ions, copper(II) ions will migrate to the cathode, accept 2 electrons and deposit out as copper metal. Accompanying this reduction will be some type of oxidation at the anode. This might be oxidation of a metal to its cation (as in today's experiment where copper is oxidized to copper(II) ions at the anode while other copper(II) ions are reduced to copper at the cathode), or another reaction such as production of chlorine from chloride.

If we measure the number of coulombs necessary for deposition (at the cathode) or loss (at the anode) of some weighable amount of copper and use the known charge on an electron, it is possible to calculate the value of Avogadro's number. The coulombs can be determined by monitoring the current in amperes (coulombs/sec) and multiplying by the time in seconds.

$$\left(\frac{\text{coulombs}}{\text{sec.}}\right)\left(\frac{\text{sec.}}{\text{grams Cu}}\right)\left(\frac{\text{1 electron}}{1.6\text{x}10^{-19}\text{ coul.}}\right)\left(\frac{\text{1 atom Cu}}{\text{2 electrons}}\right)\left(\frac{\text{63.54 g Cu}}{\text{1 mole Cu}}\right) = \text{Avogadro's \#}$$

Procedure

A. Preparation of battery. Two half cells will be set up and connected by a salt bridge to complete the circuit. Insert a 1 cm x 10 cm strip of zinc foil into a 1 M zinc sulfate solution in a 100 mL beaker. In a second 100 mL beaker, insert a 1 cm x 10 cm strip of copper foil into a 1 M copper sulfate solution. Roll a 12.5 cm piece of filter paper into a tight cylinder and insert it into a beaker of 1 M potassium nitrate solution. Remove it after it is thoroughly soaked and bend it in the middle. Place the two 100 mL beakers right next to each other and insert the salt bridge. Be sure the paper is at least 1 cm beneath the surface of each solution. Connect the zinc and copper electrodes with wires and a voltmeter (see *Figure 33-1*). After reading the voltmeter, try replacing the voltmeter with an LED. Is there enough current to light the LED?

Fig. 33-1

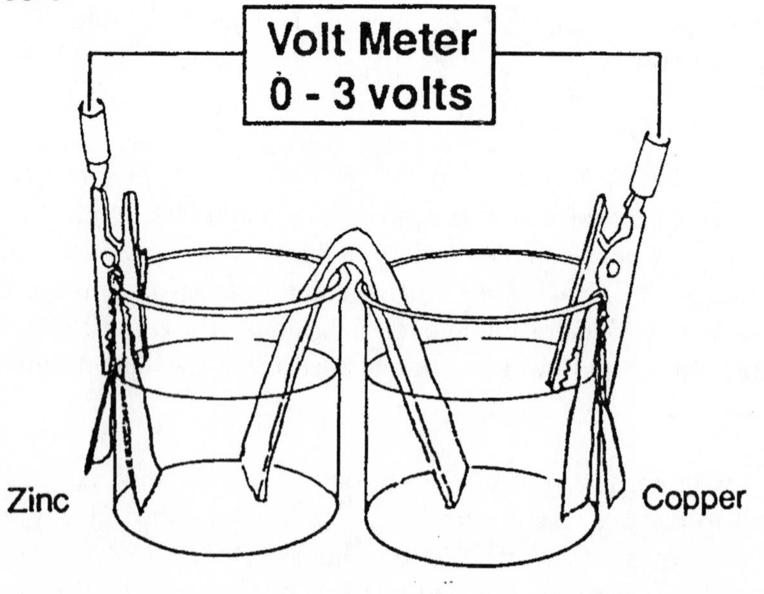

B. Determination of Avogadro's number using electrolysis. Polish two 1 cm x 10 cm copper foil strips with steel wool and weigh them to the nearest milligram. Insert the strips into a 100 mL beaker along opposite sides and clamp the strips to the beaker with alligator clips connected to wires (see *Figure 33-2*). Add about 60 mL of 1 M copper(II) sulfate to the beaker. Insert a small magnetic stirring bar into the beaker and place the beaker on a magnetic stirrer unit. Set the stirring rate to a moderate speed. Connect the lead from one electrode to a DC power supply (preferably rated about 6 volts and 3 amps). Connect the other lead to an ammeter (the optimum range of the ammeter depends on the power supply specifications and the electrode size but will probably be about 0 - 2 amperes). Be sure to use the proper polarities when hooking up the meter.

Simultaneously start a timer and the power supply. Read the ammeter immediately. Read the current again right before turning off the power supply. The proper length of time for a run will be determined by the properties of your system. If your current is about 1.5 amperes, turn off the power supply after 5 minutes. If the current is lower, the appropriate time will be longer and vice versa. Observe the electrodes, dry them with a paper towel and weigh them. Repeat the electrolysis run at least once more. Using the experimental value of 1.60×10^{-19} coulombs/electron, calculate Avogadro's number and the value of the Faraday.

Fig. 33-2

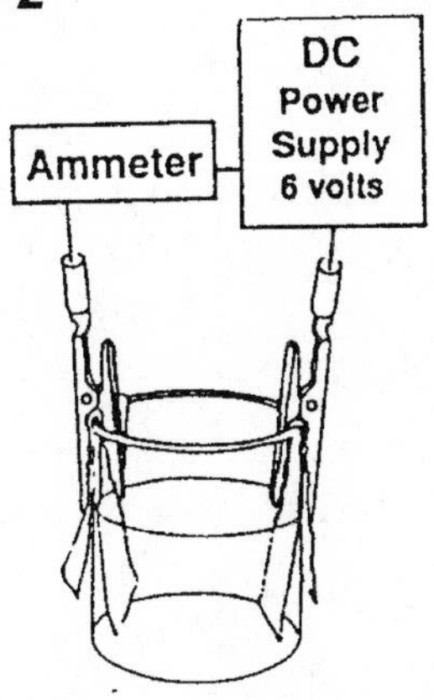

C. Options for further experimentation.

1. Refer to a table of standard reduction potentials and try to design a battery that can be constructed with materials available from your stockroom. Discuss your design with your instructor and with his/her approval, construct and test your battery.

2. Try the electrolysis experiment with a zinc strip for the anode and a copper strip for the cathode in zinc sulfate, observe the electrodes and determine Avogadro's number as in Part B above.

Name________________________________Date__________Lab Section_________

Prelaboratory Problems - *Experiment 33* - Electrochemistry

The solutions to the starred problems are in ***Appendix 4***.

Given the following electrochemical reduction potentials:

$Ag^{+} + e^{-} = Ag(s)$ 1.98 volts

$Pb^{2+} + 2e^{-} = Pb(s)$ -0.126 volts

1.* Write the net ionic equation for the cell $Pb \mid Pb^{2+} \parallel Ag^{+} \mid Ag$.

2.* Calculate the electrochemical potential for the reaction in #1. __________

3.* A $Pb \mid Pb^{2+} \parallel Ag^{+} \mid Ag$ galvanic cell generates a current of 1.5×10^{-2} amperes for 1.75 hours and, in the process, the cathode increases in mass by 0.108 g. Calculate the values of Avogadro's number and the Faraday, assuming the charge on an electron is 1.602×10^{-19} coulombs/electron.

4.* Calculate the mass the lead anode should have lost during the process in #3.

Given the following electrochemical reduction potentials:

$Mg^{2+} + 2e^- = Mg(s)$ -2.363 volts

$2H^{2+} + 2e^- = H_2(s)$ 0.000 volts

5. Write the net ionic equation for the cell $Mg|Mg^{2+}||2H^+|H_2$.

6. Calculate the electrochemical potential for the reaction in #5. ________

7. A $Mg|Mg^{2+}||2H^+|H_2$ galvanic cell generates a current of 2.15×10^{-2} amperes for 205 minutes and, in the process, 29.9 mL of hydrogen at STP are collected. Calculate the values of Avogadro's number and the Faraday, assuming the charge on an electron is 1.602×10^{-19} coulombs/electron.

8. Calculate the mass the magnesium anode should have lost during the process in #7.

Name__________________________________Date__________Lab Section__________

Results and Discussion - *Experiment 33* - Electrochemistry

A. Preparation of battery.

1. Voltage __________

2. Write the half reactions and overall reaction for the $Zn|Zn^{2+}||Cu^{2+}|Cu$ cell and calculate the cell potential from standard reduction potentials. __________

B. Determination of Avogadro's number using electrolysis.

	run 1	run 2	run 3
1. Mass of copper anode			
2. Mass of copper cathode			
3. Initial current reading			
4. Final current reading			
5. Average current reading			
6. Time for run in seconds			
7. Final mass of anode			
8. Final mass of cathode			
9. Mass loss of anode			
10. Mass gained by cathode			
11. Moles of Cu lost by anode			
12. Moles of Cu gained by cathode			
13. Coulombs used (use average current)			

14. # of electrons transferred ______ ______ ______

15. # of Cu atoms oxidized at anode ______ ______ ______

16. # of Cu atoms reduced at cathode ______ ______ ______

17. Avogadro's # from anode data ______ ______ ______

18. Avogadro's # from cathode data ______ ______ ______

19. Average value of Avogadro's # from anode data ______

20. Average value of Avogadro's # from cathode data ______

21. Faraday constant from anode data ______ ______ ______

22. Faraday constant from cathode data ______ ______ ______

23. Calculate the percent difference between the average value of Avogadro's # determined from the anode data and the cathode data ______

24. Calculate the percent difference between the average value of Avogadro's # determined from the cathode data and the literature value ______

25. Report your observations on the general changes of the electrodes with each electrolysis.

    a. anode

    b. cathode

C. Options for further experimentation.

1. Describe your experiments and observations on an additional sheet of paper.

2. Some of the *Learning Objectives* of this experiment are listed on the first page of this experiment. Did you achieve the *Learning Objectives*? Explain your answer.

Experiment 34

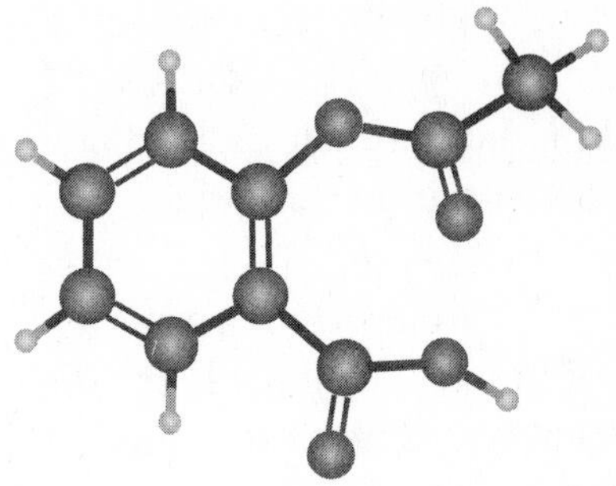

SPECTROSCOPIC ANALYSIS OF ASPIRIN

Learning Objectives

Upon completion of this experiment, students will have experienced:
1. The spectroscopic study of a complex.
2. The determination of the amount of acetylsalicylic acid in an aspirin tablet.
3. The saponification of an ester.
4. (Optional) The determination of the formation constant of a complex.

Text Topics

Spectroscopy, complex ions, organic reactions. (for text correlation, see page ix).

Notes to Students and Instructor

This experiment includes concepts presented in many earlier experiments including spectroscopy, complex ion formation, synthesis and equilibrium constants.

Discussion

Have you ever questioned or perhaps even doubted the accuracy of the information printed on the container of a product? The ultimate goal of this experiment is to determine the amount of acetylsalicylic acid in an aspirin tablet and compare it to the label value. A spectroscopic technique very similar to the procedure used to analyze the amount of cobalt ion in an unknown in ***Experiment 14*** will be utilized. Review the discussion for ***Experiment 14*** before continuing.

Acetylsalicylic acid undergoes a saponification reaction in basic solution that is typical of esters. For example, ethyl acetate when heated in a base solution saponifies to form sodium acetate and ethanol.

$$CH_3COOCH_2CH_3 + NaOH \rightarrow CH_3COO^-Na^+ + CH_3CH_2OH$$

Acetylsalicylic acid is similar to ethyl acetate except that the ethyl has been replace by a salicylic acid group (CH_3COOSA where SA represents salicylic acid). Acetylsalicylic acid will be saponified and the salicylic acid that is quantitatively released by this reaction will be analyzed by a spectroscopic technique.

Iron(III) ion forms a 1:1 complex with the "dianion form" of salicylic acid in 0.05 M acid solution. (Technically the dianion is not the form of salicylic acid that would exist in 0.05 M acid solution. However, the iron(III) complex of the dianion is apparently stable under these conditions.) Sufficient iron(III) will be used to complex virtually all of the salicylic acid present. A series of known concentrations of the complex will be prepared. The solutions will be used to determine the absorption spectrum of the complex and a Beer's Law graph will be used to determine ϵb for the complex. An aspirin tablet will be saponified, the salicylic acid complexed with iron(III) and the absorption measured to determine the amount of acetylsalicylic acid in the tablet.

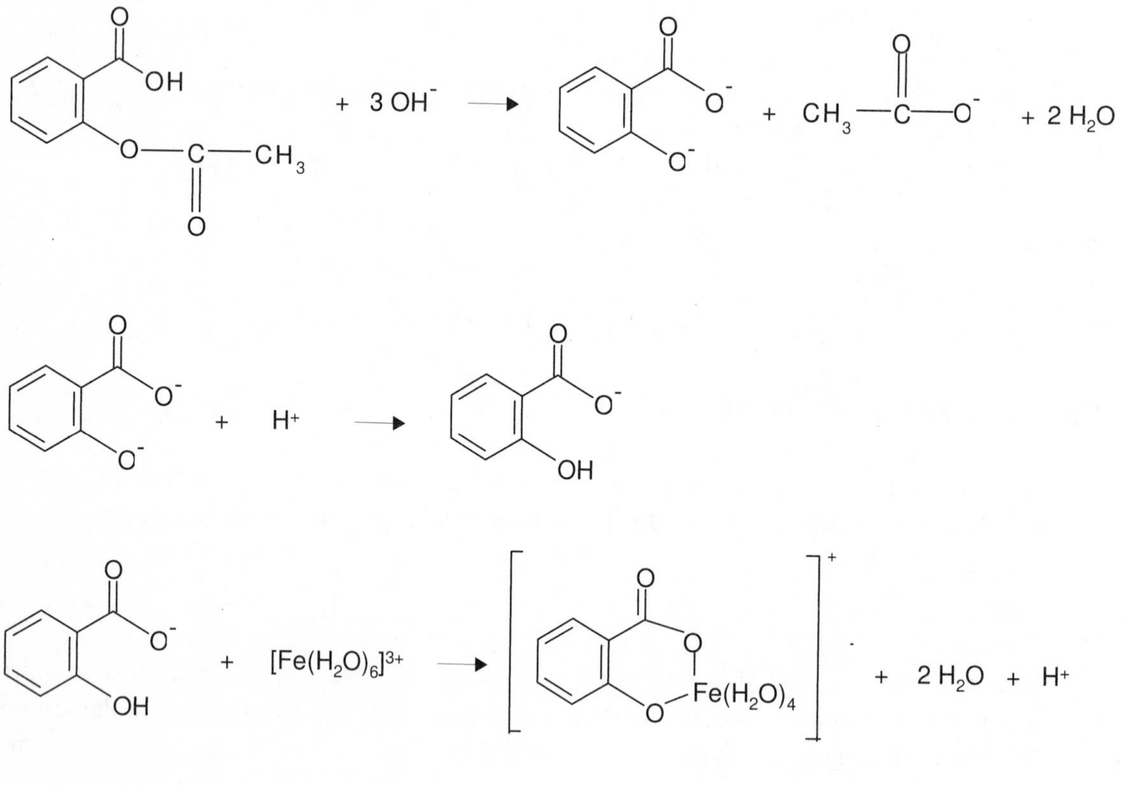

Procedure

A. Saponification of acetylsalicylic acid in aspirin. Weigh an aspirin tablet to the nearest 0.001 g. Grind it up and transfer all of it to a 125 mL Erlenmeyer flask. Add 10 mL of 1 M sodium hydroxide to the flask and heat the mixture to boiling. Be sure to rinse any solid that spatters on to the side of flask back down into the flask with deionized water.

Using a long stemmed funnel, transfer the saponified mixture to a 250 mL volumetric flask (be sure to rinse the Erlenmeyer flask at least twice and add the washings to the volumetric flask). Dilute to the mark with deionized water. Thoroughly mix the contents of the flask and then pipet 5.00 mL of the solution into a 100 mL volumetric flask. Dilute to the 100 mL mark with 0.020 M iron(III) chloride in 0.050 M HCl and mix.

B. Preparation of sodium salicylate solutions. Weigh into a beaker, to at least the nearest 0.001 g, approximately 0.4 g of sodium salicylate (formula mass = 160.11 g/mol). Add about 50 mL of water and dissolve the sodium salicylate. Using a funnel, transfer the contents of the beaker to a 250 mL volumetric flask. Rinse the beaker three times with water and add the washings to the flask. Dilute to the mark and thoroughly shake the mixture.

Rinse and fill a buret with the sodium salicylate solution. Deliver 5.00 mL of this solution into a 100 mL volumetric flask and dilute to the 100 mL mark with 0.020 M iron(III) chloride in 0.050 M HCl, and mix. Transfer this solution to a clean and dry 125 mL Erlenmeyer flask labeled "V". Rinse the volumetric flask with water several times and add 4.00 mL of the sodium salicylate from the buret to the flask. Again dilute to the mark with the iron(III) chloride solution. Transfer this solution to a 125 mL Erlenmeyer flask labeled "IV". Repeat this process with 3.00, 2.00 and 1.00 mL of the sodium salicylate solution and transfer the solutions to flasks labeled "III", "II" and "I". If you decide to do Part E, leave the sodium salicylate solution in the buret and save the rest.

C. Absorption spectrum of the iron salicylate complex. Use 0.020 M iron(III) chloride as a blank and determine the value of the absorption, A, at 420, 450, 480, 500, 520, 530, 540, 560, 580, 610 and 640 nm for solution V. See the instructions for your particular spectrometer. Every time the wavelength is set, the blank must be inserted and the absorption set at 0.00 (or %T at 100%). Plot the absorption on the y axis and the wavelength (λ) on the x axis. Set the instrument on the wavelength of maximum absorption and reset the 0 and 100% transmission values. Proceed to Part D.

D. Beer's law plot for iron salicylate complex. Rinse a cuvette with solution I and add solution I to about the ½ way point of the tube. Insert the tube into the instrument and read and record the absorption value. Repeat this procedure for solutions II - V and the saponified acetylsalicylic acid in the 100 mL volumetric flask. Plot the absorption (y axis) versus the concentration of the iron salicylate complex (x axis). Determine the slope of the line and calculate the concentration of iron(III) salicylate in the 100 mL volumetric flask (from aspirin solution).

E. (Optional) Determination of the formation constant for the iron(III) salicylate complex. The determination of the formation constant for the complex must take into account the two acid ionization constants of salicylic acid ($K_{a1} = 1.1x10^{-3}$, $K_{a2} = 4x10^{-14}$). Follow the procedure for the preparation of the solutions given below and measure their absorptions at the wavelength selected in Part C above using deionized water as a blank.

a. Add 1.00 mL of the sodium salicylate solution to a 100 mL volumetric flask. Using a second buret, add 5.00 mL of 0.020 M iron(III) chloride containing 0.050 M HCl to the volumetric flask, add 47.5 mL of 0.10 M HCl and dilute to the mark with water. Mix and transfer the solution to a clean, dry 125 mL Erlenmeyer flask.

b. Repeat the above procedure with 3.00 mL of sodium salicylate solution, 3.00 mL of iron(III) chloride solution and 48.5 mL of 0.10 M HCl followed by dilution to the mark with water.

c. Repeat the above procedure with 5.00 mL of sodium salicylate solution, 1.00 mL of iron(III) chloride solution and 49.5 mL of 0.10 M HCl followed by dilution to the mark with water.

d. Measure the absorptions of the three solutions and determine the formation constant of the iron(III) salicylate complex. Be sure to take into account the acid dissociation constants.

Name________________________________ Date____________ Lab Section__________

Prelaboratory Problems - *Experiment 34* - Spectroscopic Analysis of Aspirin.
The solutions to the starred problems are in ***Appendix 4***.

1. Write an equation for the saponification of methyl benzoate ($C_6H_5COOCH_3$).

2. a.* 50.0 mL of 0.15 M cobalt(II) nitrate is diluted to 250 mL with water. What is the final concentration of cobalt(II) nitrate?

   b.* If the extinction coefficient of cobalt(II) ion is 5.0, what would the absorption of the solution in *2-a* be in a 1.00 cm cell?

   c. 25.0 mL of 0.12 M cobalt(II) nitrate is diluted to 350 mL with water. What is the final concentration of cobalt(II) nitrate?

d. If the extinction coefficient of cobalt(II) ion is 5.0, what would the absorption of the solution in *2-c* be in a 1.00 cm cell?

Name__________________________________Date__________Lab Section__________

Results and Discussion - *Experiment 34* - Spectroscopic Analysis of Aspirin

A. Saponification of acetylsalicylic acid in aspirin.

1. Brand of aspirin __________
2. Mass of aspirin tablet __________
3. Grains of aspirin according to manufacturer __________
4. Unit conversion (grams/grains) __________
5. Grams of aspirin (calculated from label) __________

B. Preparation of sodium salicylate solutions.

1. Mass of beaker + sodium salicylate __________
2. Mass of beaker __________
3. Mass of sodium salicylate __________
4. Formula mass of sodium salicylate __________
5. Moles of sodium salicylate __________
6. Molarity of sodium salicylate (250 mL flask) __________

C. Absorption spectrum of the iron salicylate complex.

Wavelength (nm)	Absorption	Wavelength (nm)	Absorption	Wavelength (nm)	Absorption
420	______	520	______	580	______
450	______	530	______	610	______
480	______	540	______	640	______
500	______	560	______		

Plot absorption (y axis) vs wavelength (x axis) and determine the wavelength of maximum absorption. __________

D. Beer's law plot for iron salicylate complex.

1. Molarity and absorption of iron salicylate complex in solutions I - V and aspirin solution. Calculate the concentrations by considering dilution factor of B-6 above.

	molarity (moles/L)	absorption
a. Solution I	________	________
b. Solution II	________	________
c. Solution III	________	________
d. Solution IV	________	________
e. Solution V	________	________
f. Diluted aspirin solution		________

2. Plot the absorption (y axis) vs the concentration (x axis) and determine the slope of the line (ϵb). ________

3. Assuming that b is 1.0 cm and the literature value for ϵ is 1.8×10^3, what is the percent difference between your value and the literature value? ________

4. Concentration of complex in diluted aspirin solution (100 mL volumetric flask). ________

5. Concentration of acetylsalicylic acid in original aspirin solution (250 mL flask). ________

6. Moles of acetylsalicylic acid in original solution and in aspirin tablet. ________

7. Molecular mass of acetylsalicylic acid ________

8. Mass of acetylsalicylic acid in tablet from experimental results. ________

9. Percent difference between experimental result and label value (A-5). ________

10. Why is a large excess of iron chloride used for the preparation of the solutions of the complex.

E. (Optional) Determination of the formation constant for the iron(III) salicylate complex.
($K_{a1} = 1.1\text{x}10^{-3}$, $K_{a2} = 4\text{x}10^{-14}$ for the acid dissociation constants of salicylic acid)

1. Show that $K_f K_{a1} K_{a2} = \dfrac{[Fe\mathit{SA}^+][H^+]^2}{[H_2\mathit{SA}][Fe^{3+}]}$

where

K_f	=	formation constant for complex ion between iron(III) and dianion of salicylic acid
K_{a1} and K_{a2}	=	first and second acid ionization constants of salicylic acid
$[Fe\mathit{SA}^+]$	=	concentration of complex between iron(III) and dianion of salicylic acid
$[H^+]$	=	concentration of hydrogen ion
$[H_2\mathit{SA}]$	=	concentration of salicylic acid
$[Fe^{3+}]$	=	concentration of iron(III) ion

2. Data and calculations

	Trial 1	Trial 2	Trial 3
a. Concentration of stock solution of sodium salicylate	______	______	______
b. Concentration of stock $FeCl_3$ soln.	______	______	______
c. Volume of sodium salicylate	1.00 mL	3.00 mL	5.00 mL
d. Volume of $FeCl_3$	5.00 mL	3.00 mL	1.00 mL
e. Initial diluted concentration of sodium salicylate $[NaHSA]_o$	______	______	______
f. Initial diluted concentration of iron(III) ion $[Fe^{3+}]_o$	______	______	______
g. Absorption at λ_{max}	______	______	______
h. ϵb from *D-2*	______	______	______
i. $[FeSA^+] = A/\epsilon b$	______	______	______
j. $[H_2SA] = [NaHSA]_o - [FeSA^+]$	______	______	______
k. $[Fe^{3+}] = [Fe^{3+}]_o - [FeSA^+]$	______	______	______
l. $[H^+]$	______	______	______
m. $[H^+]^2$	______	______	______
n. $K_f K_{a1} K_{a2}$	______	______	______
o. K_f	______	______	______
p. Average value of K_f			______
q. Deviation of each K_f from average value of K_f	______	______	______
r. Average deviation			______
s. A typical literature value from *Stability Constants of Metal Ion Complexes*, London, The Chemical Society, p. 534 (1964) for $\log K_f$ is 16.4. What is the percent difference between the literature value and your experimental value?			______

3. Some of the *Learning Objectives* of this experiment are listed on the first page of this experiment. Did you achieve the *Learning Objectives*? Explain your answer.

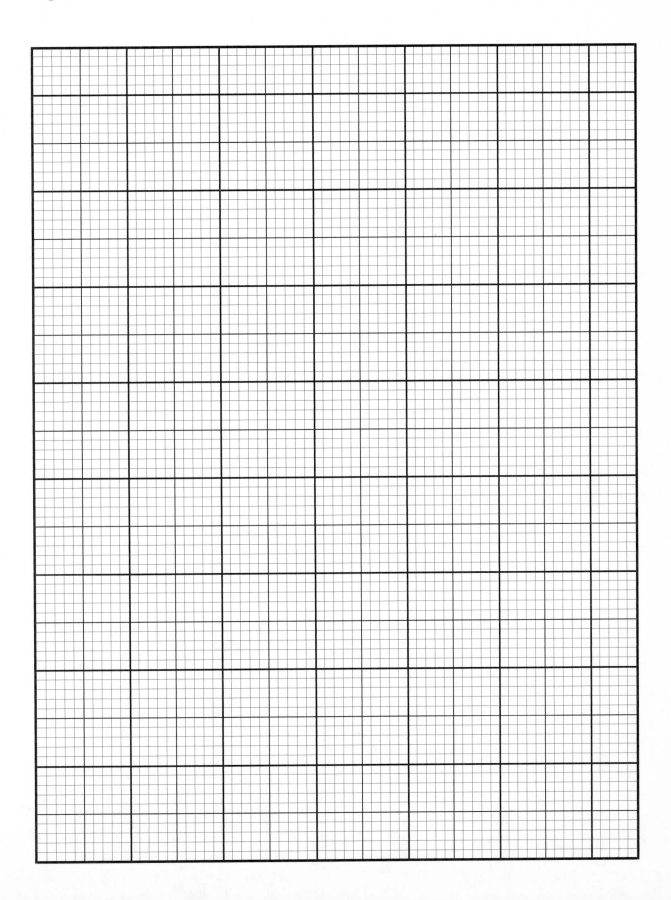

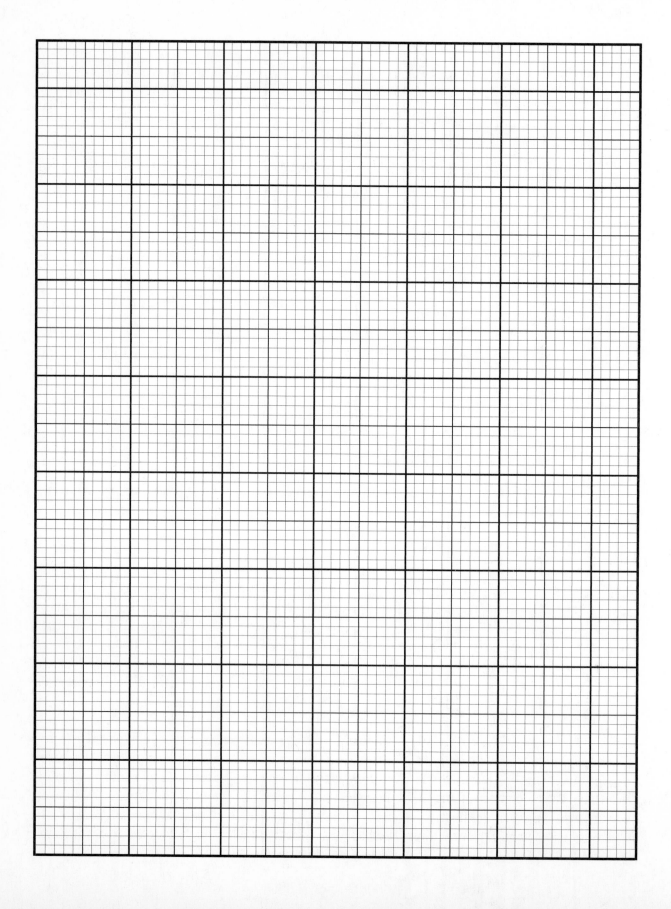

Experiment 35

POLYMERS, MATERIAL SELECTION USING THE INTERNET AND SOME FUN

Learning Objectives

Upon completion of this experiment, students will have experienced:
1. The preparation of crosslinked polymers.
2. Some interesting chemical demonstrations.

Text Topics

Polymers (for correlation with some textbooks, see page ix).

Notes to Students and Instructor

This experiment is intended to promote interest in polymer chemistry but presents a limited perspective of the very broad and very important field of polymer chemistry.

Discussion

Take a look at everything around you. From the paper of this book, the plastic of your pen, the fibers of your clothes to your hair and skin, you are surrounded by a world made from polymers. Find a bunch of paper clips and connect them together to make a chain. You have just constructed a model of a polymer. The linking of molecules (or monomers) with chemical bonds end to end hundreds of times results in the formation of a polymer. The structural material of plants, cellulose, is a polymer of the sugar, glucose. The structural material of animals, protein, is a polymer made from amino acids.

A significant percentage of chemists employed by industry work on polymer related research. Chemists are always looking for plastics with better or different melting points, strengths, flexibilities, clarities, flammabilities, durabilities and so on. A major field of biochemistry involves the study and alteration of the DNA polymer that carries all genetic information. Despite its importance, polymer chemistry in most schools is relegated to a rather small part of the undergraduate chemistry curriculum. Perhaps this is because a thorough knowledge of monomer chemistry is essential for an understanding of polymer chemistry.

Today's experiment is not intended to cover the field of polymer chemistry. It will only give you a brief excursion into the field of polymers including the changing of the properties of two polymers and the use of the Internet to select polymers for selected applications.

There are two common classes of polymers. Polystyrene, polyethylene, polyvinylchloride and teflon are all examples of addition polymers. For example, styrene can be polymerized by a free radical initiator (a source of a chemical having an unpaired electron). The free radical adds to the styrene creating a new free radical which then adds to another styrene and the process continues until several hundred styrenes have been added to a chain.

free radical initiator X- X → 2 X·, Ph = phenyl (benzene ring)

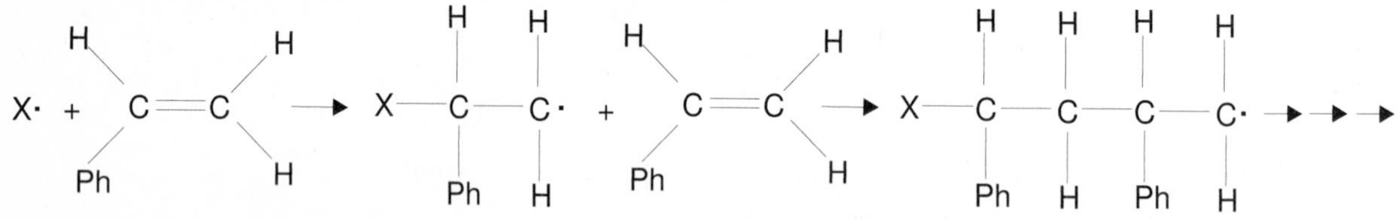

Polyesters and nylon belong to another class called condensation polymers. Nylon 66 can be made in the laboratory by reacting adipoyl chloride with 1,6-diaminohexane. Notice that after one molecule of each reacts, the two ends are still reactive and the left end will react with another 1,6-diaminohexane while the right reacts with adipoyl chloride. The process can continue hundreds of times. Your instructor will demonstrate this reaction to you.

$$Cl\text{–}\overset{\overset{O}{\|}}{C}(CH_2)_4\overset{\overset{O}{\|}}{C}\text{–}Cl + H_2N(CH_2)_6NH_2 \rightarrow Cl\text{–}\overset{\overset{O}{\|}}{C}(CH_2)_4\overset{\overset{O}{\|}}{C}\text{–}NH(CH_2)_6NH_2 \rightarrow\rightarrow\rightarrow\rightarrow$$

One way of changing the properties of a polymer is to chemically connect adjacent strands. This process is called crosslinking and is the topic of two parts of today's experiment. In one case, the crosslinking of the polymer in guar gum (or polyvinyl alcohol may be substituted), will result in a fluid with unusual properties that you should investigate.

Procedure

A. Polymer models. Link together at least 8 paper clips and explain the analogy to a polymer. Make another chain of at least 8 paper clips and lay the two chains down side by side. Link the two chains together with single paper clips at two different sites. How does this "cross linking" affect the behavior of the chains when they are moved?

Catalysts are often needed to cause polymerizations to occur. To model the use of a catalyst, fold a dollar bill into thirds like a fan. Place two paper clips on the dollar bill as illustrated in *Figure 35-1* with one paper clip clipping the first two thirds and the second clipping the last two thirds. Grab the two ends of the dollar bill with your left and right hands and quickly pull in opposite directions. Explain how the dollar bill in this "magic trick" serves as a model for a catalyst and comment on the quality of the analogy. (I want to thank Dr. Alan McCormack of California State University at San Diego for sharing this analogy with me.)

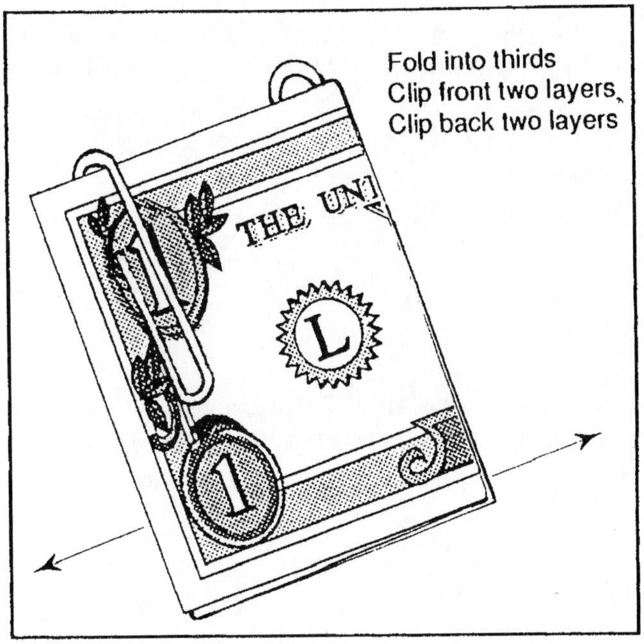

B. Demonstrations.

It is recommended that the instructor perform these experiments as demonstrations as some of the chemicals are potential irritants and/or corrosive. This also minimizes the amount of waste disposal necessary.

1. Nylon synthesis

   a. Prepare a solution containing 0.5 M NaOH and 0.5 M 1,6-diaminohexane. Add about 2 mL of the solution to a watch glass or evaporating dish.
   b. Put a small loop in the end of a 4 inch piece of copper wire.
   c. Add about 4 mL of 0.25 M adipoyl chloride in cyclohexane to the watch glass, insert the loop into the solution and slowly pull a nylon string out of the solution.

2. Disappearing polystyrene.

   Pour about 100 mL of acetone into a 600 mL beaker. Have students line up and have each one add a handful of polystyrene peanuts to the beaker. Stir between each addition.

C. Preparation of *glurp, glurp'* and *glorp*.

A 4% borax solution (sodium borate decahydrate - $Na_2B_4O_7 \cdot 10H_2O$ - 20 Mule Team Borax works fine) should be available in the laboratory.

Glurp. Prepare 100 mL of a 1% guar gum dietary fiber solution (or 50 mL of 4% polyvinyl alcohol - dissolve by using a microwave) by adding about 1 gram of guar gum to 100 mL of water in a 400 mL beaker followed by stirring for 2 minutes. Guar gum is a vegetable gum polymer or polysaccharide composed of mannose and galactose. Add 5 mL of the borax solution to the guar gum solution and stir for several minutes. Try the following experiments with your *glurp*:

a. Pull the *glurp* slowly and record your observations.
b. Pull the *glurp* quickly and record your observations.
c. Mount a funnel in a ring on a ringstand. Put some *glurp* into the funnel and push it through the funnel. Record your observations as the *glurp* comes out of the funnel.
d. Try some other safe experiments, describe them, and record your observations.

Glurp ′[see David A. Katz, *J. Chem Ed.*, **71**, 891 (1994)]. Place a 20 cm by 20 cm piece of plastic cut from a melt-away plastic bag in a 200 mL beaker containing 25 mL of water. Stir vigorously until no further change is observed. Add 5 mL of the 4% borax solution and stir vigorously again. Test the *glurp* ′just as you tested the *glurp* above.

Glorp. Add 20 mL of water to 20 grams of Elmer's glue. Add 20 mL of borax solution and stir. Try some safe experiments on your *glorp.*

D. The seesaw reaction.

The only connection of this experiment to polymers is that the polymer, starch, is used as a complexing agent. The reason the seesaw reaction is included here is that this is the last formal experiment of this book and to some, including this author, this is one of the most intriguing chemical reactions. The three solutions indicated below will be available. Add 10 mL portions of each to a 125 mL Erlenmeyer flask, swirl and transfer quickly to a 50 mL graduated cylinder and allow to stand. Report all of your observations.

Solution A: 11.7 g malonic acid and 2.5 g manganese sulfate monohydrate diluted to 750 mL with 0.4% aqueous starch solution.

Solution B: 206 mL 30% hydrogen peroxide diluted to 500 mL with water **[Note: This solution is corrosive and should be handled with care]**.

Solution C: 21.4 g potassium iodate diluted to 500 mL with water containing 2.2 mL of concentrated sulfuric acid.

E. Material Selection Using the Internet.

The very interesting and increasingly important field of material science involves determination of the properties of a material needed for an application and then selection of the best material. A very useful site for selecting polymers and metals for particular applications is:

http://www.matweb.com/

The answer section for this exercise guides you through a way to use this site to find a good polymer to use for the coating of magnets for use as magnetic stirrers. Then you will be asked to perform similar property and material searches for some additional applications.

Name________________________________ Date__________ Lab Section__________

Results and Discussion - *Experiment 35* - Polymers and Some Fun

A. Polymer models.

1. How does "crosslinking" affect the movement of the model polymer chains?

2. Explain how this dollar-paper clip "magic trick" models a catalyst and comment on the quality of the analogy.

B. Demonstrations.

1. The 1,6-diaminohexane is on the bottom in the water phase and the adipoyl chloride is in the top phase in cyclohexane. Give and account for your observations as the copper loop was lifted from the mixture.

2. Report and explain your observations as the polystyrene peanuts are added to the acetone.

3. Write an equation for the polymerization of chloroethene (vinyl chloride - CH_2=CHCl) to polyvinylchloride assuming that a free radical initiator (X·) is present.

C. Preparation of *glurp*, *glurp'* and *glorp*.

1. *Glurp*

   a. Describe your observations when the *glurp* was pulled slowly.

   b. Describe your observations when the *glurp* was pulled quickly.

   c. Describe your observations on the funnel experiment.

2. *Glurp′*

   a. Describe your observations when the *glurp′* was pulled slowly.

   b. Describe your observations when the *glurp′* was pulled quickly.

   c. Describe your observations on the funnel experiment.

3. Test the *glorp* in as many ways as you can think of. Report your tests and observations.

D. The seesaw reaction. Report your observations on this system. If you desire, try varying some parameter and run the reaction again. For additional information on the oscillating clock reaction, see the following internet sites and references therein.
http://chemlearn.chem.indiana.edu/demos/TheOsci.htm
http://chemistry.about.com/cs/demonstrations/a/aa050204a.htm
http://www.math.udel.edu/~rossi/Math512/br5.pdf
http://www.cci.unl.edu/Teacher/NSF/C10/C10Links/chemistry.about.com/library/weekly/aa100499a.htm

E. Material Selection Using the Internet. The very interesting and increasingly important field of material science involves determination of the properties of a material needed for an application and then selection of the best material. A very useful site for selecting polymers and metals for particular applications is:

http://www.matweb.com/

Before searching for a material, you should familiarize yourself with the site, its capabilities and some of the properties in its database. To begin this process, the exercises below will help you learn how to navigate around the site.

1. Determination of some properties of aluminum. At the site, click on "searchable data base of material data sheets and then select "Material Type/Category Search." Scroll down to "Pure Metallic Elements" and click "Find". Now scroll down (they are not alphabetical to aluminum and click on it and fill in the data below.

    density (g/cm^3) ________________

    electrical resistivity (ohm-cm) ________________

    melting point (°C) ________________

2. Determination of some properties of Plexiglas VO52. Now back up to the "Material Type/Category Search." Under "Thermoplastic polymers" select "acrylic" and click on "Find". Locate Plexiglas VO52 and fill in the data below.

    Plexiglas is a polymer of ________________

    water absorption (%) ________________

    melting point (°C) ________________

    transmission visible (%) ________________

As an example of a search for a material, consider the selection of a polymer for use as a coating on a magnetic stirring bar. Some of the properties that are needed are: high melting point [The stirring bar is sometimes used in solvents at their boiling points and melting would be very undesirable. As some solvents boil above 200°C [for solvent properties, see Appendix 1 and/or

http://virtual.yosemite.cc.ca.us/smurov/orgsoltab.htm]

a good starting lower limit for the melting point would be 300°C.], a low coefficient of friction ($\leq$0.1) will facilitate spinning and low water absorption ($\leq$.1%) should inhibit water penetration to the magnet. At the original site, scroll down and select "Physical Properties, Metric". Next scroll down and select "polymer. In the boxes below, set melting point at a minimum of 300°C and the coefficient of friction and water absorption at maxima of 0.1 and 0.1 (for 0.1%) respectively. Clicking on "Find" should result in three different forms of teflon or PTFE. The molded form is probably most appropriate for use as a coating for the magnet. The exercises that follow will encourage you to navigate yourself around the site to determine suitable plastics and metals and/or metal alloys for specific applications. If too may possibilities result from a search, put more severe restrictions on the properties.

3. Material selection - Polymers

   a. 2 L bottle

      desirable properties ______________________

      possible plastics ______________________

   b. frying pan coating

      desirable properties ______________________

      possible plastics ______________________

   c. telephone case

      desirable properties ______________________

      possible plastics ______________________

   d. food storage bags

      desirable properties ______________________

      possible plastics ______________________

   e. tires

      desirable properties ______________________

      possible plastics ______________________

4. Material selection - Metals and/or Metal Alloys

   a. bicycle frame

      desirable properties ______________________

      possible metals ______________________

   b. solder

      desirable properties ______________________

      possible metals ______________________

   c. electrical wire

      desirable properties ______________________

      possible metals ______________________

   d. nails

      desirable properties ______________________

      possible metals ______________________

5. Some of the *Learning Objectives* of this experiment are listed on the first page of this experiment. Did you achieve the *Learning Objectives*? Explain your answer.

REVIEW EXERCISES

Since you are reading this, you are probably nearing the end of a one year course that has hopefully continued to expand your experience with chemistry. "Continued" not because you took a previous course in high school or college but because chemistry is a significant part of your essence. From the time you were first conceived, chemical reactions have been at the center of all the important processes that govern your life including the vision process that enables you to see and read these words. By studying chemistry we hope to be able to understand the chemistry of our lives. Hopefully, you will take many more chemistry courses and continue to explore the mysteries and patterns of our universe. As stated in the **Preface**, one of the joys of life is "the pleasure of finding things out."

Before proceeding on however, it is important to review and reflect upon the year and to integrate the skills and insights you have acquired with rest of your experiences and knowledge. You now have new tools that can be used to understand and solve the changing problems that continually confront us but also provide exciting challenges and adventures. The focus of this course has been a study of the properties of matter and in particular a search for order in matter. In the laboratory, the emphasis has been on learning the manipulations, the applicability of the techniques chemists use to solve problems and experiences with the pleasure of finding things out.

At the heart of chemistry is synthesis or the preparation of chemicals. During this course, you prepared and purified calcium carbonate, zinc iodide and copper(II) glycinate monohydrate. If you continue into the especially creative field of organic chemistry, you will discover that the emphasis of your studies will be on the synthesis of many organic compounds. When the synthesis is complete or when natural products are found, the next steps are usually separation, purification and identification. You have performed many separation and purification techniques in this course including filtration, extraction, recrystallization, distillation and chromatography. Next, for previously characterized compounds, the chemist has to verify the identity of the compound. Or, for new compounds, the chemist must determine the composition and structure of the compound. Measurements such as density, melting and boiling point, chromatographic parameters (such as R_f values), spectra and molecular mass (from titration, spectroscopy, quantitative precipitation and freezing point depression) are useful for verification. Spectra and molecular mass along with chemical properties are useful for the determination of the structures of new compounds.

Finally, the accumulated information is applied to new situations. It might be used to modify synthetic procedures to improve yields or for selection of a new material for a particular application.

During the course, you probably encountered some questions in this text (particularly those in the ***Postlaboratory Problems***) that were intended to help you understand the reasons the experiment had been performed, the applicability of the technique and how it connected to other experiments and the principles introduced in the lecture portion of the course. These ***Review Exercises*** have been designed to help you organize the content of this course and reflect once more upon the applicability and relationship of the techniques of synthesis, separation, purification and identification. There are three parts to the ***Review Exercises***. The first section includes questions similar to those in the ***Postlaboratory Problems*** sections. The second section of the ***Review Exercises*** asks you to apply your experience and knowledge to the selection of materials for specific applications. The final section suggests several topics for research papers that are of extreme societal importance. Society must make informed, wise decisions on each of these topics or we will continue to jeopardize the future of life on earth. Hopefully you are going to realize that you are now able to understand these issues better than when the course started and that you might want to try to influence the direction society takes on these issues.

A. Applicability of Methods

1. Chemistry has been called "the central science." Give reasons for this terminology and discuss its appropriateness.

2. In the introduction to the ***Review Exercises*** above, it was stated that chemistry is a search for order in matter. Comment and critically evaluate this statement.

3. How would you purify the following:

   a. 2 grams of a solid compound that contains about 5% impurity.

   b. 10 mL of a liquid (b.p. = 110°C) that has two impurities present (5% A, b.p. = 50°C, 5% B, b.p. = 200°C).

   c. A solid mixture of approximately equal quantities of magnesium sulfate and barium sulfate.

4. a. What criteria should be used for the selection of a solvent for the recrystallization of a solid compound?

   b. Give the typical steps of a recrystallization.

   c. Why is vacuum filtration usually used to recover the product of recrystallization rather than gravity filtration?

5. For the collection of CuO in ***Experiment 7***, why was gravity filtration used rather than vacuum filtration?

6. Give at least three uses of melting range determinations.

7. What measurements would you make to identify an unknown pure liquid?

8. How would you distinguish between the following substances?

   a. iron and zinc

   b. methanol and ethanol

   c. ethanol and cyclohexane

9. The titration curve of ascorbic acid is on page 276.

   a. Give a detailed description of the technique you would use to determine the amount of ascorbic acid in a vitamin pill (assume no other acids are present).

   b. Determine the pK_a of ascorbic acid from the graph.

10. How would you determine the purity of each of the following solids (assume several grams of each is available)?

    a. K_2CO_3

    b. $MgSO_4$

    c. $Ca(OCl)_2$

    d. $CuCl_2 \cdot 2H_2O$ (3 ways)

    e. $Zn(NO_3)_2 \cdot 6H_2O$

    f. $K_4Fe(CN)_6 \cdot 3H_2O$

11. How would you determine the solubility of the following two compounds in water?

    a. $Sr(OH)_2$

    b. $Ca(IO_3)_2$

12. How would you distinguish between the following ions?

    a. Li^+, K^+ (2 ways)

    b. Ba^{2+}, Sr^{2+}

    c. Ba^{2+}, Mg^{2+} (2 ways)

    d. Cu^{2+}, Ca^{2+} (2 ways)

    e. Pb^{2+}, Zn^{2+}

    f. NH_4^+, K^+

13. As you have discovered, graphs and graphing techniques are very important in chemistry. Using the data in ***Appendix 1***, for the elements Al, Cr, Cu, Au, Fe, Pb, Mg, Hg, Pt, Ag, Sn, Ti, W, and Zn, graph the specific heat vs the atomic mass and vs the inverse of the atomic mass. Is the Dulong and Petit empirical relationship, $M_mC_p = 25$, correct? Explain your answer. Graph paper is available on pages 433 and 434.

B. Material Selection. ***Experiments 13*** and ***35***, included questions that dealt with the selection of materials for specific applications. In ***Experiment 13***, choices were limited to elements and ***Experiment 35*** suggested the use of polymers for one section and metals for the other. This time, without any restrictions, state the criteria you would use to select a material and suggest possible materials for the applications listed below.

1. antifreeze solution for cars
2. paint pigments
3. food coloring
4. packing material
5. containers for chemicals and reactions
6. glasses (for vision)
7. soda can material
8. filament wire for incandescent lamp

C. Societal Issues. Each of the following is a controversial issue currently confronting society. Give the arguments pro and con on each side of the issue and state your preference.

1. Should metropolitan water supplies be fluoridated?

   (Note that numbers 2 and 3 are connected and perhaps should be treated together)
2. Should more nuclear fission power plants be constructed?
3. Should we continue to expand the world's consumption of fossil fuels or should we begin a phase out of fossil fuels?
4. Should we invest heavily in nuclear fusion research?
5. Should we resume funding for the supercollider?
6. Should the NASA budget be increased or decreased?
7. Should we continue to do research in biotechnology?
8. Should we invest heavily in the human genome project?
9. Should people take vitamin C on a daily basis?
10. Should the ban on freons be reversed?
11. Should a college cafeteria use paper plates, Styrofoam plates or reusable dishware?
12. Should nitrites be added to bacon, hot dogs, bologna, etc.?
13. Should dentists use silver amalgams to fill cavities?
14. Should we invest heavily in solar voltaic cell research?
15. Should insecticides and herbicides be used in agriculture?
16. Should we be alarmed by widespread exposure to phthalates and perchlorates?.

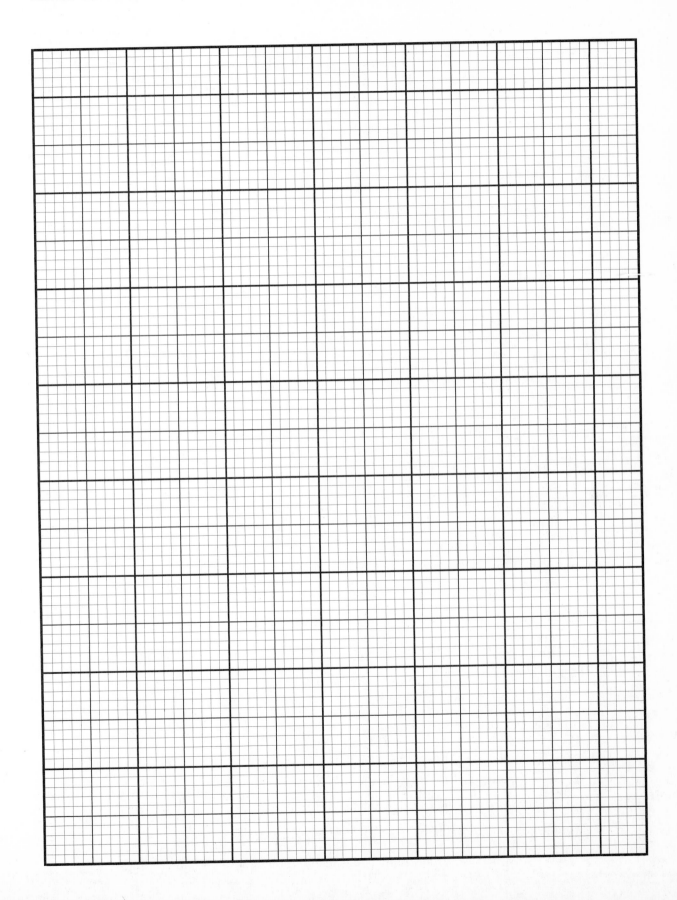

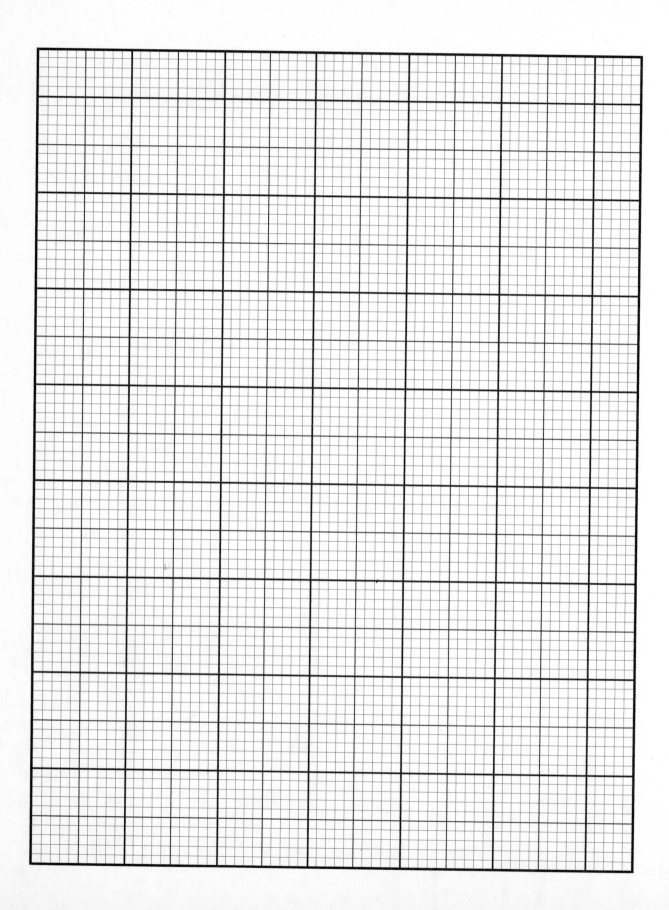

Appendix 1

PROPERTIES OF SUBSTANCES

Substance	Formula	Formula Mass (g/mol)	Melting Point (°C)	Boiling Point (°C)	Density[1] (g/cm^3)	Specific Heat J/g °	Price[2] ($/kg)
acetic acid	$C_2H_4O_2$	60.05	16.6	117.9	1.049	2.05	24
acetone	C_3H_6O	58.08	-95.3	56.2	0.790	2.18	20
acetonitrile	C_2H_3N	41.05	-45.7	81.6	0.786	2.26	36
acetophenone	C_8H_8O	120.16	20.5	202.0	1.028	1.98	26
aluminum	Al	26.98	660.2	2467	2.702	0.900	50
aluminum chloride	$AlCl_3$	133.34	183	190	2.44	0.787	60
ammonia	NH_3	17.03	-77.7	-33.3		2.19	
benzene	C_6H_6	78.12	5.5	80.1	0.879	1.74	34
benzoic acid	$C_7H_6O_2$	122.13	122.4	249	1.26	1.20	32
1-butanol	$C_4H_{10}O$	74.12	-89.5	117.2	0.810	2.3	25
2-butanol	$C_4H_{10}O$	74.12	-115	99.5	0.808		24
i-butyl alcohol	$C_4H_{10}O$	74.12	-108	108	0.803		30
t-butyl alcohol	$C_4H_{10}O$	74.12	25.5	82.2	0.789	3.0	36
calcium chloride	$CaCl_2$	110.99	772		2.15	0.686	63
carbon tetrachloride	CCl_4	153.82	-23.0	76.5	1.594	0.86	161
chloroform	$CHCl_3$	119.38	-63.5	61.7	1.483	0.97	26
chromium	Cr	52.00	1890	2482	7.20	0.449	200
copper	Cu	63.55	1083	2595	8.92	0.385	61
cyclohexane	C_6H_{12}	84.16	6.55	80.7	0.779	1.8	30
cyclohexanol	$C_6H_{12}O$	100.16	25.1	161.1	0.962	1.74	20
ethanol	C_2H_6O	46.07	-117	78.5	0.789	2.45	30
ethyl acetate	$C_4H_8O_2$	88.12	-83.6	77.1	0.900	1.93	30
ethyl ether	$C_4H_{10}O$	74.12	-116.2	34.5	0.714	2.32	36
glucose	$C_6H_{12}O_6$	180.16	146d		1.544	1.15	29
gold	Au	196.97	1064	2807	19.3	0.129	100000
gold(III) chloride	$AuCl_3$	303.33		229	3.9		80000
hexane	C_6H_{14}	86.18	-95	69.0	0.660	2.26	44
hydrogen chloride	HCl	36.46	-114.8	-84.9	1×10^{-3}	0.81	

Substance	Formula	Formula Mass (g/mol)	Melting Point (°C)	Boiling Point (°C)	Density[1] (g/cm^3)	Specific Heat J/g °	Price[2] ($/kg)
iodine	I_2	253.81	113.5	184.4	4.93	0.22	180
iron	Fe	55.85	1535	3000	7.86	0.451	64
lauric acid	$C_{12}H_{24}O_2$	200.33	44	131	0.869		20
lead	Pb	207.19	327.5	1744	11.29	0.129	36
magnesium	Mg	24.31	651	1107	1.74	1.02	90
maple wood	$(C_6H_{12}O_6)_n$			0.68			
mercury	Hg	200.59	-38.9	357	13.59	0.139	113
methanol	CH_4O	32.04	-93.9	65.0	0.791	2.51	22
methylene chloride	CH_2Cl_2	84.93	-95.1	40	1.327	1.20	27
pentane	C_5H_{12}	72.15	-129.7	36.1	0.626		46
phenyl carbonate	$C_{13}H_{10}O_3$	214.2	83	306			47
platinum	Pt	195.09	1772	3827	21.4	0.133	100000
potassium	K	39.10	63.6	774	0.86	0.753	1200
potassium chloride	KCl	74.56	776		1.490	0.68	56
1-propanol	C_3H_8O	60.11	-126.5	97.4	0.804	2.39	28
2-propanol	C_3H_8O	60.11	-89.5	82.4	0.786	2.58	22
salicylic acid	$C_7H_6O_3$	138.12	159	211	1.443		80
silver	Ag	107.87	960.8	2212	10.5	0.236	2400
silver chloride	AgCl	143.32	455	1550	5.56	0.36	2200
sodium	Na	22.99	97.8	883	0.97	1.23	78
sodium chloride	NaCl	58.44	801	1413	2.165	0.866	32
sodium hydroxide	NaOH	40.00	318.4	1390	2.130		58
tin	Sn	118.69	231.9	2260	7.30	0.22	180
titanium	Ti	47.90	1675	3260	4.5	0.523	121
toluene	C_7H_8	92.15	-95	110.6	0.867	1.69	23
tungsten	W	183.85	3410	5660	19.4	0.135	120
vanillin	$C_8H_8O_3$	152.16	81	285	1.056		84
water	H_2O	18.02	0.00	100.0	0.998	4.184	21 (HPLC)
zinc	Zn	65.37	419.4	907	7.14	0.387	56

[1] 20°C

[2] very approximate price from 2004 *Aldrich Catalog* (grade and purity dependent).

Appendix 2

COMMON IONS BY CHARGE

A. Positive Ions

1+

Ammonium	NH_4^+	Mercury(I) (mercurous)	Hg_2^{2+}
Copper(I) (cuprous)	Cu^+	Potassium	K^+
Hydrogen	H^+	Silver	Ag^+
Hydronium	H_3O^+	Sodium	Na^+
Lithium	Li^+		

2+

Barium	Ba^{2+}	Magnesium	Mg^{2+}
Cadmium	Cd^{2+}	Manganese(II) (manganous)	Mn^{2+}
Calcium	Ca^{2+}	Mercury(II) (mercuric)	Hg^{2+}
Cobalt(II) (cobaltous)	Co^{2+}	Nickel(II) (nickelous)	Ni^{2+}
Copper(II) (cupric)	Cu^{2+}	Strontium	Sr^{2+}
Iron(II) (ferrous)	Fe^{2+}	Tin(II) (stannous)	Sn^{2+}
Lead(II) (plumbous)	Pb^{2+}	Zinc	Zn^{2+}

3+

Aluminum	Al^{3+}	Cerium(III) (cerous)	Ce^{3+}
Antimony(III)	Sb^{3+}	Chromium(III) (chromic)	Cr^{3+}
Arsenic(III)	As^{3+}	Iron(III) (ferric)	Fe^{3+}
Bismuth(III)	Bi^{3+}		

4+

Lead(IV) (plumbic)	Pb^{4+}	Tin(IV) (stannic)	Sn^{4+}

5+

Antimony(V)	Sb^{5+}	Bismuth(V)	Bi^{5+}
Arsenic(V)	As^{5+}		

COMMON IONS BY CHARGE continued

B. Negative ions

1-

Acetate	$C_2H_3O_2^-$	Hydrogen sulfate (bisulfate)	HSO_4^-
Bromate	BrO_3^-	Hydrogen sulfite (bisulfite)	HSO_3^-
Bromide	Br^-	Hydroxide	OH^-
Chlorate	ClO_3^-	Hypochlorite	ClO^-
Chloride	Cl^-	Iodate	IO_3^-
Chlorite	ClO_2^-	Iodide	I^-
Cyanate	NCO^-	Nitrate	NO_3^-
Cyanide	CN^-	Nitrite	NO_2^-
Fluoride	F^-	Perchlorate	ClO_4^-
Hydride	H^-	Permanganate	MnO_4^-
Hydrogen carbonate (bicarbonate)	HCO_3^-	Thiocyanate	SCN^-

2-

Carbonate	CO_3^{2-}	Sulfate	SO_4^{2-}
Chromate	CrO_4^{2-}	Sulfide	S^{2-}
Dichromate	$Cr_2O_7^{2-}$	Sulfite	SO_3^{2-}
Oxalate	$C_2O_4^{2-}$	Tetrathionate	$S_4O_6^{2-}$
Oxide	O^{2-}	Thiosulfate	$S_2O_3^{2-}$
Persulfate	$S_2O_8^{2-}$		

3-

Ferricyanide	$Fe(CN)_6^{3-}$	Phosphate	PO_4^{3-}

4-

Ferrocyanide	$Fe(CN)_6^{4-}$

Appendix 3

SOLUBILITIES OF IONIC COMPOUNDS - APPROXIMATE # OF GRAMS OF SOLUTE PER 100 GRAMS OF SOLUTION

	$C_2H_3O_2^-$	Br^-	CO_3^{2-}	Cl^-	CrO_4^{2-}	$Fe(CN)_6^{3-}$	$Fe(CN)_6^{4-}$	OH^-	IO_3^-	I^-	NO_3^-	$C_2O_4^{2-}$	PO_4^{3-}	SO_4^{2-}	S^{2-}	SCN^-
Al^{3+}	ss	s		31			ss	$1x10^{-4}$		s,d	42	i	i	27	d	
NH_4^+	60	43	50	27	25	vs	s	47	2	63	66	4	26	43	vs	63
Ba^{2+}	42	51	$2x10^{-3}$	26	$4x10^{-4}$		.1	4	.02	68	8	$2x10^{-2}$	i	$2x10^{-4}$	d	26
Ca^{2+}	26	59	$6x10^{-3}$	43	14		36	.16	.3	68	56	$7x10^{-4}$	$2x10^{-3}$	.2	.02	s
Ce^{3+}	20	3	i	50				i	.1	s	64	$4x10^{-5}$	i	9	i	
Co^{2+}	s	54	i	35	i	i	i	$3x10^{-4}$	1	65	50	$3x10^{-3}$	i	26	$4x10^{-4}$	51
Cu^{2+}	7	56	i	42	i	i	i	$3x10^{-4}$	.1	1.1	55	$2x10^{-3}$	i	17	$2x10^{-4}$	d
Fe^{3+}		s		70			i	$1x10^{-5}$	.04		46		i	ss	$3x10^{-17}$	vs
Pb^{2+}	31	.8	$1x10^{-4}$	1	$7x10^{-6}$	ss	i	.02	$2x10^{-3}$	.07	35	$1x10^{-4}$	$1x10^{-5}$	$4x10^{-3}$	$9x10^{-5}$	.05
Li^+	31	62	1.3	45	50			11	45	62	43	7	.03	26	vs	vs
Mg^{2+}	40	50	.07	35	42		25	$2x10^{-3}$	8	58	41	.03	.02	28	d	
K^+	70	40	52	25	39	31	23	53	7.5	59	27	26	47	10	s	67
Ag^+	1.0	$8x10^{-6}$	$3x10^{-3}$	$2x10^{-4}$	$4x10^{-3}$	$7x10^{-5}$	i		$4x10^{-3}$	$3x10^{-6}$	70	$3x10^{-3}$	$6x10^{-4}$	.8	$7x10^{-13}$	$2x10^{-5}$
Na^+	32	48	22	26.4	47	23	15	52	8	64	47	3.3	11	20	16	58
Sr^{2+}	27	51	$1x10^{-3}$	35	.12		33	1	.03	64	42	$5x10^{-3}$	i	.01	s,d	vs
Zn^{2+}	25	82	$2x10^{-2}$	79	i		i	$4x10^{-4}$	.9	83	56	$7x10^{-4}$	i	30	10^{-8}	s

An arbitrary standard for solubility is that a compound is called soluble if at least 1 gram dissolves in 100 mL of solution.

When quantitative data could not be located, the symbols below were used:

vs = very soluble, s = soluble, ss = slightly soluble, i = insoluble, d = decomposes

Appendix 4

Solutions to Starred Prelaboratory Problems

Experiment 1

3. Vinegar is a polar liquid and oil is nonpolar. The attractions between molecules of each liquid are much stronger than the attractions to molecules of the other liquid thus they would rather keep to themselves than mix in this case.

5. The small diameter neck minimizes the error caused by a slight missetting of the water level.

7. a. substance, element
   b. homogeneous mixture, unsaturated solution
   c. heterogeneous mixture
   d. extensive physical property
   e. Intensive physical property
   f. homogeneous mixture, unsaturated solution

Experiment 2

1. a. $\left(\frac{55.428 - 54.730}{61.945 - 54.730}\right)(100\%) = 9.67\%$

   b. $\left(\frac{0.33}{10.00}\right)(100\%) = 3.3\%$

   c. Neither filtration or recrystallization would give any information about the mass percent of dissolved potassium chloride in the solution.

   d. As saturation is attained at about 25% by mass KCl and the solution was only 9.67%, the solution was not saturated.

Experiment 3

1. a. 4, 3, 4, 2, 3

3. a. Salt lowers the freezing point of the now salt water. Therefore addition of salt to the roads often lowers the freezing point of the salt water to a temperature below prevailing temperatures thus preventing the freezing of the salt water solution.

4. a. $\left(\frac{1.8}{2.5}\right)(100\%) = 72\%$

   b. The melting range is broad and depressed considerably from the literature value indicating the presence of significant amounts of impurity (~4%).

   c. As the literature value for the melting point of naphthalene is 80.55°C and the recrystallized sample melted over a narrow range almost identical to the literature value, the recrystallization substantially purified the sample.

Experiment 3 (continued)

6. $V = (4/3)\pi r^3 = (4/3)(3.1416)(1.85/2)^3\ cm^3 = 3.32\ cm^3$

$$d = \frac{15.00\ g}{3.31\ cm^3} = 4.52\ g/cm^3$$

This value is very close to the literature value for titanium.

Experiment 4

1. a. $\underline{\text{balance 1}}$ average = (45.747 + 45.745 + 45.748)/3 = 45.747 g
average deviation = (0.000 + 0.002 + 0.001)/3 = $\underline{0.001}$

2. a. $$\frac{64.859\ g - 37.234\ g}{25.00\ mL} = 1.105\ g/mL$$

Experiment 5

1. a. $2\ Al(s) + 3/2\ O_2(g) = Al_2O_3(s)$

b. $$(5.0\ g\ Al)\left(\frac{1\ mol\ Al}{27.0\ g\ Al}\right)\left(\frac{1\ mol\ Al_2O_3}{2\ mol\ Al}\right)\left(\frac{102.0\ g\ Al_2O_3}{1\ mol\ Al_2O_3}\right) = 9.44\ g\ Al_2O_3$$

$$(5.0\ g\ O_2)\left(\frac{1\ mol\ O_2}{32.0\ g\ O_2}\right)\left(\frac{1\ mol\ Al_2O_3}{3/2\ mol\ O_2}\right)\left(\frac{102.0\ g\ Al_2O_3}{1\ mol\ Al_2O_3}\right) = 10.6\ g\ Al_2O_3$$

∴ Al is limiting reagent and 9.44 g of Al_2O_3 will be formed.

$$(9.44\ g\ Al_2O_3)\left(\frac{1\ mol\ Al_2O_3}{102\ g\ Al_2O_3}\right)\left(\frac{3/2\ mol\ O_2}{1\ mol\ Al_2O_3}\right)\left(\frac{32.0\ g\ O_2}{1\ mol\ O_2}\right) = 4.44\ g\ O_2 \text{ reacted and } 0.56\ g\ O_2 \text{ are left}$$

3. a. $BaCl_2{\cdot}2H_2O(s) \xrightarrow{\Delta} BaCl_2(s) + 2\ H_2O(l)$

4. a. $$\left(\frac{2\ mol\ H_2O}{1\ mol\ BaCl_2{\cdot}2H_2O}\right)\left(\frac{18.02\ g\ H_2O}{1\ mol\ H_2O}\right)\left(\frac{1\ mol\ BaCl_2{\cdot}2H_2O}{244.2\ g\ BaCl_2{\cdot}2H_2O}\right)(100\%) = 14.76\%$$

5. Assume 100.0 g of sample

$$(8.74\ g\ C)\left(\frac{1\ mole}{12.011\ g}\right) = 0.727\ moles\ C$$

$$(77.73\ g\ Cl)\left(\frac{1\ mole}{35.453\ g}\right) = 2.192\ moles\ Cl$$

$$(13.83\ g\ F)\left(\frac{1\ mole}{19.00\ g}\right) = 0.728\ moles\ F$$

empirical formula = CCl_3F 137/137 = 1 molecular formula = CCl_3F

8. $$\left(\frac{0.793\ g\ H_2O}{4.00\ g}\right)(100\%) = 19.8\%\ water$$

$$(0.793\ g\ H_2O)\left(\frac{1\ mol\ H_2O}{18.02\ g\ H_2O}\right) = 0.0441\ mol\ H_2O$$

$$[(4.00 - 0.793)\ g\ NiBr_2]\left(\frac{1\ mole\ NiBr_2}{218.5\ g\ NiBr_2}\right) = 0.0147\ mol\ NiBr_2$$

$$\left(\frac{0.0441\ mol\ H_2O}{0.0147\ mol\ NiBr_2}\right) = 3.00\ moles\ H_2O/1\ mole\ NiBr_2$$

Therefore the formula is $NiBr_2 \cdot 3H_2O$.

Experiment 6

1. **reaction** — **classification**

	reaction	classification
1. a.	$Mg(s) + ZnCl_2(aq) = MgCl_2(aq) + Zn(s)$	SR
b.	$2\,AgNO_3(aq) + CaCl_2(aq) = 2\,AgCl(s) + Ca(NO_3)_2(aq)$	DR
c.	$C_2H_6(g) + 7/2\,O_2(g) = 2\,CO_2(g) + 3\,H_2O(g)$	CU
d.	$Na_2O(s) + H_2O(l) = 2\,NaOH(aq)$	CA
e.	$KClO_3(s) = KCl(s) + 3/2\,O_2(g)$	D
2. a.	$BaCl_2(aq) + Na_2SO_4(aq) = BaSO_4(s) + 2\,NaCl(aq)$	DR
b.	$2\,Fe(s) + 3\,CuCl_2(aq) = 2\,FeCl_3(aq) + 3\,Cu(s)$	SR
c.	$C_6H_6(l) + 15/2\,O_2(g) = 6\,CO_2(g) + 3\,H_2O(g)$	CU
d.	$BaCl_2(s) + 2\,H_2O(l) = BaCl_2{\cdot}2H_2O(s)$	CA

3. a. $N_2(g) + 3\,H_2(g) = 2\,NH_3(g)$

   b. $\dfrac{1\ \text{mole}N_2}{3\ \text{moles}\ H_2}$

Experiment 7

1. a. $(33\ \text{g})\left(\dfrac{1\ \text{mole}}{44.0\ \text{g}}\right) = 0.75\ \text{moles}$

2. a. $(3.45\times10^{-2}\ \text{moles})\left(\dfrac{169.9\ \text{g}}{1\ \text{mole}}\right) = 5.86\ \text{g}$

3. a. $Mg(NO_3)_2(aq) + 2\,NaOH(s) = Mg(OH)_2(s) + 2\,NaNO_3(aq)$

4. $2\,Na(s) + 2\,H_2O(l) = 2\,NaOH(aq) + H_2(g)$

$$(2.3\ \text{g Na})\left(\frac{1\ \text{mol Na}}{23.0\ \text{g Na}}\right)\left(\frac{1\ \text{mol}\ H_2}{2\ \text{mol Na}}\right)\left(\frac{2.01\ \text{g}\ H_2}{1\ \text{mol}\ H_2}\right) = 0.10\ \text{g}\ H_2\ \text{(theoretical yield)}$$

$$\left(\frac{0.080\ \text{g}}{0.10\ \text{g}}\right)(100\%) = 80\%\ \text{(percent yield)}$$

Experiment 9

1. Barium sulfate is very insoluble in water and not enough is absorbed into the body to cause a problem.

3. a. If sodium sulfate is used in the first step, marginally soluble silver sulfate might precipitate along with the barium sulfate.

Experiment 10

1. b. copper(II) hydroxide

2. a. K>Al>Pb>Ag

   b. $Al(s) + 3\,Ag^+ = Al^{3+} + 3\,Ag(s)$

2. a. $Cu(s) + 2\,Ag^+ = Cu^{2+} + 2\,Ag(s)$

   b. As the copper replaces the silver, it is more active than silver.

   c. Copper loses 2 electrons in the process (its oxidation number increases by 2) therefore it is oxidized.

Experiment 11

3. a. $$(5.00\text{x}10^{-2}\text{ L } CuSO_4{\cdot}5H_2O)\left(\frac{0.75\text{ moles } CuSO_4{\cdot}5H_2O}{1\text{ L } CuSO_4{\cdot}5H_2O}\right)\left(\frac{249.68\text{ g } CuSO_4{\cdot}5H_2O}{1\text{ mole } CuSO_4{\cdot}5H_2O}\right) = 9.36\text{ g } \underline{CuSO_4{\cdot}5H_2O}$$

4. $AgNO_3(aq) + HCl(aq) = HNO_3(aq) + AgCl(s)$

$$\left(\frac{8.50\text{x}10^{-3}\text{ L HCl}}{1.000\text{x}10^{-2}\text{ L } AgNO_3}\right)\left(\frac{0.1100\text{ mol HCl}}{1\text{ L solution}}\right)\left(\frac{1\text{ mol } AgNO_3}{1\text{ mol HCl}}\right) = 0.0935\text{ M } AgNO_3$$

7. The indicator is selected so that there is a visual change as close as possible to the point where stoichiometric amounts of the two reactants have been used.

Experiment 12

1. b. The temperature change extrapolated to 3.5 min. is 32.7 °C − 18.2 °C = 14.5 °C.

2. $$C_u = -\frac{(25.00\text{ g } H_2O)(4.184\text{ J/g deg})(5.5\text{ deg})}{(15.00\text{ g unk})(-72.5\text{ deg})} = 0.529\text{ J/g deg}$$

   atomic mass = 25/0.529 = 47.2 g/mol This value indicates that the unknown is titanium.

Experiment 14

1. As $E = hc/\lambda$, the longer the wavelength, the lower the energy. Microwaves with $\lambda \approx 0.01$ cm have a much longer wavelength than visible light with wavelengths between $4\text{x}10^{-5}$ and $7\text{x}10^{-5}$ cm.

2. a. $(5.00\text{x}10^{-2} \text{ L } CuSO_4{\cdot}5H_2O)\left(\frac{0.30 \text{ moles } CuSO_4{\cdot}5H_2O}{1 \text{ L } CuSO_4{\cdot}5H_2O}\right)\left(\frac{249.68 \text{ g } CuSO_4{\cdot}5H_2O}{1 \text{ mole } CuSO_4{\cdot}5H_2O}\right) = 3.8 \text{ g } \underline{CuSO_4{\cdot}5H_2O}$

   b. $M_1V_1 = M_2V_2 \quad 2/5(0.30) = 0.12 \text{ M}$

4. a. $-\log_{10}T = A = \epsilon bc \qquad -\log_{10}(0.59) = \epsilon(1)(0.25)\ \epsilon = 0.92$

5. Absorption and wavelength are not linearly related and a plot of absorption vs wavelength usually gives bell shaped curves if an absorption occurs in the wavelength region.

   Absorption at one wavelength is related to the concentration of the absorbing species by Beer's law, $A = \epsilon bc$. As ϵ and b are constants for a given absorbing species at a given wavelength, A is proportional to the concentration and a plot of A vs c should yield a straight line.

Experiment 16

1. a. HF because it has hydrogen bonding. Both have same number of electrons.

   d. SiH_4 has more electrons and at the same temperature, a lower molecular velocity.

2. a. NaCl is an ionic compound and is more soluble in the polar solvent, water.

Experiment 17

2. a. $V = \frac{nRT}{P} = \frac{(1 \text{ mol})(0.08206 \text{ L·atm/mol·K})(273 \text{ K})}{1 \text{ atm}} = 22.4 \text{ L/mol}$

5. $Mg(s) + 2\,HCl(aq) = MgCl_2(aq) + H_2(g)$

   $(0.525 \text{ g Mg})\left(\frac{1 \text{ mol Mg}}{24.31 \text{ g Mg}}\right)\left(\frac{1 \text{ mol } H_2}{1 \text{ mol Mg}}\right) = 2.16\text{x}10^{-2} \text{ moles } H_2$

   $V = \frac{nRT}{P} = \frac{(2.16\text{x}10^{-2})(0.08206)(300)}{752/760} = 0.537 \text{ L } H_2$

Experiment 18

2. a. $\Delta T_{emp} = mK_f \quad K_f = \frac{\Delta T_{emp}}{m} = \frac{(3.4 \text{ deg})(5.00\text{x}10^{-3} \text{ kg})}{(0.38 \text{ g}/154.2 \text{ g·mol}^{-1})} = 6.9 \frac{\text{deg·kg}}{\text{mol}}$

   b. $\Delta T_{emp} = mK_f = \left(\frac{0.150 \text{ g } C_{14}H_{10}}{3.00\times10^{-3} \text{ kg}}\right)\left(\frac{1 \text{ mole } C_{14}H_{10}}{178.24 \text{ g } C_{14}H_{10}}\right)\left(\frac{6.9 \text{ deg·kg}}{\text{mol}}\right) = 1.94\ ^{\circ}\text{C}$

   $216^{\circ}\text{C} - 1.9^{\circ}\text{C} = 214^{\circ}\text{C}$

Experiment 18 (continued)

4. $$V = \left(\frac{1\ cm^3}{8.90\ g}\right)\left(\frac{58.7\ g}{1\ mole}\right)\left(\frac{1\ mole}{6.022x10^{23}\ atoms}\right)\left(\frac{4\ atoms}{unit\ cell}\right) = 4.38x10^{-23}\ cm^3$$

The edge of the unit cell (a) is the cube root of the volume $[V^{1/3}]$

$$a = V^{1/3} = (4.38x10^{-23}\ cm^3)^{1/3} = (43.8x10^{-24}\ cm^3)^{1/3} = 3.52x10^{-8}\ cm$$

$$atomic\ radius = r = a \times 2^{1/2}/4 = (3.52x10^{-8}\ cm)(0.354) = 1.24x10^{-8}\ cm$$

Experiment 19

3. $$\left(\frac{5.00x10^{-3}\ moles}{1\ L}\right)\left(\frac{1\ L}{10^3\ mL}\right)\left(\frac{mL\ EDTA}{mL\ H_2O}\right)\left(\frac{1\ mol\ Ca^{2+}}{1\ mol\ EDTA}\right)\left(\frac{100\ g\ CaCO_3}{1\ mol\ CaCO_3}\right)(10^6\ ppm) = \frac{500(mL\ EDTA)}{mL\ H_2O}$$

Experiment 21

1. b. $$(1.1\ g\ KHP)\left(\frac{1\ mole\ KHP}{204.22\ g\ KHP}\right)\left(\frac{1\ mole\ NaOH}{1\ mole\ KHP}\right)\left(\frac{1\ L\ NaOH}{0.25\ moles}\right) = 0.022\ L = 22\ mL\ NaOH\ soln.$$

2. $$\left(\frac{0.4904\ g\ KHP}{0.02382\ L\ NaOH}\right)\left(\frac{1\ mol\ KHP}{204.2\ g\ KHP}\right)\left(\frac{1\ mol\ NaOH}{1\ mol\ KHP}\right) = 0.1008\ moles\ NaOH/L\ soln.$$

Experiment 22

1. $$\left(\frac{0.03122\ L\ NaOH}{0.02500\ L\ H_2SO_4}\right)\left(\frac{0.1234\ mol\ NaOH}{1\ L\ NaOH}\right)\left(\frac{1\ mol\ H_2SO_4}{2\ mol\ NaOH}\right) = 7.705\times10^{-2}\ M\ H_2SO_4$$

2. $$\left(\frac{0.01628\ L\ NaOH}{0.01000\ L\ vin.}\right)\left(\frac{0.5120\ mol\ NaOH}{1\ L}\right)\left(\frac{1\ mol\ acetic\ acid}{1\ mol\ NaOH}\right) = 0.8335\ M\ acetic\ acid$$

$$\left(\frac{0.8335\ mol\ acetic\ acid}{1\ L\ soln.}\right)\left(\frac{0.01000\ L\ soln.}{10.05\ g\ soln.}\right)\left(\frac{60.05\ g\ acetic\ acid}{1\ mol\ acetic\ acid}\right)(100\%) = 4.98\%\ acetic\ acid$$

3. $(.01556\ L\ NaOH)(0.1020\ mol\ NaOH/L)(1\ mol\ HX/1\ mol\ NaOH) = 1.587x10^{-3}\ mol\ HX$

$0.1936\ g/1.587x10^{-3}\ mol = 122.0\ g/mol$

Experiment 23

1.

	pH	$[H^+]$ (M)	$[OH^-]$ (M)	pOH
a.	4.7	$2x10^{-5}$	$5x10^{-10}$	9.3
b.	2.11	$7.7x10^{-3}$	$1.3x10^{-12}$	11.89

6. $H_2S(aq) = H^+ + SH^-$

$[H^+] = [SH^-] = 1.0x10^{-4}\ M,\ [H_2S] = 1.0x10^{-1} - 1.0x10^{-4} = 1.0x10^{-1}\ M$

$$K_a = \frac{[H^+][SH^-]}{[H_2S]} = \frac{(1.0x10^{-4})^2}{0.10} = 1.0x10^{-7}$$

Experiment 24

3. $(0.300 \text{ mol HCl/L})(0.0250 \text{ L}) = 7.50\text{x}10^{-3}$ moles HCl (total)

$(0.200 \text{ mol NaOH/L})(0.0248 \text{ L}) = 4.96\text{x}10^{-3}$ moles NaOH

$(4.96\text{x}10^{-3} \text{ moles NaOH})(1 \text{ mol HCl}/1 \text{ mol NaOH}) = 4.96\text{x}10^{-3}$ moles HCl (excess)

$7.50\text{x}10^{-3} - 4.96\text{x}10^{-3} = 2.54\text{x}10^{-3}$ moles HCl consumed by carbonate

$$\left(\frac{1 \text{ mol BaCO}_3}{2 \text{ mol HCl}}\right)(2.54\text{x}10^{-3} \text{ mol HCl}) = 1.27\text{x}10^{-3} \text{ moles BaCO}_3$$

$0.250 \text{ g}/1.27\text{x}10^{-3} \text{ mol} = 197 \text{ g/mol}$

Use of the periodic table to determine the formula mass of $BaCO_3$ yields 197.3 g/mol. The agreement well within experimental error strongly supports the conclusion that the product was barium carbonate.

Experiment 26

1. a. $\Delta G^\circ = \Delta H^\circ - T\Delta S^\circ = 25.7 \text{ kJ/mol} - (298 \text{ K})(0.1087 \text{ kJ/mol·K}) = -6.69 \text{ kJ/mol}$

$\Delta G^\circ = -RT\ln K$

$$\ln K = -\Delta G^\circ/RT = \frac{-(-6700 \text{ J/mol})}{(8.313 \text{ J/mol·K})(298 \text{ K})} = 2.70$$

$K = 14.9$

b. Endothermic

c. The positive ΔS is consistent with the increase in disorder that occurs when a solid dissolves in a liquid.

Experiment 27

1. a. $K_{sp} = [Ag^+][Br^-] = 5\text{x}10^{-13}$

$$K_f = \frac{[Ag(NH_3)_2^+]}{[Ag^+][NH_3]^2} = 1.7\text{x}10^7$$

$$K_fK_{sp} = \frac{[Ag(NH_3)_2^+][Br^-]}{[NH_3]^2} = 8.5\text{x}10^{-6} = \frac{(x)(0.01)}{(6.0)^2}$$

x = 0.031 M As the complex concentration comes out greater than 0.01 M, the silver bromide would dissolve.

Experiment 28

1. a. $\frac{\Delta[A]}{\Delta t} = \frac{2\Delta[P]}{\Delta t}$

2. a. $\frac{\Delta[P]}{\Delta t} = k[A][B]$ slow step $A + B \rightarrow I_{ntermediate}$
   fast step $I + B \rightarrow P$

Experiment 29

1. $$(12\text{ g SA})\left(\frac{1\text{ mol SA}}{138\text{ g SA}}\right)\left(\frac{1\text{ mol Asp}}{1\text{ mol SA}}\right)\left(\frac{180\text{ g Asp}}{1\text{ mol Asp}}\right) = 15.7\text{ g Aspirin (theoretical yield)}$$

   (11/15.7)(100%) = 70% (percent yield)

3. c. $Ca(NO_3)_2(aq) + K_2C_2O_4(aq) = CaC_2O_4(s) + 2\,KNO_3(aq)$

   As calcium oxalate is insoluble, equal volumes of only approximately equal molar solutions of calcium nitrate and potassium oxalate should be mixed and the product collected by vacuum filtration.

Experiment 30

1. a. 2, 4, 7

2. a. $Al(s) = Al^{3+} + 3\,e^-$

   $3\,H^+ + 3\,e^- = 3/2\,H_2(g)$

   $Al(s) + 3\,H^+ = Al^{3+} + 3/2\,H_2(g)$

5. $$\left(\frac{0.800\text{ g }K_3Fe(CN)_6}{1.90\text{x}10^{-2}\text{ L }Na_2S_2O_3}\right)\left(\frac{1\text{ mol }K_3Fe(CN)_6}{329.3\text{ g }K_3Fe(CN)_6}\right)\left(\frac{1\text{ mol }Na_2S_2O_3}{1\text{ mol }K_3Fe(CN)_6}\right) = 0.128\text{ M }NaS_2O_3$$

Experiment 31

1. a. $C_2O_4^{2-} = 2\,CO_2(g) + 2\,e^-$

$MnO_4^- + 8\,H^+ + 5\,e^- = Mn^{2+} + 4\,H_2O(g)$

$2\,MnO_4^- + 16\,H^+ + 5\,C_2O_4^{2-} = 2\,Mn^{2+} + 8\,H_2O(g) + 10\,CO_2(g)$

b. $$\left(\frac{0.2600\text{ g }Na_2C_2O_4}{2.250\text{x}10^{-2}\text{ L }KMnO_4}\right)\left(\frac{1\text{ mol }Na_2C_2O_4}{134.00\text{ g }Na_2C_2O_4}\right)\left(\frac{2\text{ mol }KMnO_4}{5\text{ mol }Na_2C_2O_4}\right) = 0.03449\text{ M }KMnO_4$$

c. Permanganate is purple colored. During the titration, the permanganate reacts leaving the solution with little or no color. However, the slightest amount of permanganate beyond the stoichiometric amount will turn the solution purple.

3. $$\left(\frac{0.02112\text{ L }NaS_2O_3}{0.00100\text{ L "Cl"}}\right)\left(\frac{0.150\text{ mol }NaS_2O_3}{1\text{ L }NaS_2O_3}\right)\left(\frac{1\text{ mol NaClO}}{2\text{ mol }NaS_2O_3}\right) = 1.584\text{ M NaClO}$$

$$\left(\frac{1.584\text{ mol}}{\text{L soln.}}\right)\left(\frac{74.4\text{ g NaClO}}{1\text{ mol NaClO}}\right)\left(\frac{1\text{ L soln.}}{1180\text{ g soln}}\right)(100\%) = 9.99\%$$

Experiment 32

1. a. $Cu(s) + Sn^{4+} = Cu^{2+} + Sn^{2+}$ $E^o = -0.19$ volts

---

Experiment 33

1. $Pb(s) + 2\,Ag^+ = Pb^{2+} + 2\,Ag(s)$

2. 2.11 volts

3. $(1.5\text{x}10^{-2}\text{ coulombs/sec})(1.75\text{ hr})(3600\text{ sec/hr}) = 94.5\text{ coulombs}$

$$\left(\frac{94.5\text{ coul}}{0.108\text{ g Ag}}\right)\left(\frac{1\text{ electron}}{1.602\text{x}10^{-19}\text{ coul}}\right)\left(\frac{107.9\text{ g Ag}}{1\text{ mol Ag}}\right)\left(\frac{1\text{ atom Ag}}{1\text{ electron}}\right) = 5.89\text{x}10^{23}\ \frac{\text{atoms Ag}}{\text{mol Ag}}$$

$$\left(\frac{94.5\text{ coulombs}}{0.108\text{ g Ag}}\right)\left(\frac{107.9\text{ g Ag}}{1\text{ mol Ag}}\right) = 9.4\text{x}10^4\text{ coulombs/mol}$$

4. $$(0.108\text{ g Ag})\left(\frac{1\text{ mol Ag}}{107.9\text{ g Ag}}\right)\left(\frac{1\text{ mol Pb}}{2\text{ mol Ag}}\right)\left(\frac{207.2\text{ g Pb}}{1\text{ mol Pb}}\right) = 0.104\text{ g Pb}$$

Experiment 34

2. a. $(0.15\text{ M})(50/250) = 0.030\text{ M}$

b. $A = \epsilon bc = (5.0)(1.00)(0.030) = 0.15$

PERIODIC TABLE OF THE ELEMENTS

s		d										p					
1	2	3	4	5	6	7	8	9	10	11	12	13	14	15	16	17	18
1A	*2A*	*3B*	*4B*	*5B*	*6B*	*7B*	*8B*	*8B*	*8B*	*1B*	*2B*	*3A*	*4A*	*5A*	*6A*	*7A*	*8A*
1 ***H*** 1.00794																	2 ***He*** 4.0026
3 **Li** 6.941	4 **Be** 9.012											5 **B** 10.81	6 **C** 12.011	7 *N* 14.007	8 ***O*** 15.999	9 ***F*** 18.998	10 ***Ne*** 20.180
11 **Na** 22.990	12 **Mg** 24.305											13 **Al** 26.982	14 **Si** 28.086	15 **P** 30.974	16 **S** 32.066	17 *Cl* 35.453	18 ***Ar*** 39.948
19 **K** 39.098	20 **Ca** 40.078	21 **Sc** 44.956	22 **Ti** 47.867	23 **V** 50.942	24 **Cr** 51.996	25 **Mn** 54.938	26 **Fe** 55.847	27 **Co** 58.933	28 **Ni** 58.693	29 **Cu** 63.546	30 **Zn** 65.39	31 **Ga** 69.723	32 **Ge** 72.61	33 **As** 74.922	34 **Se** 78.96	35 Br 79.904	36 ***Kr*** 83.798
37 **Rb** 85.468	38 **Sr** 87.62	39 **Y** 88.906	40 **Zr** 91.224	41 **Nb** 92.906	42 **Mo** 95.94	43 Tc (98)	44 **Ru** 101.07	45 **Rh** 102.91	46 **Pd** 106.42	47 **Ag** 107.87	48 **Cd** 112.41	49 **In** 114.82	50 **Sn** 118.71	51 **Sb** 121.76	52 **Te** 127.60	53 **I** 126.90	54 ***Xe*** 131.29
55 **Cs** 132.91	56 **Ba** 137.33	57 **La** 138.91	72 **Hf** 178.49	73 **Ta** 180.95	74 **W** 183.85	75 **Re** 186.21	76 **Os** 190.23	77 **Ir** 192.22	78 **Pt** 195.08	79 **Au** 196.97	80 Hg 200.59	81 **Tl** 204.38	82 **Pb** 207.2	83 **Bi** 208.98	84 Po (209)	85 At (210)	86 *Rn* (222)
87 Fr (223)	88 Ra 226.03	89 Ac 227.03	104 Rf (261)	105 Db (262)	106 Sg (263)	107 Bh (262)	108 Hs (265)	109 Mt (266)	110 Ds (269)	111 Rg (272)	112 1996 (277)	113 2003 (284)	114 1998 (289)	115 2003 (288)	116 2000 (292)		

f

58 **Ce** 140.12	59 **Pr** 140.91	60 **Nd** 144.24	61 Pm (145)	62 **Sm** 150.36	63 **Eu** 151.96	64 **Gd** 157.25	65 **Tb** 158.93	66 **Dy** 162.50	67 **Ho** 164.93	68 **Er** 167.26	69 **Tm** 168.93	70 **Yb** 173.04	71 **Lu** 174.97
90 Th 232.04	91 Pa 231.04	92 U 238.03	93 Np 237.05	94 Pu (244)	95 Am (243)	96 Cm (247)	97 Bk (247)	98 Cf (251)	99 Es (252)	100 Fm (257)	101 Md (258)	102 No (259)	103 Lr (260)

key: bold or normal italics - gas, shadow - liquid, bold or normal - solid, normal print - all known isotopes are radioactive